TODAY'S TECHNICIAN ™

CLASSROOM MANUAL

For Automotive Engine Repair & Rebuilding

SIXTH EDITION

TODAY'S TECHNICIAN™

CLASSROOM MANUAL

For Automotive Engine Repair & Rebuilding

SIXTH EDITION

Chris Hadfield
Director, Minnesota Transportation Center of Excellence

Randy Nussler
South Puget Sound Community College & New Market Skills Center

CENGAGE
Learning·

Australia · Brazil · Japan · Korea · Mexico · Singapore · Spain · United Kingdom · United States

Today's Technician: Automotive Engine Repair & Rebuilding, Sixth Edition

Chris Hadfield

Randy Nussler

SVP, GM Skills & Global Product Management: Jonathan Lau

Product Director: Matthew Seeley

Senior Product Manager: Katie McGuire

Senior Director, Development: Marah Bellegarde

Senior Product Development Manager: Larry Main

Senior Content Developer: Mary Clyne

Product Assistant: Mara Ciacelli

Vice President, Marketing Services: Jennifer Ann Baker

Associate Marketing Manager: Andrew Ouimet

Senior Production Director: Wendy Troeger

Production Director: Andrew Crouth

Senior Content Project Manager: Cheri Plasse

Senior Art Director: Jack Pendleton

Production Service/Composition: SPi Global

Cover image(s): Umberto Shtanzman/ Shutterstock.com

Library of Congress Control Number: 2017930371

Classroom Manual ISBN: 978-1-305-95811-1
Package ISBN: 978-1-305-95813-5

Cengage Learning
20 Channel Center Street
Boston, MA 02210
USA

Cengage Learning is a leading provider of customized learning solutions with employees residing in nearly 40 different countries and sales in more than 125 countries around the world. Find your local representative at **www.cengage.com**.

Cengage Learning products are represented in Canada by Nelson Education, Ltd.

To learn more about Cengage Learning, visit **www.cengage.com**

Purchase any of our products at your local college store or at our preferred online store **www.cengagebrain.com**

Printed in the United States of America
Print Number: 02   Print Year: 2017

CONTENTS

Thanks to the support the *Today's Technician*™ series has received from those who teach automotive technology, Cengage Learning, the leader in automotive-related textbooks, is able to live up to its promise to provide new editions of the series every few years. By revising this series on a regular basis, we can respond to changes in the industry, changes in technology, changes in the certification process, and to the ever-changing needs of those who teach automotive technology.

The *Today's Technician*™ series features textbooks and digital learning solutions that cover all mechanical and electrical systems of automobiles and light trucks. The individual titles correspond to the ASE (National Institute for Automotive Service Excellence) certification areas and are specifically correlated to the 2016 standards for Automotive Service Technicians (AST), Master Automotive Service Technicians (MAST), and Maintenance and Light Repair (MLR).

Additional titles include remedial skills and theories common to all of the certification areas and advanced or specific subject areas that reflect the latest technological trends, *such as this updated title on engine repair.*

Today's Technician: Automotive Engine Repair and Rebuilding, 6th edition, is designed to give students a chance to develop the same skills and gain the same knowledge that today's successful technicians have. This edition also reflects the changes in the guidelines established by the National Automotive Technicians Education Foundation (NATEF).

The purpose of NATEF is to evaluate technician training programs against standards developed by the automotive industry and recommend qualifying programs for certification (accreditation) by ASE. Programs can earn ASE certification upon NATEF's recommendation. NATEF's national standards reflect the skills that students must master. ASE certification through NATEF evaluation ensures that certified training programs meet or exceed industry-recognized, uniform standards of excellence.

The technician of today and for the future must know the underlying theory of all automotive systems and be able to service and maintain those systems. Dividing the material into two volumes, a Classroom Manual and a Shop Manual, provides the reader with the information needed to begin a successful career as an automotive technician without interrupting the learning process by mixing cognitive and performance learning objectives into one volume.

The design of Cengage's *Today's Technician*™ series was based on features that are known to promote improved student learning. The design was further enhanced by a careful study of survey results, in which the respondents were asked to value particular features. Some of these features can be found in other textbooks, while others are unique to this series.

Each Classroom Manual contains the principles of operation for each system and subsystem. The Classroom Manual also contains discussions on design variations of key components used by the different vehicle manufacturers. It also looks into emerging technologies that will be standard or optional features in the near future. This volume is organized to build upon basic facts and theories. The primary objective of this volume is to allow the reader to gain an understanding of how each system and subsystem operates. This understanding is necessary to diagnose the complex automobiles of today and tomorrow. Although the basics contained in the Classroom Manual provide the knowledge needed for diagnostics, diagnostic procedures appear only in the Shop Manual. An understanding of the underlying theories is also a requirement for competence in the skill areas covered in the Shop Manual.

A spiral-bound Shop Manual delivers hands-on learning experiences with step-by-step instructions for diagnostic and repair procedures. Photo Sequences are used to illustrate some of the common service procedures. Other common procedures are listed and are accompanied with fine line drawings and photos that allow the reader to visualize and conceptualize the finest details of the procedure. This volume also contains the reasons for performing the procedures, as well as when that particular service is appropriate.

The two volumes are designed to be used together and are arranged in corresponding chapters. Not only are the chapters in the volumes linked together, the contents of the chapters are also linked. The linked content is indicated by marginal callouts that refer the reader to the chapter and page where the same topic is addressed in the companion volume. This valuable feature saves users the time and trouble of searching the index or table of contents to locate supporting information in the other volume. Instructors will find this feature especially helpful when planning the presentation of material and when making reading assignments.

Both volumes contain clear and thoughtfully selected illustrations, many of which are original drawings or photos specially prepared for inclusion in this series. This means that the art is a vital part of each textbook and not merely inserted to increase the number of illustrations.

The layout of *Automotive Engine Repair & Rebuilding*, 6th edition, is easy to follow and consistent with the *Today's Technician*™ series. Complex systems are broken into easier-to-understand explanations. Industry standardized terms and vernacular are used and explained in the text.

Jack Erjavec

HIGHLIGHTS OF THE NEW EDITION—CLASSROOM MANUAL

The Classroom Manual for this edition of *Today's Technician: Automotive Engine Repair and Rebuilding* includes updated coverage of the NATEF AST, MAST, and MLR tasks for engine repair and rebuilding. In addition to updated coverage of industry trends, new information has been added on the following:

- 0w16 oil
- Engine design changes for gas direct injection (GDI)
- EPDM belts
- Stretch belts
- Wet timing belts
- Flat plane crankshafts
- Cam-phaser design, operation, and service
- Variable valve timing
- Variable lift
- Active fuel management
- Variable cylinder displacement

HIGHLIGHTS OF THE NEW EDITION—SHOP MANUAL

Like all textbooks in the *Today's Technician*™ series, the understanding acquired by reading the Classroom Manual is required for competence in the skill areas covered in the Shop Manual. Service information related to the topics covered in the Classroom Manual is included in this manual. In addition, several photo sequences are used to highlight typical service procedures and provide the student the opportunity to get a realistic idea of a procedure. The purpose of these detailed photo sequences is to show students what to expect when they perform the same procedure. They can also provide a student with familiarity of a system or type of equipment they may not be able to perform at their school.

To stress the importance of safe work habits, Chapter 1 covers safety issues and has been updated to include hybrid vehicle high-voltage safety. Included in this chapter are common shop hazards, safe shop practices, safety equipment, and the legislation concerning and the safe handling of hazardous materials and wastes.

Chapter 2 covers special tools and procedures. Procedures include the use of engine condition and diagnostic test equipment, precision engine measuring tools and specialty measuring tools, along with engine reconditioning tools and equipment.

The subsequent Shop Manual chapters synch up with those in the Classroom Manual, and the related content of each manual's chapters is linked by use of page references in the margins. This allows the student to quickly cross-reference the theory with the practical. Redundancy between the Classroom Manual and the Shop Manual has been kept to a minimum; the only time theory is discussed again is if it is necessary to explain the diagnostic results or as an explanation of the symptom. Currently accepted service procedures are used as examples throughout the text. These procedures also served as the basis for the job sheets that are included in the textbook at the end of each chapter. Updated coverage in the Shop Manual addresses:

- Engine pre-oiling
- Engine break-in
- 500-mile service for newly rebuilt engines
- HEV service and safety
- Concerns related to improper oil service on hydraulically controlled systems

CLASSROOM MANUAL

Features of the Classroom Manual include the following:

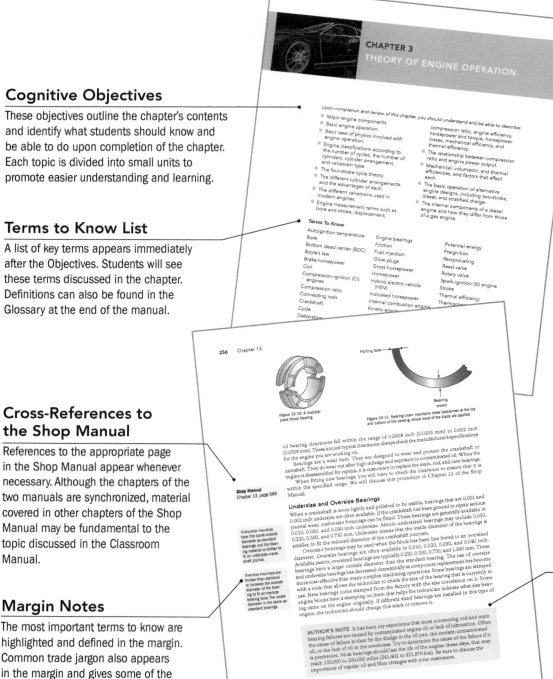

Cognitive Objectives

These objectives outline the chapter's contents and identify what students should know and be able to do upon completion of the chapter. Each topic is divided into small units to promote easier understanding and learning.

Terms to Know List

A list of key terms appears immediately after the Objectives. Students will see these terms discussed in the chapter. Definitions can also be found in the Glossary at the end of the manual.

Cross-References to the Shop Manual

References to the appropriate page in the Shop Manual appear whenever necessary. Although the chapters of the two manuals are synchronized, material covered in other chapters of the Shop Manual may be fundamental to the topic discussed in the Classroom Manual.

Margin Notes

The most important terms to know are highlighted and defined in the margin. Common trade jargon also appears in the margin and gives some of the common terms used for components. This helps students understand and speak the language of the trade, especially when conversing with an experienced technician.

Author's Note

This feature includes simple explanations, stories, or examples of complex topics. These are included to help students understand difficult concepts.

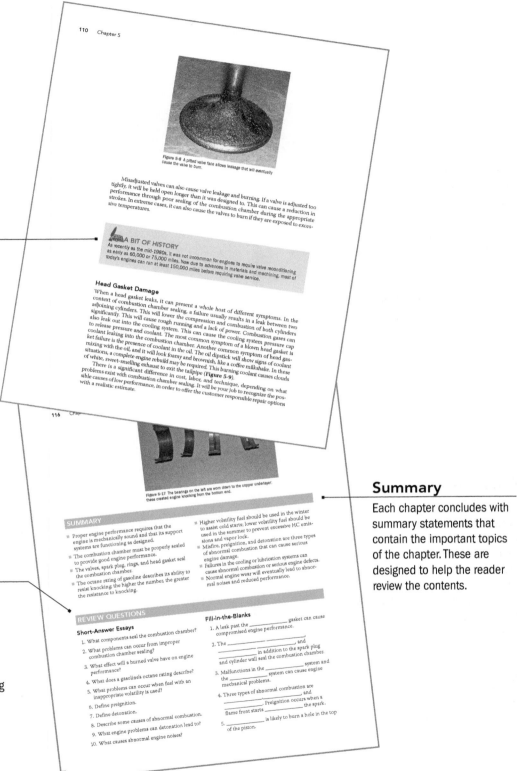

A Bit of History

This feature gives the student a sense of the evolution of the automobile. This feature not only contains nice-to-know information, but also should spark some interest in the subject matter.

Review Questions

Short-answer essays, fill-in-the-blanks, and multiple-choice questions follow each chapter. These questions are designed to accurately assess the student's competence in the stated objectives at the beginning of the chapter.

Summary

Each chapter concludes with summary statements that contain the important topics of the chapter. These are designed to help the reader review the contents.

SHOP MANUAL

To stress the importance of safe work habits, the Shop Manual dedicates one full chapter to safety. Other important features of this manual include:

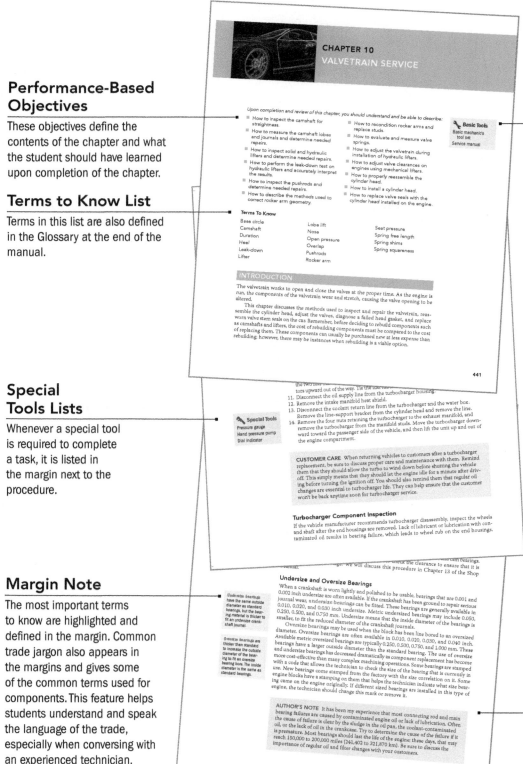

Performance-Based Objectives

These objectives define the contents of the chapter and what the student should have learned upon completion of the chapter.

Terms to Know List

Terms in this list are also defined in the Glossary at the end of the manual.

Special Tools Lists

Whenever a special tool is required to complete a task, it is listed in the margin next to the procedure.

Margin Note

The most important terms to know are highlighted and defined in the margin. Common trade jargon also appears in the margins and gives some of the common terms used for components. This feature helps students understand and speak the language of the trade, especially when conversing with an experienced technician.

Basic Tools Lists

Each chapter begins with a list of the basic tools needed to perform the tasks included in the chapter.

Author's Note

This feature includes simple explanations, stories, or examples of complex topics. These are included to help students understand difficult concepts.

Photo Sequences

Many procedures are illustrated in detailed Photo Sequences. These photographs show the students what to expect when they perform particular procedures. They also familiarize students with a system or type of equipment that the school might not have.

Service Tips

Whenever a shortcut or special procedure is appropriate, it is described in the text. Generally, these tips describe common procedures used by experienced technicians.

Warnings and Cautions

Cautions appear throughout the text to alert readers to potentially hazardous materials or unsafe conditions. Warnings advise the student of things that can go wrong if instructions are not followed or if an incorrect part or tool is used.

References to the Classroom Manual

References to the appropriate page in the Classroom Manual appear whenever necessary. Although the chapters of the two manuals are synchronized, material covered in other chapters of the Classroom Manual may be fundamental to the topic discussed in the Shop Manual.

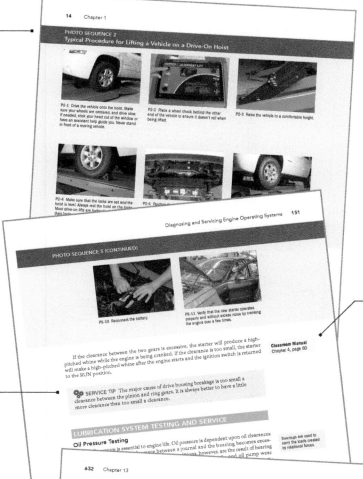

Case Studies

Each chapter ends with a Case Study describing a particular vehicle problem and the logical steps a technician might use to solve the problem. These studies focus on system diagnosis skills and help students gain familiarity with the process.

ASE-Style Review Questions

Each chapter contains ASE-Style Review Questions that reflect the performance objectives listed at the beginning of the chapter. These questions can be used to review the chapter as well as to prepare for the ASE certification exam.

Job Sheets

Located at the end of each chapter, the Job Sheets provide a format for students to perform procedures covered in the chapter. A reference to the NATEF Tasks addressed by the procedure is included on the Job Sheet.

ASE Challenge Questions

Each technical chapter ends with five ASE challenge questions. These are not more review questions; rather, they test the students' ability to apply general knowledge to the contents of the chapter.

---

570   Chapter 12

CASE STUDY

Inspection of a V8 engine block indicated that one cylinder was excessively scuffed and damaged. The technician researched the options available and discussed them with the customer. It was decided that the cost of a new block was not feasible and having to bore each cylinder would require the purchase of new pistons. In this instance, it was decided that the most cost-effective method of repairing the block was the installation of a sleeve. After completing all the required machining operations on the engine block, the technician was ready to turn her attention to the crankshaft. The inspection notes taken indicated that one of the main-bearing journals required grinding to restore its surface finish. All other main-bearing journals were good. Realizing it is unusual to have only one bearing fail in this manner, the technician inspected the old bearing very closely and discovered that the original bearing was undersize even though the journal was not ground. Someone had attempted to remove an engine noise by simply installing a thicker bearing to take up clearance. The extra friction caused the journal to score. To maintain proper crankshaft position in the block, all main-bearing journals were ground to the next standard undersize, new bearings were installed, and oil clearances were checked. By taking the time to find what is in the best interest of the customer, your reputation as an honest technician will grow.

ASE-STYLE REVIEW QUESTIONS

1. *Technician A* says to reverse the tightening sequence when loosening the cylinder head.
Technician B says to loosen the main caps starting at the front of the engine and moving toward the rear.
Who is correct?
A. A only
B. B only
C. Both A and B
D. Neither A nor B

2. *Technician A* says that you can pry most harmonic balancers off with two big pry bars.
Technician B says that you can damage the crankshaft if you do not protect the threads while using a puller.
Who is correct?
A. A only
B. B only
C. Both A and B
D. Neither A nor B

4. *Technician A* says that the pistons should come out the top of the block.
Technician B says to drive on the edge of the piston skirt with a punch to remove the pistons.
Who is correct?
A. A only
B. B only
C. Both A and B
D. Neither A nor B

5. The crankshaft has been removed for inspection.
Technician A says the area around the fillet is a common location for stress cracks.
Technician B says a crack near the number 1 piston connecting-rod journal may indicate a faulty vibration damper.
Who is correct?
A. A only
B. B only
C. Both A and B
D. Neither A nor B

6. The cylinder block is ready for inspection.
Technician A says that deck warpage can be checked using a precision straightedge and feeler gauge.

Name _____   Date _____

PERFORMING AN OIL AND FILTER CHANGE

Upon completion of this job sheet, you should be able to properly perform an oil and filter change.

JOB SHEET 20

NATEF Correlation

This job sheet addresses the following **AST/MAST** task for Engine Repair:

D.10.   Perform engine oil and filter change.

This job sheet addresses the following **MLR** task for Engine Repair:

D.5.   Perform engine oil and filter change.

Tools and Materials
• Technician's tool set
• Shop rag
• Correct type of engine oil
• Service manual
• Oil filter wrench
• Used oil container
• Torque wrench

Diagnosing and Servicing Engine Operating Systems   189

ASE CHALLENGE QUESTIONS

1. A starter makes a grinding noise when it engages the flywheel teeth.
Technician A says that it could be shimmed improperly.
Technician B says that the starter drive gear could be damaged.
Who is correct?
A. A only
B. B only
C. Both A and B
D. Neither A nor B

2. A technician is performing a starter current draw test because the engine will not turn over. The test results indicate that the current draw on the starter is higher than the specification. This means that the:
A. starter solenoid is bad.
B. battery needs charging.
C. ignition switch is bad.
D. engine may be hydrostatically locked.

3. A cooling system is pressurized with a pressure tester to locate a leak. After 15 minutes, the tester gauge has dropped from 15 psi to 5 psi, and there are no visible leaks in the engine compartment.
Technician A says that the engine may have an internal head gasket leak.

Technician B says that the heater core may be leaking.
Who is correct?
A. A only
B. B only
C. Both A and B
D. Neither A nor B

4. A customer says that his oil pressure warning light comes on while the car is idling. An oil pressure test shows low oil pressure.
Technician A says that the engine bearings may be worn.
Technician B says that the oil pressure relief valve may be stuck closed.
Who is correct?
A. A only
B. B only
C. Both A and B
D. Neither A nor B

5. A coolant temperature gauge does not move from its lowest reading when the vehicle is driven.
Technician A says that the coolant temperature sensor wires may be disconnected.
Technician B says that the thermostat could be stuck open.
Who is correct?
A. A only
B. B only
C. Both A and B
D. Neither A nor B

SUPPLEMENTS

Instructor Resources

The *Today's Technician*™ series offers a robust set of instructor resources, available online at Cengage's Instructor Resource Center and on DVD. The following tools have been provided to meet any instructor's classroom preparation needs:

- An Instructor's Guide including lecture outlines, teaching tips, and complete answers to end-of-chapter questions.
- PowerPoint presentations with images and animations that coincide with each chapter's content coverage.
- Cengage Learning Testing Powered by Cognero® provides hundreds of test questions in a flexible, online system. You can choose to author, edit, and manage Test Bank content from multiple Cengage Learning solutions and deliver tests from your LMS, or you can simply download editable Word documents from the DVD or Instructor Resource Center.
- An Image Gallery includes photos and illustrations from the text.
- The Job Sheets from the Shop Manual are provided in Word format.
- End-of-chapter Review Questions are provided in Word format, with a separate set of text rejoinders available for instructors' reference.
- To complete this powerful suite of planning tools, correlation guides are provided to the NATEF tasks and to the previous edition.

REVIEWERS

The author and publisher would like to extend special thanks to the following instructors for reviewing this material:

Matthew Bockenfeld
Vatterott College—Joplin Joplin, MO

David Chavez
Austin Community College Austin, TX

Joseph Cortez
Reynolds Community College Richmond, VA

Jason S. Grice
Black Hawk College Galva, IL

Todd Mikonis
Manchester Community College Manchester, NH

Vincenzo Rigaglia
Bronx Community College Bronx, NY

Ronald Strzalkowski
Baker College of Flint Flint, MI

Michael White
The University of Northwestern Ohio Lima, OH

CHAPTER 1
AUTOMOTIVE ENGINES

Upon completion and review of this chapter, you should understand and be able to describe:

- The purpose of the automotive engine.
- The basic operation of the internal combustion four-stroke engine.
- The basic function of the cooling and lubrication systems.
- The basic function of the intake and exhaust systems.
- The basics of the cylinder head and its related components.
- The engine block and its related components.
- Many of the major components of the engine.

Terms To Know

Camshaft
Combustion
Combustion chamber
Cylinder
Cylinder head
Engine block
Exhaust manifold

Intake manifold
Internal combustion engine (ICE)
Journals
Overhead camshaft (OHC) engine
Piston

Powertrain
Powertrain control module (PCM)
Pushrod
Thermostat
Valve timing
Valvetrain

INTRODUCTION

The automotive engine is the power source that drives the automobile. Today's engines are capable of providing good performance and smooth operation in a variety of ambient pressures and temperatures while accelerating, decelerating, cruising at high speeds, or at idle. They also achieve good fuel economy and low toxic emissions. While the basic operation of the four-stroke gasoline **internal combustion engine (ICE)** has not changed in more than 100 years, efficiency and power ratings have increased more than 1,000 percent. Henry Ford's Model T sported a 2.7-L engine producing about 20 horsepower (hp). Today we commonly see more than 300 hp out of the same size engine.

Refinements to the base engine and its support systems have allowed the gains in performance, control of fuel consumption, and emissions. Lighter materials within the engine have allowed weight reductions. The precision and accuracy of the manufacturing process of modern engines have increased, allowing tighter engine internal clearances. The durability of engines has increased also due to improvements in the process of manufacturing materials and metals. With proper maintenance, most engines should provide more than 150,000 miles of trouble-free service. Improved parts, new component designs, advanced fluids, and electronic controls have allowed for fewer maintenance

> **Internal combustion engine (ICE)** burns its fuels within the engine. The power that is used as a result of burning that fuel is also developed inside the engine. In comparison, in an external combustion engine the burning of fuel occurs in an external source, or tank. The heat is then transferred to a separate component where it can be used to power the engine and move parts. Examples of external combustion engines would be steam locomotives and the Stirling engine.

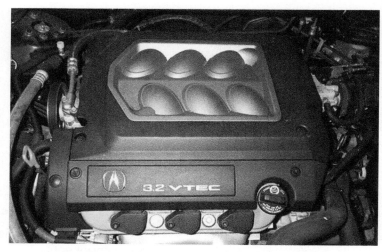

Figure 1-1 Today's smaller engines can produce some hefty horsepower.

A **powertrain** includes the engine and all the components that deliver that force to the road, including the transmission and axles.

The **combustion chamber** is a sealed area in the engine where the burning (combustion) of the air and fuel mixture takes place.

The **piston** is a round-shaped part that is driven up and down in the engine cylinder bore.

An engine **cylinder** is a round hole bored into the cylinder block. The pistons are housed in the cylinders.

The **powertrain control module (PCM)** is an onboard computer that controls functions related to the powertrain, such as fuel delivery, spark timing, temperature, and shift points.

Combustion is the controlled burn created by the spark igniting the hot, compressed air and fuel mixture in the combustion chamber.

requirements. **Powertrains** are now more than 95 percent cleaner in terms of toxic emissions than those of the 1960s, while the number of vehicles on the road has more than doubled.

This text will focus on the operation and service of these powerful and efficient new gasoline engines (**Figure 1-1**). You must thoroughly understand the operation of the engine and the functions of its components to become a skilled automotive diagnostic and repair technician. To provide good customer service and ensure proper engine performance, you must accurately diagnose faults and perform precise repairs of the engine and its supporting systems. This chapter will highlight the contents of the classroom and shop manuals. Each chapter in the classroom manual will explain the theory and operation of a system. The corresponding shop manual chapter will describe the diagnostic and repair procedures and strategies for those systems.

BASIC ENGINE OPERATION

The gas engine is often described as an air pump because it uses a tremendous amount of air. But to make power, it also uses a small percentage of gasoline. The engine pulls air into a sealed **combustion chamber** through the intake system. The combustion chamber is sealed on the bottom by the **piston** and its rings (**Figure 1-2**) and on the top by the combustion chamber formed in the cylinder head (**Figure 1-3**).

Air enters into the **cylinders** as the pistons move downward. This creates a difference in pressure between the cylinder and the atmosphere. The lower pressure area is in the cylinder. When the intake valve opens, the atmospheric pressurized (fresh) air rushes into the cylinder via the intake manifold. Fuel is injected into the airstream near (or sometimes in) the combustion chambers. The timing and quantity of fuel delivery are precisely controlled by the **powertrain control module (PCM)**.

The air and fuel mixture is then compressed to make it more combustible as the piston moves toward the top of its travel. At the optimum instant, the PCM initiates ignition that will deliver a spark to the spark plug. When the spark jumps across the gap of the spark plug, the air and fuel mixture is ignited and burns rapidly, and the engine harnesses some of the heat energy available in the fuel. This rapid, but controlled, burning is called **combustion**. The power of the expanding gases pushes down with a force comparable to an elephant standing on top of the piston. The piston is connected to the crankshaft by a

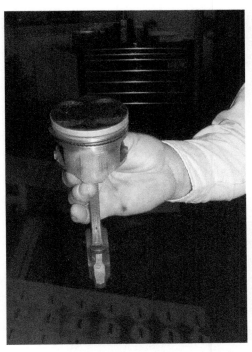

Figure 1-3 The bottom of this cylinder head (shown) shows the four combustion chambers sealed by the valves.

Figure 1-2 This piston and rings, with the connecting rod attached, fit closely in the engine cylinder to seal the bottom of the combustion chamber.

connecting rod (**Figure 1-4**). The crankshaft causes the engine to turn. Then the spent gases are allowed to exit the combustion chamber through the exhaust system. This process is repeated in each of the engine cylinders to keep the crankshaft spinning.

The engine valves allow airflow into and out of the combustion chamber. There are typically one or two intake valves per cylinder and one or two exhaust valves per cylinder (**Figure 1-5**). Refer to Figure 1-3 to see the heads of the valves within the combustion chamber. The valves are opened by the **camshaft** (**Figure 1-6**). The camshaft is a gear, belt, or chain driven by the crankshaft. The relationship between the camshaft and crankshaft rotation creates the proper **valve timing**. It is carefully timed to ensure that the valves open at the correct time. The camshaft has egg-shaped (or cam-shaped) lobes on it to force the opening of the valves. Valve springs close the valves.

The crankshaft and connecting rod work together to change the reciprocating and linear motion of the piston into the rotary motion of the crankshaft.

The **camshaft** is a shaft with eccentrically shaped lobes on it to force the opening of the valves.

Valve timing refers to the relationship between the rotation of the crankshaft and camshaft. That correct relationship ensures that the valves open at the correct time.

Connecting rod

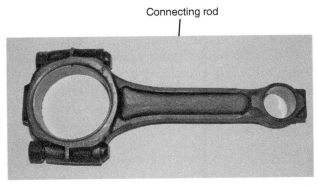

Figure 1-4 The connecting rod connects the piston to the crankshaft.

Figure 1-5 An engine valve.

Figure 1-6 The camshaft has large, round journals that fit into bearings in the block and small, eccentrically shaped lobes that open and close the valves.

Figure 1-7 This transversely mounted engine sits sideways in the engine compartment with the transaxle bolted on its end.

Engine Location

Most engines are installed in the front of the vehicle. They may be installed longitudinally, with the front of the engine toward the bumper and the rear toward the passenger compartment, or transversely, with the engine placed in the engine compartment sideways. Trucks and rear-wheel drive (RWD) vehicles most commonly use a longitudinally mounted engine. Many passenger vehicles and most front-wheel drive (FWD) vehicles use a transversely mounted engine (**Figure 1-7**). This design allows a shorter hood line, desired in today's vehicles to provide greater passenger space. A few vehicles use a mid-engine design, where the engine is placed between the rear axle and the driver. Often, this well-balanced configuration is associated with modern sports cars.

Engines are held in place by motor mounts. These are attaching pieces with a flexible rubber portion that isolates much of the engine vibration from the automobile frame and passenger compartment. The mounts connect the engine to the vehicle chassis (**Figure 1-8**).

Engine Function

The engine is attached to the transmission through the flywheel and clutch on a manual transmission (**Figure 1-9**) or through the flexplate and torque converter on an automatic transmission. The engine delivers rotational speed and a twisting force

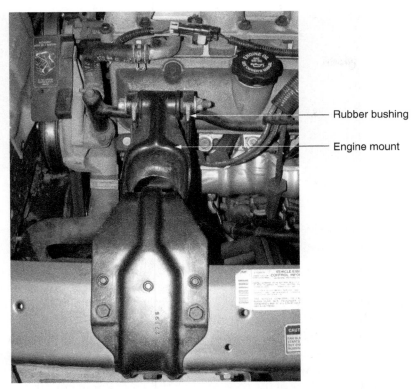

— Rubber bushing

— Engine mount

Figure 1-8 The engine mount has a rubber bushing to isolate the vibration of the engine from the chassis.

Figure 1-9 The flywheel bolts to the crankshaft and is attached to the transmission through the clutch on a manual transmission.

to spin the transmission components. The twisting force is called torque, and the speed that develops is called horsepower. The transmission uses this torque and horsepower and manipulates them to provide the appropriate level of torque and speed to drive the wheels and propel the vehicle. The transmission multiplies the torque of the engine through gear sets in the transmission and differential to allow for good acceleration. In higher gear(s), torque is reduced and speed is increased to improve fuel economy.

Figure 1-10 A thermostat.

Figure 1-11 The head gasket seals between the block and the cylinder head.

COOLING SYSTEM

Combustion develops a tremendous amount of heat in the engine. The expansion of gases from the heat pushes the piston down, but some of the heat is absorbed in the engine components. A liquid cooling system is used on all current gas automobile engines to remove excess heat from the engine. An engine belt or gear drives a water pump that circulates coolant throughout the engine to keep all areas at an acceptable temperature, around 200°F (110°C). When the coolant reaches a certain temperature, a **thermostat** (**Figure 1-10**) opens to allow coolant to flow through the radiator to cool it down. Air flowing across the radiator lowers the temperature of the coolant; then the water pump pulls the coolant back into the engine and recirculates it. The cooling system must retain heat to speed the warm-up period to enhance combustion and lower emissions after cold starts. It must also maintain an even engine operating temperature for better engine efficiency. To prevent the meltdown of components, the cooling system must remove excess heat.

Cooling system problems can cause very serious damage to an engine. Heat expansion of parts can cause the engine to seize. When an engine overheats, it is likely to blow the seal between the cylinder head and the cylinder block. That seal is the head gasket, and replacement can cost several hundred dollars or more and many hours of work (**Figure 1-11**).

> A **thermostat** is a mechanical component that blocks or allows coolant flow to the radiator.

LUBRICATION SYSTEM

The lubrication system is critical to reduce the tremendous friction created between the moving components in the engine. This also reduces engine heat and wear. The oil pump is mounted on the bottom of the engine or on the crankshaft to circulate engine oil under pressure to all friction areas in the engine (**Figure 1-12**). The oil is then filtered to remove particles of metal or debris that could damage finely machined engine components. The oil travels through oil passages drilled throughout the engine to reach the crankshaft, the camshaft, the cylinder walls, and other key friction areas. Just a few minutes of engine operation without adequate lubrication can turn an engine into irreparable scrap metal. Proper lubrication system maintenance is essential to the longevity of any engine.

ENGINE BREATHING

> The **intake manifold** connects the air inlet tubes to the cylinder head ports to equally distribute air to each cylinder.

Engine power is dependent on how well the engine can breathe. The intake and exhaust systems provide the breathing tubes for the engine. The **intake manifold** connects the air intake ductwork to the cylinder head (**Figure 1-13**). The intake manifold distributes an equal amount of air to each cylinder. Ports into the combustion chamber allow fresh

Figure 1-12 This oil pump is driven off of the front of the crankshaft.

Figure 1-13 The intake manifold distributes air equally to the cylinders. Note that this is for an eight-cylinder engine.

air to flow past the intake valves when they are open. The **exhaust manifold** connects the exhaust ports on the cylinder head to the rest of the exhaust system. To a major degree, the more air an engine can take in, the more power it can put out. The intake and exhaust systems must flow freely to allow good airflow. Turbocharging and supercharging are systems that increase airflow by forcing air under pressure into the intake. These systems dramatically increase engine power (**Figure 1-14**).

Problems with the engine's ability to breathe significantly affect engine performance. Something as simple as a restricted air filter can cause the engine to barely run or even to not start. You will learn simple maintenance procedures for the intake and exhaust systems, as well as more sophisticated tests to diagnose tricky problems.

> The **exhaust manifold** connects to the cylinder head and acts as a chamber to let exhaust gases exit the engine in an efficient and safe manner.

ENGINE PERFORMANCE

An engine has many required conditions to function as designed and deliver good performance. An engine needs to have the combustion chamber well sealed to develop compression of the air and fuel mixture. The engine and intake system must be large enough to

Figure 1-14 This supercharger creates powerful boost and acceleration.

accommodate and sealed tightly enough to maintain the low pressure needed for the exchange of exhaust gases and fresh air in each cylinder for the engine's given displacement. The lubrication and cooling systems must be operating properly. Each engine component must be within its operating tolerance to allow the engine as a whole to hold together and run well (**Figure 1-15**). Problems in how the engine performs will keep you busy as an engine technician.

To recommend repairs, you will need to learn how to evaluate the engine's mechanical condition. Vacuum tests and compression tests as well as listening and smelling will be some of the weapons in your arsenal to correctly diagnose the cause of engine problems (**Figure 1-16**). You must test and cross-check before you recommend major engine repairs, trying to gather as much information as possible. You'll need a detailed diagnosis to be able to offer the customer an accurate estimate and to

Figure 1-15 This severely worn lifter would cause poor engine performance.

Figure 1-16 You will learn many different ways to analyze the engine.

be sure you correct the problem on the first try. Engine repairs can become very costly; therefore, it is crucial that the diagnosis and repairs are performed with care and attention to detail.

CYLINDER HEAD

The **cylinder head** houses the valves and often the camshaft. In this case it is called an **overhead camshaft (OHC) engine** (**Figure 1-17**). In cases where the camshaft is in the block, the cylinder head still holds many of the **valvetrain** components. The pushrods, lifters, rocker arms, and springs are all parts of the valvetrain because they work to open and close the valves. When the camshaft is in the block, as the high point of the camshaft lobe comes up, it pushes on a lifter, which acts like a plunger, to move the **pushrod**. The pushrod is a simple, hollow rod that transfers the motion from the lifter to the rocker arm. The rocker arm rides on a pivot. When the pushrod raises one end of the rocker, the other end pushes down on the tip of the valve to open it, much like a typical first-class lever (**Figure 1-18**). In later chapters you will discover there are many different combinations that the camshaft can be used in.

Problems within the cylinder head and valvetrain can be quite serious. A valve that is not adjusted properly, for example, can cause that cylinder to drop in performance, become noisy, and cause the engine to run rough. During an engine overhaul, it is typical to perform a valve job. This involves refinishing all the valves and their seats and replacing the valve seals. This requires complete disassembly of the cylinder head and some precision machining and repair (**Figure 1-19**).

A **cylinder head** is the large object that houses most of the valve train and covers the combustion chamber.

If a cylinder head contains the camshaft, it is called an **overhead camshaft (OHC) engine**. The camshaft can also be located in the cylinder block.

The **valvetrain** is made up of the components that open and close the valves.

A **pushrod** is a long tube that connects that camshaft to the valve train components in the cylinder head. This is typically not used on an OHC engine.

Rocker arms

Figure 1-17 This OHC engine uses rocker arms to open the valves.

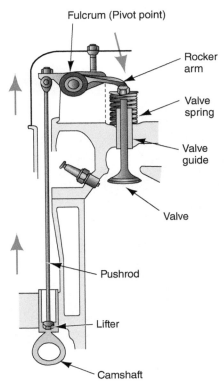

Fulcrum (Pivot point)

Rocker arm

Valve spring

Valve guide

Valve

Pushrod

Lifter

Camshaft

Figure 1-18 The valvetrain components open the valve in response to the camshaft lobe.

Figure 1-19 You will learn to use specialized equipment used in engine overhauling, such as this valve grinder.

TIMING MECHANISM

As discussed earlier, proper valve timing is necessary to keep the valves opening and closing at the correct times. This is achieved by the timing mechanism. A sprocket or gear on the crankshaft is attached by a gear, belt, or chain to a sprocket or gear on the camshaft (**Figure 1-20**). Timing belts require periodic maintenance. When that is neglected, the timing belt can snap, and serious engine damage can result. In some cases, the valves contact the top of the pistons and bend, and a new set of valves is required (**Figure 1-21**).

Figure 1-20 The lower crankshaft gear turns at twice the speed of the larger camshaft gear.

Figure 1-21 Valve timing failures can be catastrophic.

In the worst cases, those valves damage the piston tops, and the whole engine needs to be overhauled. Timing chains also wear out over time, and replacing them requires skill and attention to detail. Variable valve timing systems increase performance and minimize emissions. More and more vehicles employ these systems to vary the times when the valves can be opened. This makes timing mechanism theory and service a new and more complex challenge.

ENGINE BLOCK

The **engine block** is the main supporting structure of the engine (**Figure 1-22**). It holds the pistons, connecting rods, crankshaft, and sometimes the camshaft. The cylinder head bolts to the top of the block. The block is bored (drilled) to create cylinders. An eight-cylinder engine has eight-cylinder bores (**Figure 1-23**). The pistons are pushed up and down in the cylinders by the crankshaft. On the power stroke, the piston is actually pushed down by the force of combustion to turn the crankshaft. Rings on the pistons seal the small clearance between the cylinder walls. This prevents the hot, expanding gases from combustion within the combustion chamber, which lets all the force act to push the piston down. The rings also scrape excess oil from the cylinder walls so the oil does not burn inside the combustion chamber. When oil is allowed to burn during combustion, it forms blue smoke that exits the tailpipe. The block also holds the crankshaft in its main bore. The soft main bearings are placed within the main bore to provide a place for the finely machined crankshaft **journals** to ride. The journals are the machined round areas of the crankshaft that allow the bearings and bearing caps to bolt around them and hold the crankshaft in the engine block. The clearance between the journals and the bearings must be just right: enough to allow adequate oil, but not so much that precious oil pressure leaks through excessive clearances. Crankshaft bearings of the friction type are softer than the metal they support. Bearings protect the crankshaft from damage, but they also wear. When they wear excessively, they cause low oil pressure and engine knocking.

> The **engine block** is the main structure of the engine that forms the combustion chamber and houses the pistons, crankshaft, and connecting rods.

> Main **journals** are round machined portions on the centerline of the crankshaft where the crankshaft is held in the main bore.

Figure 1-22 An eight-cylinder engine block.

Figure 1-23 The cylinder bores are finely finished to provide a good sealing surface for the rings.

Figure 1-24 This carriage is powered by one of Daimler's first engines.

✒️📖 A BIT OF HISTORY

Jean-Joseph Lenoir created the first workable internal combustion engine in 1860. Nicholas Otto was credited with creating the first four-stroke internal combustion engine back in 1876.

All previous internal combustion engines did not compress the air-fuel mixture. They attempted to draw the air-fuel mixture in during a downward movement of the piston and then ignite the mixture. The expansion of gases would force the piston down the rest of its travel. This design was used to make the piston forceful on each downward stroke. The four-stroke engine, however, proved much more efficient and powerful. Gottlieb Daimler received the first patent for the internal combustion engine in 1885. The basic operation of today's engines is still similar to those first four-stroke engines. **Figure 1-24** shows an early working model from Daimler's workshop, which is now a museum.

An engine overhaul involves rebuilding the engine to nearly new condition. Each step of engine repair requires careful inspection, measurement, and judgment. Together the *Today's Technician Engine Repair & Rebuilding Classroom* and *Shop Manuals* provide the information needed to learn about component theory, operation, diagnosis, and the repair procedures of modern engines. The text covers information on principles, safety, the engine repair industry, ancillary systems, diagnosis, measuring and testing, removal, service, installation, along with alternative fuel vehicle operation and service.

SUMMARY

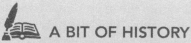

- Today's smaller engines make more power with longer life, fewer emissions, and better fuel economy than their older counterparts.
- The powertrain control module controls the engine's operation.
- The engine ignites a compressed air and fuel mixture in a confined chamber to harness the heat energy of expanding gases.

- The power of combustion pushes down on the piston. The piston is connected to the crankshaft by the connecting rod.
- The intake and exhaust systems allow the engine to take in and expel the great quantity of air that the engine needs to make power.
- The cooling system prevents the meltdown of engine components.

- The lubrication system reduces temperature and friction in the engine.
- The cylinder head houses the valvetrain that opens and closes the engine valves.
- The camshaft controls the operation and timing of the valves. It can be used in different configurations.
- The cylinder block holds the pistons, connecting rods, and crankshaft and provides the main structure of the engine.
- All of the engine systems and components must work together as designed to create a smooth-running engine that has sufficient power and produces low emissions.

REVIEW QUESTIONS

Short-Answer Essays

1. Explain the benefits of today's modern engines.

2. Explain what the combustion chamber is.

3. How does fresh air enter the cylinders?

4. What controls fuel injection and spark timing?

5. What is the function of the intake manifold?

6. Where can an oil pump be mounted?

7. How does a turbocharger create more power?

8. What is valve timing?

9. Name three components located in the engine block.

10. What forces the crankshaft to turn?

Fill-in-the-Blanks

1. The _____ of modern engines have decreased more than 90 percent from earlier models.

2. A difference in air pressure produces the movement of fresh air into the _____ through the intake system.

3. The PCM controls _____, _____, and _____, _____ among other functions.

4. The _____ _____ distributes air equally to each cylinder.

5. The oil pump may be driven off the front of the _____.

6. When the thermostat opens, coolant flows through the _____.

7. The _____ _____ connects the cylinder head to the exhaust system.

8. The _____ and _____ are compressed in the combustion chamber.

9. The _____ both opens and closes the valves.

10. The _____ _____ holds the crankshaft in its main bore.

Multiple Choice

1. Today's powertrains require:
   A. more maintenance.   C. more fuel.
   B. bigger engine sizes. D. less fuel.

2. The engine is often described as a(n):
   A. water pump.   C. impeller.
   B. air pump.   D. gear driver.

3. *Technician A* says that the more air an engine can take in, the more power it can put out.
   Technician B says that the exhaust can affect how well an engine can breathe.
   Who is correct?
   A. A only   C. Both A and B
   B. B only   D. Neither A nor B

4. *Technician A* says that the piston is connected to the crankshaft by the rocker arm.
   Technician B says that the connecting rod converts the reciprocating motion of the piston to the rotary motion of the crankshaft.
   Who is correct?
   A. A only   C. Both A and B
   B. B only   D. Neither A nor B

5. The cylinder head forms the top of the:
   A. cylinder.
   B. combustion chamber.
   C. intake manifold.
   D. throttle bore.

6. When a spark ignites the air and fuel mixture, _____ occurs.

   A. an explosion
   B. a blast
   C. conflagration
   D. combustion

7. The _____ is the main supporting structure of the engine.

   A. motor
   B. block
   C. cylinder head
   D. main bore

8. *Technician A* says that the main bearings protect the crankshaft journals.
   Technician B says that the camshaft lobes lift the pistons.
   Who is correct?

   A. A only
   B. B only
   C. Both A and B
   D. Neither A nor B

9. Which of the following is true?

   A. The crankshaft rotates at the same speed as the camshaft.
   B. The camshaft rotates twice as fast as the crankshaft.
   C. The camshaft rotates three times as fast as the crankshaft.
   D. The crankshaft rotates twice as fast as the camshaft.

10. Valve timing is:

    A. when the spark occurs.
    B. when the pistons rotate.
    C. when the valves open and close.
    D. how long combustion takes.

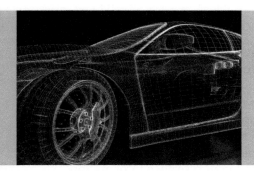

CHAPTER 2
ENGINE REPAIR AND REBUILDING INDUSTRY

Upon completion and review of this chapter, you should understand and be able to describe:

- What types of repair are performed at a full-service repair facility.
- The reasons for the existence of specialty shops and repair facilities.
- What changes are occurring in the engine repair and rebuilding industry.
- The skills needed for employment in various settings.
- How specialized repair facilities operate.
- The basic engine rebuilding process at a remanufacturing facility.

Terms To Know

Computer numerically controlled (CNC)
Core engine

INTRODUCTION

The engine repair and rebuilding industry has changed significantly in the past decade. This chapter was designed to give you a brief introduction to the current industry and the changes that have been made to machining equipment, repair facilities, and the daily operations.

In the past, engines and vehicles were not built to the high standards they are now. If an engine and vehicle drove past 100,000 miles without a major repair, you were considered lucky and running on borrowed time. Today, consumers expect their engines and vehicles to give them reliable service for many more miles and years. The engines in the past were not built to today's high standards and repairs such as valve jobs and rebuilds were common. Today, things have changed significantly. There are very few shops that will rebuild an engine in-house.

When you are employed in the field, one of your jobs may be to help the customers choose what option is best for them. The information in this book, your experiences, and the shop you work in will all help you inform the customer as to what action should be taken.

If repairing and rebuilding engines is what you have chosen to do as your career, you will have to know what options you will have and what types of operations the particular places of employment may do.

FULL-SERVICE REPAIR FACILITIES

There are a lot of full-service repair shops across the nation. The full-service repair shops include independently owned, franchised, and new and used dealership facilities. The majority of these shops are willing to work on anything that comes in the door. Full-service repair shops will typically perform oil changes and other small repairs at a lower advertised cost to drive customers to their door. This technique is common and allows the customer to become aware of the amount and type of work that is performed there. It also allows the shop to inspect the vehicle and make the customer aware of any problems with the vehicle. Once the customer's trust is earned, there is a good chance that they will come back for a different repair and become long-term customers.

The average full-service repair facility will have several bays and hoists to work with, along with computer terminals to write repair orders and look up service information. Rebuilding an engine in this type of repair facility is not nearly as common as it once was. Many independent shops used to own their own machine equipment, such as valve grinding machines. Since today's engines need rebuilding less often, those same machines collect more dust and the shop owner or manager has to consider whether it is worth still keeping those machines around (and paying to maintain them) or to sell them (**Figure 2-1**). The space that a machine takes up on a floor must be usable if a shop is to remain profitable and stay open. New shops often do not purchase engine machining equipment unless they plan on specializing in engine work. If a shop does specialize in engine work, they often have a relationship with many neighboring repair facilities that send engine work to them. When an extensive engine repair is determined to be the course of action, many shops will subcontract the job to a specialist.

Today, most independent shops perform engine repairs instead of rebuilds. These repairs may include, but are not limited to, head gaskets, engine seals, timing belts and chains, and oil leaks. When a major engine repair is needed, they may often choose to either install a crate engine or component, or send the engine parts out to an engine machine shop to be machined and then put the engine back together at the independent shop. One example is if an engine were to overheat, blow a cylinder head gasket, and crack the cylinder head. If this were to happen, the full-service repair shop will estimate the costs of replacing the head versus sending it out to a machine shop for machining and head reassembly. The shop in turn will charge the customer for the sublet service. Depending on the vehicle and engine, it may be more cost-effective to purchase crate components, especially if it is a popular component where there is a lot of supply and the cost is low.

Most shops now prefer to install remanufactured components and engines because the company they purchase from is mostly responsible for the warranty of their product. But their preferred installation choice may be limited by the customer choosing a lower cost alternative.

Figure 2-1 This valve machine has not been used for the last few years.

Engine repair work at a full-service shop does not usually require a technician to obtain training beyond what this book provides, but some shops prefer to have the lead technician or a specialized technician perform the work. Sometimes this technician is chosen because of the technician's interest, experience, background, or specialized training.

MACHINE SHOP AND ENGINE REBUILD FACILITIES

One-third of the full-service shops surveyed in the past few years indicate that they still own engine machining equipment, but most of them are now planning on selling them. Machine shops and engine rebuild facilities have been around for a long time. They exist as part of another facility, independent, or connected to a parts supply store. Many high-performance facilities will also perform repairs and rebuilding of normal engines. Most of these repair facilities have not changed their type of business much over the years but have found themselves focusing on the changes in technology and how it affects their business model.

Machine shops will take in either complete engine repair and rebuilding or component repairs and rebuilding. Most machine shops survive and exist because they have a relationship with the local surrounding independent shops, franchises, and dealers. These supporters will send work out to the machine shop. Often the vehicle will not be housed at the machine shop, but there are a few exceptions.

Some of the most common repairs at a machine shop are cylinder head remanufacturing and repairs, cylinder block remanufacturing, repairs and resizing, piston pressing, and component assembly. The extent of these repairs will vary. Machine shops have a significant advantage over purchasing crate engines. Both shipping costs and local support are two things that machine shops offer to distinguish themselves from large nationwide rebuilders.

Often, machine shops will come across a repair for an older, antique, or high-performance engine. Each of these cases is handled separately, and the shop foreman or manager usually talks directly with the customer. Some machine shops have started to accept more high-performance work because of the instability of other work. Machine shops used to rely on cylinder heads and internal components failing at high enough rates to keep them in business. Modern engines have lower failure rates that keep them from needing extensive service frequently.

Machine shops usually have a very costly inventory of machines and equipment (**Figure 2-2**). These machines not only have a high initial purchase, but also must be periodically maintained, calibrated, cleaned, and updated. **Computer numerically controlled (CNC)** equipment is sometimes used in a machine shop (**Figure 2-3**). This type of equipment requires special training, as do many of the machines and equipment

A **computer numerically controlled (CNC)** machine uses special computer software to machine close tolerances.

Figure 2-2 This milling machine can be used to resurface cylinder heads, block decks, or flywheels.

Figure 2-3 A CNC machine is one of the many pieces of expensive equipment at a machine shop.

Figure 2-4 Machinists need to be able to understand and apply math.

in a machine shop. A CNC machine can cut and mill engine components to a smaller tolerance (more accuracy) than most human-operated and controlled machines. This is desirable because the smaller tolerances usually mean less comebacks and a higher-quality product. Machine shop technicians are required to have a good working knowledge of engine components, precision measuring tools, math, and the ability to locate resources quickly (**Figure 2-4**). There are a few specialized schools across the United States that specialize in teaching automotive machining.

ENGINE REPAIR AND REPLACEMENT SPECIALTY FACILITIES

Engine repair and replacement has almost become a separate industry in itself. In the past, the transmission repair and rebuilding industry had separated itself from the rest of the industry by having shops offer complete transmission work at one facility. Transmissions are very complicated and can require extensive testing and highly skilled technicians to work on them. The growing complexity of modern engines requires similarly skilled technicians and extensive testing.

Today's engines are designed with tighter clearances for more efficiency. Replacing or rebuilding the engine in a modern vehicle requires a higher level of knowledge and skill. Technicians must also be able to transfer that knowledge over to a different vehicle and a different setup. The facility must be well-equipped if a profit is to be made.

The amount and cost of the equipment and technician training are significant factors to consider when offering engine rebuilding and repair at a facility. It is for these reasons that specialized shops exist. These shops offer engine and related work. They can usually diagnose and repair the cause in-house, saving money for the customer and keeping the profit in one place. Sometimes, a component replacement is all that is needed.

Specialized shops rely on a good source of parts and equipment to operate. If cost is a strong customer consideration, then installing a good, used engine may be an option to consider with the customer.

Some of the specialized shops offer and advertise high-performance and restoration work. This may account for a significant amount of their business in some cases. Specialized shops may also perform custom machining, cylinder porting, and engine tuning.

ENGINE AND COMPONENT REMANUFACTURING FACILITIES

As mentioned earlier, today's engines are lasting longer, and repairs are less frequent. Because the bodies and other mechanical components do not rust and wear out as frequently either, many consumers are repairing their high-mileage engines that need replacement, because it costs less than purchasing another vehicle.

For many years, manufacturers have been placing the same engine in several models. For example, the Chevrolet 5.7 L (350 cubic inch displacement) engine was used in many light truck and car applications for decades. Many of the components are interchangeable from one to the next. Because of the volume and number of components out there, the cost of the parts and crate components (and engines) is significantly lower than many other engine families. This is because of the volume of parts and components produced for this engine. Many manufacturers beside General Motors have adapted this strategy. Many domestic vehicles are being built on a "global" platform.

An example of engine production in volume is the Global Engine Manufacturing Alliance LLC (GEMA) World Engine, which until 2012 had a joint venture with Chrysler. The World Engine had five factories worldwide and produced about 2 million engines per year. Approximately 20 different vehicles by Chrysler, Mitsubishi, Hyundai, and Fiat used the 1.8 L, 2.0 L, and 2.4 L World Engines, each slightly different for its own application. The World Engine was used in vehicles all over the globe.

With more manufacturers producing similar products, another segment of engine rebuilding and remanufacturing was created because of the volume. There are only a few engine and component remanufacturing facilities in the United States. Their products can be purchased directly from them, a parts store, or an authorized dealer/installer. These facilities are large and have the capability of "assembly line production style" engine remanufacturing because of their volume.

Remanufacturing starts out with a core engine. The **core engine** is an old, used, or discarded engine that was received from a variety of sources. These sources include salvage yards and a customer's old engine that was returned. A core charge is usually placed on a new engine when it is ordered. This is an extra charge that the customer (or repair facility) must pay until the old engine that was removed is sent back in the same shipping crate. The old engine is then inventoried and completely stripped down to the bare components.

The major components, such as the cylinder head, block, and crankshaft, are checked for cracks (**Figure 2-5**) and are machined, repaired, or discarded as needed. They are then

> A **core engine** is a used engine that is disassembled for purposes of rebuilding in large volumes.

Figure 2-5 Machine shops and engine remanufacturing plants usually have expensive equipment. This crankshaft is being inspected for cracks using a magnet.

inventoried and stored for later use. The major components are then thoroughly cleaned and ready for reassembly. These engine components are also often sold as individual remanufactured components.

During reassembly, new parts are installed, and everything is measured, tested, and torqued to specification. The engines are then started and tested to ensure that the customer receives a high-quality product. Some of the main advantages of purchasing an engine from a remanufacturer are the warranty, quality control, and lower cost.

SUMMARY

- Automotive and light truck engines last longer today than they did in the past. This has led to changes in the engine repair and rebuilding industry.
- Machine shops perform the majority of engine machining and repairs today.
- A CNC machine uses special computer software to machine close tolerances.

- Specialized shops sometimes offer high-performance engine work.
- Engine remanufacturing in large volumes helps lower the price.
- A core engine is a used engine that is disassembled for purposes of rebuilding in large volumes.

REVIEW QUESTIONS

Short-Answer Essays

1. Describe why an engine remanufacturer receives core engines.

2. Describe what a CNC machine is.

3. Explain why full-service repair shops perform more engine repairs and component replacements than engine rebuilding.

4. Mention where engine remanufacturers receive their core engines from.

5. Describe the relationship between a full-service repair facility and a machine shop.

6. Explain why engine repair and rebuilding specialty shops exist.

7. Describe the type of repairs an engine repair and rebuilding specialty shop may perform.

8. List where you can purchase a remanufactured engine.

9. Explain why a shop sometimes prefers to install a remanufactured engine or component instead of rebuilding.

10. List the advantages and disadvantages of a full-service repair facility owning engine machining equipment.

Fill-in-the-Blanks

1. A new car dealership is considered a _____ service repair facility.

2. Technicians who work at a machine shop have to use _____ measuring tools as part of their daily job.

3. When an installer purchases a remanufactured engine, a _____ _____ must be paid until the old engine is returned.

4. A CNC is a_____ _____ _____ machine.

5. Engine remanufacturing in large volumes helps lower the _____.

6. Rebuilding a modern engine requires a _____ technician due to their growing complexity.

7. A machine shop's inventory or machines and equipment cost is usually very _____.

8. High-performance and restoration engine work is usually performed at a _____ repair facility.

9. A crate remanufactured engine offers a _____ with its purchase.

10. Modern engines have _____ failure rates that keep them from needing extensive service frequently.

Multiple Choice

1. Where would you typically find an "assembly line" for engine rebuilding?
   A. Specialty shop
   B. Full-service shop
   C. New car dealership
   D. Remanufacturing facility

2. The World Engine can be found in what vehicles?
   A. Chrysler
   B. Ford
   C. Chevrolet
   D. Both B and C

3. Specialized engine rebuilding shops usually work on:
   A. high-performance engines.
   B. restoration engines.
   C. race engines.
   D. all of the above.

4. *Technician A* says that a remanufactured engine that comes from an engine remanufacture facility typically has a warranty.
   Technician B says that an engine remanufacture facility will charge you a core charge if you do not send the old engine back to them.
   Who is correct?
   A. A only
   B. B only
   C. Both A and B
   D. Neither A nor B

5. *Technician A* says that a machine shop's inventory is high compared to other automotive repair shops.
   Technician B says that there are very few machines at new car dealerships.
   Who is correct?
   A. A only
   B. B only
   C. Both A and B
   D. Neither A nor B

6. *Technician A* says that a CNC machine can cut and mill engine components.
   Technician B says that a CNC machine is computer controlled.
   Who is correct?
   A. A only
   B. B only
   C. Both A and B
   D. Neither A nor B

7. *Technician A* says that a CNC machine can mill an engine part with more accuracy than a regular machine.
   Technician B says that using a CNC will usually result in more customer comebacks.
   Who is correct?
   A. A only
   B. B only
   C. Both A and B
   D. Neither A nor B

8. *Technician A* says that a remanufacturing facility may receive its core engines from salvage yards.
   Technician B says that a remanufacturing facility may receive its core engines from customers who purchased engines from them.
   Who is correct?
   A. A only
   B. B only
   C. Both A and B
   D. Neither A nor B

9. The failure rate on a modern engine is
   A. lower than an older engine.
   B. higher than an older engine.
   C. similar to that of an older engine.
   D. none of the above.

10. *Technician A* says that an engine-rebuilding technician at a machine shop must be able to make precision measurements.
    Technician B says that not much math is involved when you are working as an engine rebuilding technician at a machine shop.
    Who is correct?
    A. A only
    B. B only
    C. Both A and B
    D. Neither A nor B

CHAPTER 3
THEORY OF ENGINE OPERATION

Upon completion and review of this chapter, you should understand and be able to describe:

- Major engine components.
- Basic engine operation.
- Basic laws of physics involved with engine operation.
- Engine classifications according to the number of cycles, the number of cylinders, cylinder arrangement, and valvetrain type.
- The four-stroke cycle theory.
- The different cylinder arrangements and the advantages of each.
- The different valvetrains used in modern engines.
- Engine measurement terms such as bore and stroke, displacement,

compression ratio, engine efficiency, horsepower and torque, horsepower losses, mechanical efficiency, and thermal efficiency.

- The relationship between compression ratio and engine power output.
- Mechanical, volumetric, and thermal efficiencies, and factors that affect each.
- The basic operation of alternative engine designs, including two-stroke, diesel, and stratified charge.
- The internal components of a diesel engine and how they differ from those of a gas engine.

Terms To Know

Autoignition temperature
Bore
Bottom dead center (BDC)
Boyle's law
Brake horsepower
Coil
Compression-ignition (CI) engines
Compression ratio
Connecting rods
Crankshaft
Cycle
Detonation
Displacement
Dual overhead cam (DOHC)
Engine
Efficiency

Engine bearings
Friction
Fuel injection
Glow plugs
Gross horsepower
Horsepower
Hybrid electric vehicle (HEV)
Indicated horsepower
Internal combustion engine
Kinetic energy
Law of conservation of energy
Mechanical efficiency
Net horsepower
Overhead cam (OHC)
Overhead valve (OHV)
Piston rings

Potential energy
Preignition
Reciprocating
Reed valve
Rotary valve
Spark-ignition (SI) engine
Stroke
Thermal efficiency
Thermodynamics
Top dead center (TDC)
Torque
Transmission
Transverse-mounted engine
Vacuum
Valve overlap
Volumetric efficiency (VE)
Wrist pin

INTRODUCTION

Today's engines are designed to meet the demands of the automobile-buying public and many government-mandated emissions and fuel economy regulations. High performance, fuel economy, reduced emissions, low noise level, smooth operation, and reliability are demanded from consumers. To fulfill these consumer needs, manufacturers are producing engines using lightweight blocks and cylinder heads, and nontraditional materials such as powdered metals and composites. In the past, engines were made from heavier and lower-cost raw materials such as iron and steel.

While many of the engine's mechanical components are similar to those of over 100 years ago, refinements in design, materials, and machining help the manufacturers improve engine performance and reliability. Many engine and support system functions are now closely controlled by the powertrain control module (PCM). Fuel delivery, spark timing, and often valve timing are managed so precisely that the PCM can make adjustments within milliseconds; a millisecond is one-thousandth of a second. Today's technician is called upon to diagnose and service these advanced engines properly.

The **internal combustion engine** used in automotive applications uses several laws of physics and chemistry to operate. Although engine sizes, designs, and construction vary greatly, they all operate on the same basic principles. This chapter discusses these basic principles and engine designs.

MAJOR ENGINE COMPONENTS

The engine's cylinder block is the structure that the rest of the engine is built upon (**Figure 3-1**). The block has precisely machined holes, known as cylinder bores, in which the pistons are moved up and down.

The number of cylinder bores describes the engine: a three-, four-, five-, six-, eight-, ten-, or twelve-cylinder engine. These bores are typically between 2.5 and 4.25 inches (6.35 and 10.795 cm) on common production vehicles and light duty trucks. Lengthwise across the bottom of the block is another bore that supports the **crankshaft**. The downward linear movement of the pistons rotates the crankshaft. The pistons are connected to the crankshaft by the **connecting rods**.

> **Internal combustion engines** burn their fuels within the engine.

> A **crankshaft** is a shaft held in the engine block that converts the linear and reciprocating motion of the pistons into a nonreversing rotary motion used to turn the transmission. The crankshaft rotates in only one direction. It is not reversible.

> The **connecting rod** forms a link between the piston and the crankshaft.

Figure 3-1 This engine block is an eight cylinder. It is a V-style engine with four cylinders on each bank of the V. This photo shows four cylinders on one bank of the V.

Figure 3-2 This engine cutaway shows the major components of the engine.

The crankshaft connects to the transmission and provides the rotary motion to turn the wheels. In many cases the transmission rotates in the same direction as the engine, except when it is in the reverse gear. On top of the block covering the cylinders is the cylinder head. It houses valves that open and close to allow fresh air and fuel into the cylinders and burnt exhaust gases out. Look at **Figure 3-2** to identify these components in the engine.

The cylinder block also has oil and coolant passages drilled throughout it to circulate these vital fluids to key areas in the block and the cylinder head. The water and oil pumps are mounted directly to the block in several different ways. Sometimes these components are internal and behind a cover. In other cases they are external and visible without removing any components. Both the water and oil pumps are moving objects. They are usually driven by the movement of the crankshaft via a belt, gear, or chair, but they are sometimes electrically driven.

The crankshaft is a long iron or steel fabrication with round bearing surfaces machined onto it called journals (**Figure 3-3**). The crankshaft must be made to be very strong. The

Figure 3-3 A crankshaft for an eight-cylinder engine.

main journals form the centerline of the crankshaft. Main **engine bearings** fit into the main bores of the engine block, and the crankshaft can spin freely in this mounting. There is just enough clearance (or space) between the bearings and the journals to allow a thin film of pressurized oil to keep the journals from actually riding on the bearings. Half of the main bore is integral to the block. Main caps form the other half of the bore; these can be removed to service the crankshaft, bearings, pistons, and other components.

The crankshaft also has journals that are offset from the centerline. These are called rod journals because the connecting rods, fitted with rod bearings, bolt around these journals to attach to the crankshaft. The rod bearings are very similar to the main bearings. When the piston pushes the connecting rod down on the offset journal of the crankshaft, it forces the crankshaft to rotate. The continued motion of the crankshaft in the same direction is caused by the piston moving in the opposite direction (in the cylinder bore). The crankshaft also has counterweights or crankshaft throws that are offset around the shaft. The weight of these throws helps offset the crankshaft speed fluctuations and vibrations as the different cylinders produce power.

The pistons are slightly smaller than the cylinders, to allow them to move up and down with minimal friction. They are fitted with **piston rings** that seal the area between the piston and the cylinder wall. The piston connects to the small end of the connecting rod through a **wrist pin** (sometimes called a piston or gudgeon pin) and a bushing. The force of combustion (which is the explosion of fuel and air) on the top of the piston forces it to move with power. It is transferred to the crankshaft through the connecting rod. As the piston moves up and down in the cylinder, it rotates the crankshaft. This process converts the reciprocating motion of the piston into rotary motion (**Figure 3-4**). This rotary motion is what is needed to turn the tires and wheels after the energy is transferred through the **transmission** and differential. A flywheel or flexplate at the rear of

Engine bearings are two bearing halves that form a circle to fit around the crankshaft journals.

Piston rings are hard cut rings that fit around the piston to form a seal between the piston and the cylinder wall.

A **wrist pin** is a hardened steel pin that connects the piston to the connecting rod and allows the rod to rock back and forth as it travels with the crankshaft.

The **transmission** is a device that uses the rotary motion and power of the crankshaft to turn the differential and driveshafts. The transmission multiplies the power output and changes the speed of the driveshaft using different gear ratios.

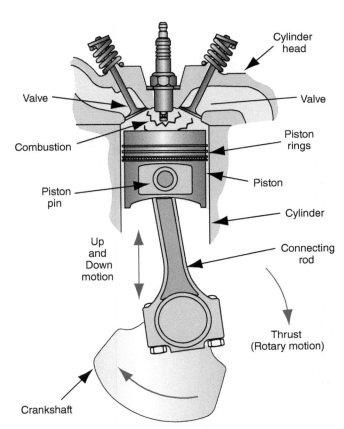

Figure 3-4 The engine components at work.

the crankshaft, along with a balancer pulley at the front, provides a large, stable mass for smoothing out the rotation and dampening the pulsations of the power stroke. Automatic transmissions use flexplates, and manual transmissions use flywheels.

The cylinder head forms a tight cover for the top of the cylinders and contains machined chambers into which the air-fuel mixture is forced and ignited (**Figure 3-5**). The void area of the cylinder head located above the cylinder bore is called the combustion chamber. It is sealed on the top by the cylinder head and valves and on the bottom by the pistons and rings. The cylinder head also has threaded holes for the spark plugs that screw right into the head so their tips protrude into the combustion chambers. Some cylinder heads have threaded holes for fuel injectors (such as diesel and gasoline direct injection engines). The valves in each cylinder are opened and closed by the action of the camshaft and related valvetrain. The camshaft is connected to the crankshaft by a gear, chain, or belt and is reduced to drive at one-half of the crankshaft's speed. In other words, for every complete revolution the crankshaft makes, the camshaft makes one-half of a revolution. This makes the speed of the crankshaft exactly half of the crankshaft. The camshaft may be mounted in the engine block itself or in the cylinder head, depending on design. Some engines use two camshafts in the head; these are called dual overhead cam engines.

The camshaft has lobes that are used to open the valves (**Figure 3-6**). A camshaft mounted in the block operates the valves remotely through pushrods and rocker arms; others act on followers placed directly on top of the valves. In some overhead camshaft (OHC) engines, the camshaft lobes contact the rocker arms that may be mounted on a rocker arm shaft.

Lubricating oil for the engine is normally stored in the oil pan (sometimes called a sump) mounted to the bottom of the engine. The oil is force-fed under pressure to almost all parts of the engine by the oil pump.

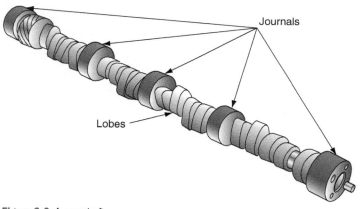

Figure 3-6 A camshaft.

Figure 3-5 The cylinder head and valves seal the top of the combustion chamber.

ENGINE OPERATING PRINCIPLES

One of the many laws of physics used within the automotive engine is **thermodynamics**. The driving force of the engine is the expansion of gases. Gasoline (a liquid fuel) will change its state to a gas if it is heated or burned. Gasoline must be mixed with oxygen before it can burn. In addition, the air-fuel mixture must be burned in a confined area to produce power. Gasoline that is burned in an open container produces very little power, but if the same amount of fuel is burned in a closed container, it will expand producing usable force. When the air-fuel mixture changes states, it also expands as the gas molecules collide with each other and bounce apart. Increasing the temperature of the gasoline molecules increases their speed of travel, causing more collisions and expansion.

Heat is generated by compressing the air-fuel mixture within the combustion chamber. Igniting the compressed mixture causes the heat, pressure, and expansion to multiply. This process releases the energy of the gasoline so it can produce work. The igniting of the mixture is more of a controlled burn, rather than an explosion. The controlled combustion releases the fuel energy at a controlled rate in the form of heat energy. The heat, and consequential expansion of molecules, dramatically increases the pressure inside the combustion chamber. Typically, the pressure works on top of a piston that is connected to the connecting rod and crankshaft. The expanding gases push the piston down with tremendous force and speed. As the piston is forced down, it causes the crankshaft to rotate. This rotation is a twisting force.

This twisting force on the crankshaft is called **torque**. Torque is applied to the drive wheels through the transmission and differential. As the engine drives the wheels to move the vehicle, a certain amount of work is done. The rate of work being performed in a certain amount of time is measured in **horsepower**.

Energy and Work

In engineering terms, in order to have work, there must be motion. Using this definition, work can be measured by combining distance and weight and is expressed as foot-pound (ft.-lb.) or newton meter (Nm). These terms describe how much weight can be moved a certain distance. A foot-pound is the amount of energy required to lift and move 1-pound of weight 1 foot in distance. The amount of work required to move a 500-pound weight 5 feet is 2,500 foot-pounds (3,390 Nm). In the metric system, the unit used to measure force is called a newton meter (Nm). A Newton is a unit measure of force, similar to a pound. Torque is measured as the amount of force in Newtons multiplied by the distance that the force acts in meters. One foot-pound is equal to 1.355 Nm. It is important to understand both systems because manufacturers will use both systems.

Energy produces work that can be measured in units, such as torque. Basically, energy is anything that is capable of resulting in motion. Common forms of energy include electrical, chemical, heat, radiant, mechanical, and thermal.

The **law of conservation of energy** states that the total amount of energy in an isolated system remains constant. Energy cannot be created nor destroyed; however, it can be stored, controlled, and changed into other forms of energy. For example, the vehicle's battery stores chemical energy that is changed into electrical energy when a load is applied. An automotive **engine** converts thermal (heat) energy into mechanical energy or power to produce force and motion. During combustion, when the gasoline burns, thermal energy is released. The engine converts this thermal energy into mechanical energy. Mechanical energy provides movement. The heat, expansion, and pressure during combustion all supply thermal energy within the combustion chamber. This thermal energy is converted to mechanical energy as the piston is pushed downward in the cylinder with great force. Often the engine under the hood is referred to as a motor. In automotive

Thermodynamics is the study of the relationship and efficiency between heat energy and mechanical energy.

Torque is a rotating force around a pivot point.

Horsepower is a measure of the rate of work.

The **law of conservation of energy** is a principal law of physics that states that the total energy of an isolated system remains constant despite any internal changes.

An **engine** is defined as a device or machine that converts thermal energy into mechanical energy.

terms, a motor is referred to as a driving and moving device that is electrically or vacuum controlled.

Types of Energy

Potential energy is energy that is not being used at a given time, but which can be used.

There are two types of energy: potential and kinetic. **Potential energy** is available to be used but is not being used. An example of this in the automobile is the chemical energy of the battery when the engine and ignition are turned off. Most of the chemical energy is stored as potential energy. (A very small percentage of the available energy is actually being used to keep computer memories alive.) When the ignition key is turned to start and the starter begins to crank the engine over, kinetic energy is being used. **Kinetic energy** is energy that is being used or working. The battery is now providing significant chemical energy to the starter so it can do work.

Kinetic energy is the amount of energy that is currently being used or is currently working.

Energy Conversion

In our example of the battery starting the engine, several energy conversions occur. Energy cannot be destroyed or eliminated, but it can be converted from one state or form to another. The starter converts the chemical energy from the battery into electrical energy to engage the starter. The starter converts this electrical energy into mechanical energy to crank the engine over. At the same time, chemical energy from the battery is being converted to electrical energy to provide power to the fuel, ignition, and computer-controlled systems so they will have the needed fuel and spark to start and run the engine. When combustion is in a rapid enough sequence, the starter is disengaged and the engine is kept running without the starter's assistance. The speed of the engine when it is running and idling is faster than its speed when it is starting. Combustion uses the potential energy in gasoline and converts it into kinetic energy by burning it. The chemical energy of the fuel is converted to thermal energy by the heat produced during combustion. The thermal energy is converted to mechanical energy by the movement of the pistons and the crankshaft. The following energy conversions are all common automotive applications:

Chemical Energy Conversion to Thermal Energy. As fuel is burned in the combustion chambers, the chemical energy in the fuel is converted to thermal energy.

Thermal Energy Conversion to Mechanical Energy. As the heat, pressure, and expansion of gases develop during combustion, thermal energy increases. This thermal energy is converted to mechanical energy to drive the vehicle through energy exerted on top of the pistons, which turns the crankshaft to drive the transmission and drive wheels.

Chemical Energy Conversion to Electrical Energy. The battery stores chemical energy and converts it to electrical energy as electrical loads such as the starter, the powertrain control module, the fuel injectors, and the radio demand electricity.

A hybrid electric vehicle (HEV) uses an electric motor in combination with a gas, diesel, or alternate fuel engine to power the vehicle.

Electrical Energy Conversion to Mechanical Energy. The starter, fuel injectors, electric cooling fan, and windshield wiper motor are just a few of the many automotive applications of conversion of electrical energy into mechanical motion. A **hybrid electric vehicle (HEV)** uses electricity to power an electric motor, which then turns the transmission with or without the engine running.

Mechanical Energy Conversion to Electrical Energy. The generator provides electrical energy to recharge the battery by using the mechanical energy of rotation through a belt connected to the rotating crankshaft. HEVs use regenerative braking during deceleration to recharge the high-voltage battery pack.

Mechanical Energy Conversion to Thermal Energy. The braking system on a traditional internal combustion engine vehicle is a fine example of this form of energy conversion. The mechanical energy of the rotating wheels is changed into thermal energy through the **friction** created between the rotating brake discs or drums and the stationary friction surfaces of the brake pads or shoes. The thermal energy is dissipated to the atmosphere.

Newton's Laws of Motion

First Law. A body in motion tends to stay in motion. A body at rest tends to stay at rest. Forces applied to the body at rest can overcome this tendency and make it move. Examples of some of the opposition forces that act on a body in motion are gravity and friction. Certainly the crankshaft, transmission, and drive wheels will not propel the car forward from a rest position unless significant force is applied to the pistons to rotate the crankshaft and overcome the opposition forces.

Second Law. A body's acceleration is directly proportional to the force applied to it. The greater the force applied to the drive wheels, the greater the acceleration of the vehicle. The greater the friction on a braking system, the greater the deceleration of the vehicle.

Third Law. For every action, there is an equal and opposite reaction. A simple application of this law on the automobile involves the suspension system. When the tire travels over a bump, the wheel moves upward with a force that opposes the spring's tension. When the wheel passes the bump, the spring exerts its opposite reaction to force the wheel back down toward the road and return it to its original position.

Inertia

Inertia is the tendency of an object at rest (static inertia) to stay at rest or an object in motion (dynamic inertia) to stay in motion. The heavy flywheel attached to the end of the crankshaft on manual transmission vehicles uses the inertia of the rotating crankshaft to help it continue to spin smoothly between the combustion events in the cylinders. The engine would not run as smooth if it had a lighter weight flywheel.

Momentum

Momentum is when force overcomes static inertia and causes movement of an object. As the object moves, it gains momentum. Momentum is calculated by multiplying the object's weight times its speed. An opposing force will reduce momentum. The larger the mass of an object, the more momentum it will have. A large truck has more momentum than a small car.

Friction

In the engine, the piston rings generate friction against the cylinder walls. Friction opposes momentum and mechanical energy. Friction creates heat and reduces the mechanical energy of the piston as it is pushed downward to turn the crankshaft. This is an example of how part of an engine's efficiency is decreased. Friction may occur in solids, liquids, and gases. The airflow over a vehicle causes friction.

    The manufacturers carefully engineer their vehicles to be aerodynamic. This means that the resistance to motion from friction between the vehicle and the air is minimized. This is carefully calculated and measured as the coefficient of drag (Cd). Unfortunately, there are certain external conditions and factors that can cause a vehicle to lose up to half of its mechanical energy. Examples of some of these factors are aerodynamic drag, excessive rolling resistance, and the environment the vehicle drives in. Manufacturers spend a lot of time and focus on trying to reduce a vehicle's coefficient of drag so that the engine does not have to produce much power to keep it moving on the road. This will help a vehicle get better fuel mileage.

> **Friction** is resistance to motion when one object moves across another object. Friction generates heat and is an excellent method of controlling mechanical action.

BEHAVIOR OF LIQUIDS AND GASES

Molecular Energy

Electrons are in constant motion around the nucleus of atoms or molecules. Kinetic energy is always occurring. The kinetic energy of liquids, gases, and solids changes in response to temperature. When temperature decreases, so does kinetic energy. As temperature increases, so does kinetic energy. Solids have the slowest reaction to temperature, while gases have the quickest response to changes in temperature. We know this theory already through our discussion of combustion. As temperature increases within the combustion chamber, the gases rapidly expand and the increasing kinetic energy is used to create thermal energy that is converted into mechanical energy.

Temperature

The volume of liquids, gases, and solids increases with temperature. It is the increased volume of gases on top of the pistons that produces the force to push the piston downward. When filling engine fluids, such as tire air pressure and coolant, we must be aware of the effects of temperature on volume. It is easy to overfill a coolant reservoir if the coolant is at ambient temperature and you top the fluid off to the maximum level. As the fluid heats up, it may expand beyond the capacity of the reservoir and develop enough pressure to blow the cap's pressure relief valve. The same is true for checking air pressure in a tire. After driving for some time, the friction of the tires will raise the temperature of the tires and then raise the temperature of the air inside the tires. This results in raising the pressure inside the tires. Manufacturers suggest that you check and adjust tire pressure when it is cold or be aware of the difference that may occur when doing it at higher temperatures.

Pressure and Compressibility

Pressure is always applied equally to all surfaces of a closed container. If there is a leak in the container, the pressure will leak out and equalize with the pressure that is outside the container. Gases and liquids both flow, so they are classified as fluids. But there are distinct and significant differences between the characteristics of gases and liquids. Gases can be compressed; the molecules squeeze tighter together and form greater pressure. You can easily fill a tire with excess pressure. Further compression of the gases in a combustion chamber will produce more power. Liquids, on the other hand, are difficult to compress. A liquid fills a chamber, and you cannot add more liquid; you can only increase pressure. Brake and clutch hydraulic systems work with liquids. When you depress the clutch or brake pedals, you move the liquid, under pressure, to exert force on a piston. This force produces mechanical work from the brake caliper piston or the clutch hydraulic lever. If there is a leak in the system, air can enter. Air is a gas and is compressible. Air in a hydraulic system will cause a lack of pressure and force.

PRESSURE AND VACUUM

Air is a gas with weight and mass. It is layered high above the surface of the earth and exerts a pressure on the earth's surface. This pressure is called atmospheric pressure. A 1-square-inch column of air extending from the earth's surface at sea level to the outer edge of the earth's atmosphere weighs 14.7 pounds. We say that atmospheric pressure is 14.7 pounds per square inch at sea level. This pressure changes as we climb in altitude. If we are 14,000 feet above sea level in the mountains, that column of air no longer weighs 14.7 pounds because the distance to the edge of the earth's outer atmosphere is less;

atmospheric pressure therefore decreases. Also, as temperature decreases, the molecules become less dense and therefore weigh less; atmospheric pressure decreases. These changes in atmospheric pressure caused by altitude and temperature are important for the PCM to know about to deliver the correct amount of fuel for the actual amount of air in the cylinder.

When air pressure is greater than atmospheric pressure, it is a positive pressure. When air pressure is lower than atmospheric pressure, it is a negative pressure, or vacuum. The engine creates a **vacuum** when the intake valves are open and the pistons move from the top of the cylinder to the bottom As the pistons reaches BDC the difference in pressure between the low pressure area created above the piston in the cylinder and the outside air cause a rush of air until the pressure has equalized Vacuum is typically not measured in pounds per square inch. It is typically measured in inches of mercury vacuum (in. Hg) or bar. When working on a vehicle, you will use a vacuum/pressure gauge. At atmospheric pressure, a vacuum gauge will read 0 in. Hg. It will also read 1 bar, 14.7 psi atmospheric, or 0 psi gauge. When there is a total absence of air or 0 absolute pressure (no atmospheric pressure at all), a perfect vacuum exists. It is measured at 29.9 in. Hg vacuum, 0 bar, or 0 psi atmospheric.

> A **vacuum** is a space in which the pressure is significantly lower than the pressure around it. In an engine, we define this as the pressure inside the engine being less than atmospheric and therefore a vacuum exists. Vacuum is considered a force and can move fluids from their rest position.

An engine develops significant vacuum during its intake stroke as the pistons move downward and the valves begin to close. Liquids, solids, and gases move from an area of high pressure to an area of low pressure because of the force caused by the state of vacuum. Automotive engines rely on engine vacuum and this principle to fill the combustion chambers with air to support combustion. When the throttle is opened, atmospheric pressure flows rapidly into the low pressure area of the cylinder to fill it with air. An engine that cannot develop adequate vacuum will not pull in enough air to make maximum power. A modern, good-running engine will typically develop about 16–18 in. Hg vacuum at idle.

It is important to understand that a vacuum does not mean that air is sucked into the space with a lower pressure. When a vacuum is created, air is forced from the area of higher pressure (typically atmospheric pressure outside of the engine and near the intake) to the area of lower pressure (typically the combustion chambers). Vacuum is also dependent on volume and temperature.

Turbocharged and supercharged engines increase this pressure difference before or after the throttle plate to help fill the combustion chambers fully under all conditions.

BOYLE'S LAW

Boyle's law states that if the temperature remains constant, the volume of a given mass of gas is inversely proportional to the absolute pressure. Boyle's law is a summary of the relationships between pressure, temperature, and volume. The fuel and air mixture enters the combustion chamber because of a vacuum. The combustion chamber still has pressure, but it is less than atmospheric pressure. This lower pressure is a direct result of the increasing volume in the combustion chamber.

ENGINE OPERATION

> The **stroke** is the amount of piston travel from TDC to BDC measured in inches or millimeters.

Most automotive and truck engines are four-stroke cycle engines. A **stroke** is the movement of the piston from one end of its travel to the other; for example, if the piston is at the top of its travel and then is moved to the bottom of its travel, one stroke has occurred. Another stroke occurs when the piston is moved from the bottom of its travel to the top again. A **cycle** is a sequence that is repeated. In the four-stroke engine, four strokes are required to complete one power-producing cycle.

> A **cycle** is a complete sequence of events.

The internal combustion engine must draw in an air-fuel mixture, compress the mixture, ignite the mixture, and then expel the exhaust. This is accomplished in four-piston strokes (**Figure 3-7**). Diesel and gasoline direct injection engines draw only air into the combustion chamber; fuel is then injected directly into the combustion chamber and mixed.

The first stroke of the cycle is the intake stroke (see Figure 3-7A). As the piston moves down from **top dead center (TDC)**, the intake valve is opened so the vaporized air-fuel mixture can be pushed into the cylinder by atmospheric pressure. As the piston moves downward in its stroke, a vacuum is created (low pressure). Since high-pressure air is forced to move toward low pressure, the air-fuel mixture is pushed past the open intake valve and into the cylinder. After the piston reaches **bottom dead center (BDC)**, the intake valve is closed, and the stroke is completed. Closing the intake valve after BDC allows an additional amount of air-fuel mixture to enter the cylinder. Even though the piston is at the end of its stroke and no more vacuum is created, the additional mixture enters the cylinder because it weighs more than air alone. The momentum of air flowing into the cylinder also helps pull more air in after the piston has reached BDC.

The compression stroke begins as the piston starts its travel back to TDC (see Figure 3-7B). The intake and exhaust valves are both closed, trapping the air-fuel mixture in the combustion chamber. The movement of the piston toward TDC compresses the mixture. When the piston reaches TDC, the mixture is fully compressed and is highly combustible because the mixture is close to its **autoignition temperature**. The ignition system induces a spark at exactly the right time to produce maximum power The actual timing of the spark on a modern engine is controlled by the PCM (powertrain control module), also referred to as ECM (engine control module) by some manufacturers, and it varies based on temperature, speed, fuel content, load, and other operating factors. This ensures that peak pressure will act on the piston when it is just headed downward and before the volume of the chamber has increased significantly.

When the spark occurs in the compressed mixture, the rapid burning causes the molecules to expand, beginning the power stroke (see Figure 3-7C). The expanding molecules create a pressure above the piston and push it downward. The downward movement of the piston in this stroke is the only time the engine is productive concerning power output. During the power stroke, the intake and exhaust valves remain closed.

Just prior to the piston reaching BDC, the exhaust valve is opened and the exhaust stroke of the cycle begins (see Figure 3-7D). The upward movement of the piston back toward TDC pushes out the exhaust gases from the cylinder past the exhaust valve and into the vehicle's exhaust system. As the piston approaches TDC, the intake valve opens and the exhaust valve is closed a few degrees after TDC. The degrees of crankshaft rotation when both the intake and exhaust valves are open are called **valve overlap**. Valve overlap uses the movement and momentum of air to help fully clean out the spent gases and fill the chamber with a fresh charge of air and fuel. The cycle is then repeated again as the piston begins the intake stroke. This cycle is occurring simultaneously in the other cylinders of the engine, though the individual strokes do not occur at the same time. For example, cylinder number one will produce its power stroke; then while its exhaust stroke occurs, cylinder number three will be on its power stroke. This keeps rotating the crankshaft around with regular impulses. Most engines in use today are referred to as **reciprocating**. Power is produced by the up-and-down movement of the piston in the cylinder. This linear motion is then converted to rotary motion by a crankshaft.

Top dead center (TDC) is a term used to indicate that the piston is at the very top of its stroke.

Bottom dead center (BDC) is a term used to indicate that the piston is at the very bottom of its stroke.

The **autoignition temperature** is the minimum temperature required to ignite a gas in air without a spark or flame to assist.

Valve overlap is the length of time, measured in degrees of crankshaft revolution, that the intake and exhaust valves of the same combustion chamber are open simultaneously. In some engines, valve timing and overlap are variable.

Reciprocating is an up-and-down or back-and-forth repetitive motion.

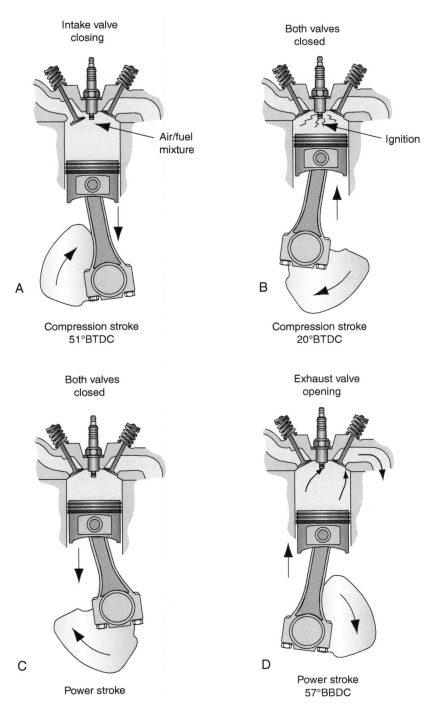

Figure 3-7 The four strokes of an automotive engine: (A) intake stroke, (B) compression stroke, (C) power stroke, (D) exhaust stroke.

 A BIT OF HISTORY

License plates made their first appearance on vehicles in the United States in 1901. In 1905, California began licensing and registering motor vehicles. By the end of 1905, California had registered 17,015 vehicles. Owners had to conspicuously display the circular or octagonal tags in the vehicle. To register the vehicle, it had to have working lamps, satisfactory brakes, and a bell or a horn.

ENGINE CLASSIFICATIONS

Engine classification is usually based on the number of cycles, the number of cylinders, cylinder arrangement, and valvetrain type. Additional descriptors include the location and number of camshafts and the type of fuel and ignition systems. Engine displacement is commonly used to identify engine size.

Sometimes different methods may be used to identify the same engine. For example, a Chrysler 2.0-liter DOHC is also identified as a 420A and D4FE. In this case, the engine has a displacement of 2.0 liters and uses dual overhead camshafts. The 420A is derived from four valves per cylinder, has 2.0 liters displacement, and is American built. The D4FE means the engine uses dual overhead camshafts with four valves per cylinder and has front exhaust (the exhaust manifold is mounted facing the front of the vehicle). This section will define many of the terms used by automotive manufacturers to classify their engines.

Number of Cylinders

One cylinder would not be able to produce sufficient power to meet the demands of today's vehicles. Most automotive and truck engines use three, four, five, six, eight, ten, or twelve cylinders. The number of cylinders used by the manufacturer is determined by the amount of work required from the engine. Generally, the greater the number of cylinders, the more power the engine can produce. Also, the more cylinders, the greater number of power pulses in one complete 720-degree cycle. This produces a smoother running engine because the vibration of power stroke pulses cancel out each other. Many Jaguars have V12 engines not just because they are powerful but because they are inherently smooth.

Vehicle manufacturers attempt to achieve a balance among power, economy, weight, and operating characteristics. An engine having more cylinders generally runs and idles more smoothly than those having three or four cylinders. This is because there is less crankshaft rotation between power strokes. There is less time between the combustion chamber explosions that cause the sudden acceleration of the piston, which is the source of vibrations. However, adding more cylinders to the engine design usually increases the weight of the assembly, the complexity of the design, and the cost of production. Over the years, manufacturers have changed their cylinder number, size, and arrangement designs to meet the ever-changing consumer demands, fuel mileage, and emissions laws.

Cylinder Arrangement

Engines are also classified by the arrangement of the cylinders. The cylinder arrangement used is determined by vehicle design and purpose. The most common engine designs are in-line and V-type.

The in-line engine places all of its cylinders in a single row (**Figure 3-8**). Advantages of this engine design include ease of manufacturing and serviceability. The disadvantage

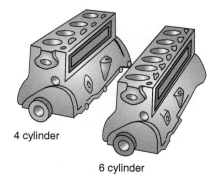

4 cylinder

6 cylinder

Figure 3-8 In-line engines are designed for ease of construction and service.

of this engine design is the block height. This causes higher hood lines on rear wheel drive vehicles resulting in less than optimum front aerodynamics. Many manufacturers overcome this disadvantage by installing a **transverse-mounted engine** in the engine compartment of front-wheel-drive vehicles. A modification of the in-line design is the slant cylinder (**Figure 3-9**). By tilting the engine slightly, the manufacturer can lower the height of the hood line for improvement in the vehicle's aerodynamics.

The V-type engine has two rows of cylinders set 60 to 90 degrees from each other (**Figure 3-10**). The V-type design allows for a shorter block height, thus easier vehicle aerodynamics. The length of the block is also shorter than in-line engines with the same number of cylinders.

Another common engine design is the horizontally opposed cylinder engine, sometimes called a "boxer" engine (**Figure 3-11**). This engine design has two rows of cylinders directly across from each other. The main advantage of this engine design is the very low vertical height. Boxer engines are commonly used by Porsche, Volkswagen, and Subaru.

A **transverse-mounted engine** faces from side to side instead of front to back. The opposed cylinder engine may be called a pancake engine.

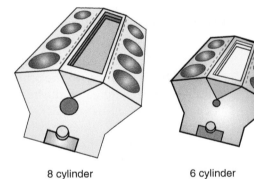

8 cylinder   6 cylinder

Figure 3-10 The V-type engine design allows for lower and shorter hood lines.

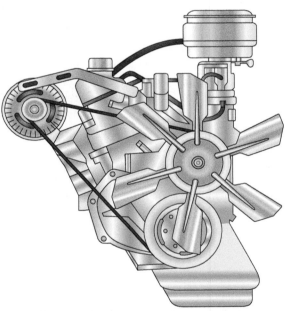

Figure 3-9 Slant cylinder design lowers the overall height of the engine.

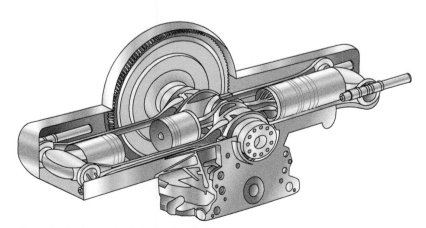

Figure 3-11 Opposed cylinder engine design.

Valvetrain Type

Engines can also be classified by their valvetrain types. The three most commonly used valvetrains are the following:

1. **Overhead valve (OHV).** The intake and exhaust valves are located in the cylinder head, while the camshaft and lifters are located in the engine block (**Figure 3-12**). Valvetrain components of this design include lifters, pushrods, and rocker arms.
2. **Overhead cam (OHC).** The intake and exhaust valves are located in the cylinder head along with the camshaft. The valves are operated directly by the camshaft and a follower, eliminating many of the moving parts required in the OHV engine. If a single camshaft is used in each cylinder head, the engine is classified as a single overhead cam (SOHC) engine. This designation is used even if the engine has two-cylinder heads with one camshaft each.
3. **Dual overhead cam (DOHC).** The DOHC uses separate camshafts for the intake and exhaust valves. A DOHC V-8 engine is equipped with a total of four camshafts (**Figure 3-13**).

Ignition Types

Most automotive engines use a mixture of gasoline and air, which is compressed and then ignited with a spark plug. This type of engine is referred to as a **spark-ignition (SI) engine**. Diesel engines, on the other hand, do not use spark plugs. The air-fuel mixture is ignited by the heat created during the compression stroke. These engines are referred to as **compression-ignition (CI) engines**.

Spark-ignition engines may be further identified by the type of ignition system used. The older, less common design is called a distributor ignition (DI). This form of ignition uses a mechanical device to distribute the spark from one **coil** to the correct cylinder. Newer ignition systems tend to use multiple coils to deliver the spark to the cylinders. These systems are termed electronic-ignition (EI) systems. A form of EI is the coil on plug (COP) ignition system, where one coil sits on top of each spark plug to deliver the spark directly (**Figure 3-14**). Another type of EI is called a waste spark-ignition system, where one coil provides the spark to two cylinders.

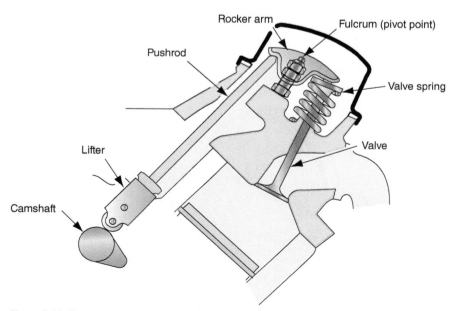

Figure 3-12 The overhead valve engine has been one of the most popular valve designs.

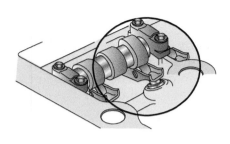

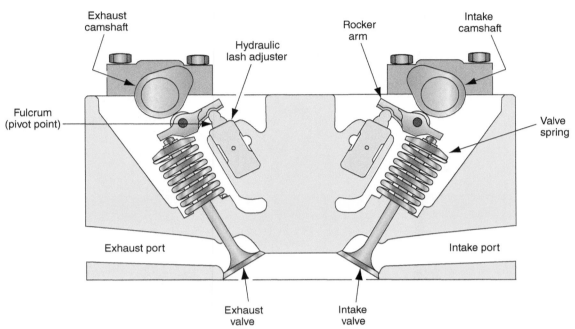

Figure 3-13 The dual overhead camshaft engine places two camshafts in the cylinder head.

Figure 3-14 This late-model V-6 engine has a coil on top of each spark plug (this is the forward cylinder head).

Fuel Injection Types

All new vehicles currently sold in the United States and Canada use a form of **fuel injection** to deliver fuel to the engine. Fuel injectors spray just the right amount of fuel for each combustion event. The required amount of fuel injected is controlled and determined by the PCM, based on inputs on air, engine speed, and temperature, and other inputs that will be affected by how much fuel is injected. Most fuel injection systems deliver fuel into the intake system very close to the intake valve. These systems use one injector per cylinder and are termed *multipoint fuel injection (MPFI)*. The fuel injector is located in the intake manifold. Many manufacturers refine this system to make sure that the injectors each fire on each cylinder's intake stroke for excellent mixing of the air and fuel. These systems are called sequential fuel injection (SFI) systems because the injectors fire in the same sequence as the cylinders. A few newer gas engines use fuel injectors that spray fuel directly into the combustion chamber and are located in the cylinder head. These are called gasoline direct injection (GDI) systems. This type of fuel injection improves combustion and reduces fuel consumption and emissions. Most compression-ignition diesel engines use a similar fuel injection method as the GDI system.

ENGINE VIBRATION

Balance Shafts

Many engine designs have inherent vibrations. The in-line four-cylinder engine is an example of this problem. Some engine manufacturers use balance shafts to counteract engine vibration. Four-cylinder engine vibration occurs at two positions of piston travel. The first is when a piston reaches TDC. The inertia of the piston moving upward creates an upward force in the engine. The second vibration occurs when all pistons are level. At this point, the downward-moving pistons have accelerated and have traveled more than half their travel distance. The engine forces are downward because the pistons have been accelerating since they have to travel a greater distance than the upward-moving pistons to reach the point at which all pistons are level.

Piston travel is farther at the top 90 degrees versus the bottom 90 degrees of crankshaft rotation. First, determine the total length. This is rod length, from the center of the big end to the center of the small end, and the crankshaft throw (stroke divided by 2). Next, determine the amount the piston drops in the first 90 degrees of crankshaft rotation (TDC to 90 degrees). Once this is known, the final amount of piston travel can be determined.

The balance shafts rotate at twice the speed of the crankshaft. This creates a force that counteracts crankshaft vibrations. At TDC, the force of the piston will exert upward; to offset the vibration, the crankshaft and the balance shaft(s) lobes have a downward force (**Figure 3-15**). At 90 degrees of crankshaft rotation, all pistons are level. Remember, the pistons traveling downward have traveled more than half of their travel distance to reach this point. The engine forces are in a downward direction since these pistons have been accelerated to reach the midpoint at the same time as the upward-moving pistons, which need to travel a lesser distance. Since the balance shaft(s) rotates at twice the crankshaft speed, the lobes of the balance shaft(s) are facing up, exerting an upward force to counteract the downward force of the pistons (**Figure 3-16**). At 180 degrees of crankshaft rotation, the upward-moving pistons reach TDC and again exert an upward force. The balance shaft(s) have their lobes facing downward to exert a downward force to counteract the upward force (Figure 3-15). Finally, at 270 degrees of crankshaft rotation, the pistons are level again. At this point, the lobes of the balance shaft(s) are exerting an upward force to counteract the downward force of the pistons (Figure 3-16). Some modern diesel engines use a pulsing fuel injection as part of a strategy to smooth out the vibrations of the engine. Diesel engines generally produce more power on the power stroke and therefore have more inherent vibration problems.

Balance shafts may also be used in V-type engines (**Figure 3-17**). Some balance shafts on OHV engines are mounted in the block V above the camshaft (**Figure 3-18**). Bearing journals on the balance shaft are mounted on bushings in the block. Some engines use two gears mounted on the front end of the camshaft to drive the balance shaft. The inner camshaft gear is meshed with the gear on the balance shaft, and the timing chain surrounds the outer camshaft gear and the crankshaft gear to drive the camshaft. On some DOHC engines, the balance shaft is mounted in the block above the crankshaft (**Figure 3-19**).

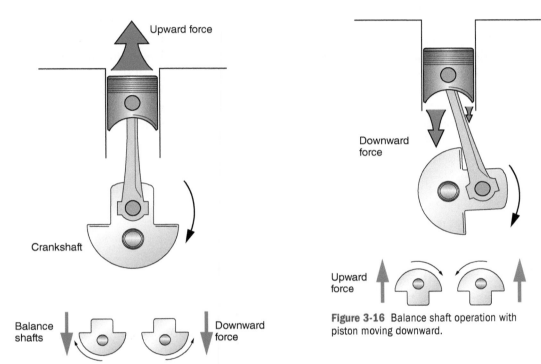

Figure 3-16 Balance shaft operation with piston moving downward.

Figure 3-15 Balance shaft operation with piston moving upward.

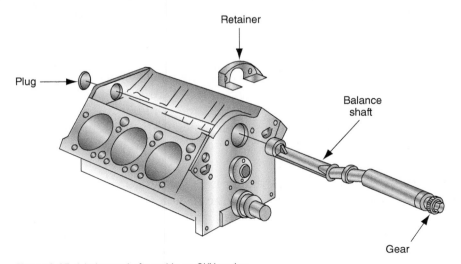

Figure 3-17 A balance shaft used in an OHV engine.

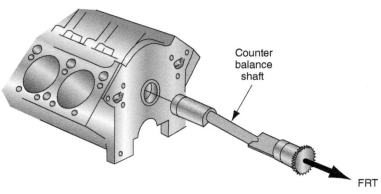

Figure 3-18 A balance shaft used in a DOHC V-6 engine.

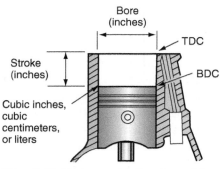

Figure 3-19 Displacement is the volume of the cylinder between TDC and BDC.

AUTHOR'S NOTE One time a student drove his freshly rebuilt engine into the shop to have me look at why it "shook" so badly. The student said the engine ran well and had plenty of power but had a very significant vibration throughout the engine rpm (revolutions per minute) range. We took a test drive to verify the concern and tested the engine and transmission mounts. During further discussion, I realized the engine had balance shafts and asked the student if he had lined them up according to the marks described in the service information. Unfortunately, he was unaware of that necessary step. Once he successfully timed his balance shafts, his "new" engine ran beautifully.

ENGINE DISPLACEMENT

Displacement is the measure of engine volume. The larger the displacement, the greater the power output.

Displacement is the volume of the cylinder between TDC and BDC measured in cubic inches, cubic centimeters, or liters (**Figure 3-19**). Most engines today are identified by their displacement in liters. Engine displacement is an indicator of engine size and power output. The more fuel (mixed with air) that can be burned, the greater the amount of energy that can be produced. For more fuel to burn, more air must also enter the cylinders. Engine displacement measurements are an indicator of the amount or mass of air that can enter the engine. The mass of air allowed to enter the cylinders is controlled by a throttle plate. If the throttle plate is closed, a smaller mass of air is allowed to enter the cylinder, and the power output is reduced. As the throttle plate is opened, the mass of air allowed to enter the cylinders is increased, resulting in greater power output. The more air that is allowed, the more fuel that can be added to it. As engine speed increases, it will eventually reach a point at which no more air can enter. Any increase in speed beyond this point causes a loss in power output. The major factor in determining the maximum mass of air that can enter the cylinders is engine displacement. The amount of displacement is determined by the number of cylinders, cylinder bore, and length of the piston stroke (**Figure 3-20**).

A **bore** is the diameter of a hole. If the bore is larger than its stroke, the engine is referred to as oversquare. If the bore is smaller than the stroke, the engine is referred to as undersquare. A square engine has the same bore size as it does stroke.

Stroke is the distance that the piston travels from TDC to BDC or vice versa. Some engine enthusiasts say "there is no replacement for displacement."

Bore and Stroke. The **bore** of the cylinder is its diameter measured in inches (in.) or millimeters (mm). The **stroke** is the amount of piston travel from TDC to BDC measured in inches or millimeters. An oversquare engine is used when high rpm (revolutions per minute) output is required. This meets the requirements of many automotive engines. Most truck engines are designed as undersquare since they deliver more low-end torque.

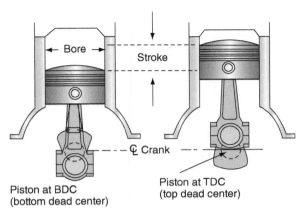

Figure 3-20 The bore and stroke determine the amount of displacement of the cylinder.

Calculating Engine Displacement

Mathematical formulas can be used to determine the displacement of an engine. As with many mathematical formulas, more than one method can be used. The first is as follows:

Engine displacement = $0.785 \times B^2 \times S \times N$
where 0.785 is a constant
  B = bore
  S = stroke
  N = number of cylinders

The second formula is:

Engine displacement = $\pi \times R^2 \times S \times N$
where $\pi = 3.1416$
  R = bore radius (diameter/2)
  S = stroke length
  N = number of cylinders

Pi times the radius squared is equal to the area of the cylinder cross section. This area is then multiplied by the stroke, and this product is multiplied by the number of cylinders.

The calculation for total cubic inch displacement (CID) of a V-6 engine with a bore size of 3.800 inches (9.65 cm) and stroke length of 3.400 inches (8.64 cm) would be as follows:

Engine displacement = $0.785 \times 3.800^2 \times 3.400 \times 6 = 231.24$ CID

or

Engine displacement = $3.1416 \times 1.9^2 \times 3.400 \times 6 = 231.35$ CID

The difference in decimal values is due to rounding pi to four digits. Round the product to the nearest whole number. In this case, the CID would be expressed as 231.

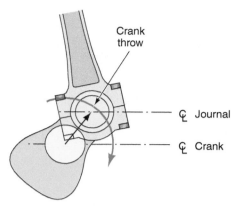

Figure 3-21 An alternate method of determining the piston stroke is to double the crankshaft throw.

Most engine manufacturers designate engine size by metric displacement. To calculate engine displacement in cubic centimeters or liters, simply use the metric measurements in the displacement formula. In the preceding example, the result would be 3,789.56 cc, which would be rounded off to 3.8 liters. Another way to convert to metric measurements is to use the CID and multiply it by .01639 to find the size of the engine in liters. For example, 231 × .01639 = 3.79, or 3.8 liters. Use a service manual to determine the bore and stroke of the engine.

An alternate method of determining the stroke is to measure the amount of crankshaft throw (**Figure 3-21**). The stroke of the engine is twice the crankshaft throw. You can measure the diameter of the cylinder to find the bore size.

If the engine is machined or modified during the engine rebuild process, the engine displacement will change. For example, during the inspection process, an engine rebuild technician may find that the cylinders need to be bored out 0.060 inch (0.1524 cm). This new cylinder size would have to be included in the new displacement calculation.

DIRECTION OF CRANKSHAFT ROTATION

The Society of Automotive Engineers (SAE) standard for engine rotation is counterclockwise when viewed from the rear of the engine (flywheel side). When viewing the engine from the front, the rotation of the crankshaft is clockwise. Most automotive engines rotate in this direction, but a few engines rotate the opposite direction. Some marine, older Volkswagen, and Honda engines are known as reverse rotation engines.

ENGINE MEASUREMENTS

As discussed previously, engine displacement is a measurement used to determine the output of the engine and can be used to identify the engine. Other engine measurements include compression ratio, horsepower, torque, and engine efficiency. Some of these are also used to identify the engine; for example, an engine with 3.8 liters displacement may be available in different horsepower and torque ratings.

Compression Ratio

The **compression ratio** expresses a comparison between the volume the air-fuel mixture is compressed into at TDC and the total volume of the cylinder (piston at BDC). The ratio is the result of dividing the total volume of the cylinder by the volume with the piston at TDC (**Figure 3-22**).

The **compression ratio** is a comparison between the volume above the piston at BDC and the volume above the piston at TDC. Compression ratio is the measure used to indicate the amount the piston compresses each intake charge.

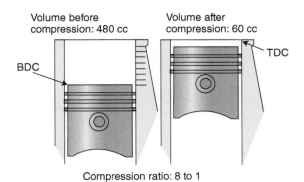

Figure 3-22 The compression ratio is a measurement of the amount the air-fuel mixture will be compressed.

Torque = 1 foot × 10 pounds = 10 foot-pounds

Torque = 2 feet × 10 pounds = 20 foot-pounds

Compression ratios affect engine efficiency. Higher compression ratios result in increased cylinder pressures, which compress the air-fuel molecules tighter. This is desirable because it causes the flame to travel faster. A higher compression ratio may be desirable because the mechanical efficiency is higher and typically more power will be produced from the same displacement.

There are some drawbacks to just having higher compression ratios, such as a higher octane fuel requirement and increased emissions. These are unwanted outcomes of higher compression ratio engines. However, some modern engines have been able to increase the compression ratio while still allowing for the use of regular octane fuel and having similar emission levels. The use of a VVT system or a variable length manifold, along with alternative fuels and advanced computer programs, have given design engineers more flexibility when designing higher compression ratio engines.

Theoretically, more power can be produced from a higher compression ratio. An increase in the compression ratio results in a higher degree of compression. Combustion occurs at a faster rate because the molecules of the air-fuel mixture are packed tighter. It is possible to change the compression ratio of an engine by performing some machining operations or changing piston designs. Increasing the compression ratio may increase the power output of the engine, but the higher compression ratio increases the compression temperature. This can cause **preignition** and detonation. To counteract this tendency, higher octane gasoline must be used. A higher octane fuel will burn slower than a lower octane fuel. The octane number of a fuel is also known as the anti-knock index; the higher the number, the more resistant to knock, or **detonation**, the fuel is.

Most gas automotive engines have a compression ratio between 8:1 (expressed as "eight to one") and 10:1 and develop dry cylinder pressures during compression of between 125 and 200 psi. Diesel engines typically have compression ratios between 17:1 and 22:1. They can produce up to 600 psi of dry compression pressure. The usable compression ratio of an engine is determined by the following factors:

- The temperature at which the fuel will ignite
- The temperature of the air charge entering the engine
- The density of the air charge
- Combustion chamber design

Preignition is an explosion in the combustion chamber resulting from the air-fuel mixture igniting prior to the spark being delivered from the ignition system.

Detonation is different from preignition. It is characterized as abnormal combustion. Detonation occurs when pressure and temperature build in gas chambers outside of the normal spark flame. This may be caused by improper octane, high engine loads, or improper combustion. Detonation is often known as "spark knock" or "pinging."

Calculating Compression Ratio

When calculating the total volume above the piston, the combustion chamber in the cylinder head must also be considered. The formula for calculating compression ratio is:

$$CR = (VBDC + VCH)/(VTDC + VCH)$$

where

CR = compression ratio

VBDC = volume above the piston at BDC

VCH = combustion chamber volume in the cylinder head

VTDC = volume above the piston at TDC

Example: VBDC = 56 cubic inches

VCH = 4.5 cubic inches

VTDC = 1.5 cubic inches

The compression ratio would be:

$$\frac{56 + 4:5}{1:5 + 4:5} = 10:1$$

During an engine rebuild, certain machining procedures may affect the compression ratio. Some examples of this would be resurfacing the cylinder head, resizing or boring the cylinders, and installing a different crankshaft. The compression ratio should always be considered when making choices during an engine rebuild.

Horsepower and Torque

Torque is a mathematical expression for rotating or twisting force around a pivot point. As the pistons are forced downward, this pressure is applied to a crankshaft that rotates. The crankshaft transmits this torque to the drivetrain and ultimately to the drive wheels. Horsepower is the rate at which an engine produces torque.

To convert terms of force applied in a straight line to force applied rotationally, the formula is:

torque = force × radius

If a 10-pound force is applied to a wrench 1 foot in length, 10 foot-pounds (ft.-lb.) of torque is produced. If the same 10-pound force is applied to a wrench 2 feet in length, the torque produced is 20 ft.-lb. (**Figure 3-23**).

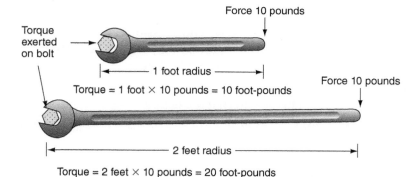

Figure 3-23 Example of how torque can be increased.

Power is the rate at which work is done. The power an engine produces is rated in horsepower. *Horsepower* is a term that can be used to measure many mechanical actions. When discussing engines, we are measuring mechanical horsepower. One horsepower is equal to 33,000 foot-pounds per minute. The horsepower of an engine can be measured on a dynamometer, which places a load on the engine and measures torque. Knowing the torque of an engine and the rpm (engine speed) at which it was measured, you can calculate horsepower by using the following equation.

$$\text{Horsepower} = \frac{\text{torque} \times \text{rpm}}{5,252}$$

Torque and horsepower will peak at some point in the rpm range (**Figure 3-24**). This graph is a mathematical representation of the relationship between torque and horsepower in one engine. The graph indicates that this engine's torque peaks at about 1,700 rpm, while brake horsepower peaks at about 3,500 rpm. The third line in the graph represents the amount of horsepower required to overcome internal resistances or friction in the engine. The amount of **brake horsepower** is less than **indicated horsepower** due to this and other factors.

The SAE standards for measuring horsepower include ratings for gross and net horsepower. **Gross horsepower** expresses the maximum power developed by the engine without additional accessories operating. These accessories include the air cleaner and filter, cooling fan, charging system, and exhaust system. **Net horsepower** expresses the amount of horsepower the engine develops as installed in the vehicle. Net horsepower is about 20 percent lower than gross horsepower. Most manufacturers express horsepower as SAE net horsepower.

For example, if the torque of the engine is measured at the following values, a graph can be plotted to show the relationship between torque and horsepower (**Figure 3-25**):

Brake horsepower is the usable power produced by an engine. It is measured on a dynamometer using a brake to load the engine.

Indicated horsepower is the amount of horsepower the engine can theoretically produce.

Gross horsepower is the maximum power that an engine can develop. It is measured without additional accessories installed (such as belt-driven items and the transmission).

Net horsepower is the amount of horsepower an engine has when it is installed in the vehicle.

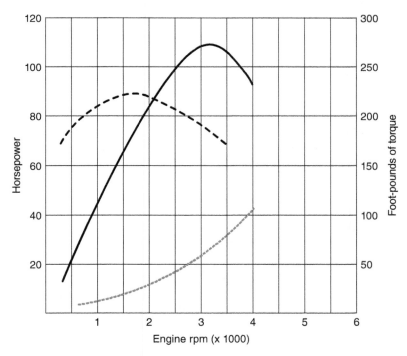

Figure 3-24 Example of torque and horsepower curves.

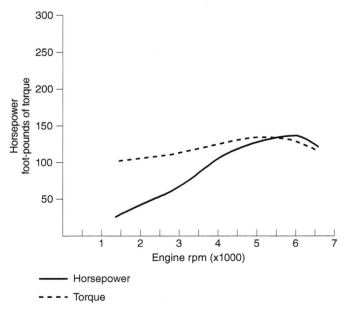

Figure 3-25 It is possible to graph the relationship between torque and horsepower. Highest torque output is usually at lower engine speeds, while peak horsepower is at higher engine speeds.

Torque output @ 1,500 rpm = 105 lb.-ft.
Torque output @ 2,800 rpm = 110 lb.-ft.
Torque output @ 3,500 rpm = 115 lb.-ft.
Torque output @ 5,000 rpm = 130 lb.-ft.
Torque output @ 6,000 rpm = 125 lb.-ft.
Torque output @ 6,500 rpm = 100 lb.-ft.

In this instance, the horsepower is as follows:

Horsepower @ 1,500 rpm = 30
Horsepower @ 2,800 rpm = 59
Horsepower @ 3,500 rpm = 77
Horsepower @ 5,000 rpm = 124
Horsepower @ 6,000 rpm = 143
Horsepower @ 6,500 rpm = 124

Engine Efficiency

Efficiency is a measure of a device's ability to convert energy into work.

Several different measurements are used to describe the **efficiency**. Efficiency is mathematically expressed as output divided by input. The terms used to define the input and output must be the same. Some of the most common efficiencies of concern are mechanical, volumetric, and thermal efficiencies.

The **mechanical efficiency** of an engine is the comparison between actual power measured at the crankshaft and the actual power developed within the cylinders. There will always be frictional losses in the crankshaft, cylinders, and connecting rods.

Mechanical efficiency is a comparison between brake horsepower and indicated horsepower; in other words, it is a ratio of the power actually delivered by the crankshaft to the power developed within the cylinders at the same rpm. The formula used to calculate mechanical efficiency is:

Mechanical efficiency = brake horsepower/indicated horsepower

Ideally, mechanical efficiency would be 100 percent. However, the power delivered will always be less due to power lost in overcoming friction between moving parts in the engine. The mechanical efficiency of an internal combustion four-stroke engine is approximately 90 percent. If the engine's brake horsepower is 120 hp and the indicated horsepower is 140 hp, the mechanical efficiency of the engine is 85.7 percent. Mechanical efficiency is a top concern for engine developers. Increasing the mechanical efficiency can increase the power output and allow for a smaller, more efficient engine to be installed into a larger vehicle. Low-friction lubricants, small manufacturing tolerances, and precision manufacturing are just a few examples of the techniques of how engine developers are increasing engine efficiency.

Volumetric efficiency (VE) is a measurement of the amount of air-fuel mixture that actually enters the combustion chamber compared to the amount that could be ingested at that speed. Volumetric efficiency describes how well an engine can breathe. The more air an engine can take in, the more power the engine will put out. As the volumetric efficiency increases, the power developed by the engine increases proportionately. The formula for determining volumetric efficiency is:

> **Volumetric efficiency** is the comparison between the amount of air that could enter the engine and the amount of air that actually enters the engine.

$$\frac{\text{Actual volume of air in cylinder}}{\text{Maximum volume of air in cylinder}}$$

An example of a calculation for volumetric efficiency is an engine with a 350 CID displacement that actually has an air-fuel volume of 341 cubic inches at a particular rpm. The actual amount of air taken in is less than the available space; this engine is 97 percent volumetrically efficient at this rpm.

A BIT OF HISTORY

The term *horsepower* was coined by James Watt (1736–1819). Mr. Watt was working with the first practical steam-powered engines. Steam engines were on the verge of replacing horses for doing some of the work. At the time, miners, farmers, and other industries were using horses to do most of their work. Mr. Watt was working with mine ponies who were used to lift coal out of a mine. He figured that on average a pony could pull 220 pounds a distance of 100 feet in one minute. He then judged that horses were 50 percent stronger than ponies, and thus he arrived at the equation 1 horsepower = 33,000 foot pounds of work per minute. James Watt and Matthew Boulton standardized this measurement in 1783.

Ideally, 100 percent volumetric efficiency is desired, but actual efficiency is reduced due to pumping losses. It must be realized that a given mass of air-fuel mixture occupies different volumes under different conditions. Atmospheric pressures and temperatures will affect the volume of the mixture. Pumping losses are the result of restrictions to the flow of air-fuel mixture into the engine. These restrictions are the result of intake manifold passages, throttle plate opening, valves, and cylinder head passages. There are some machining techniques used by builders of performance engines that will reduce these restrictions. Pumping losses are influenced by engine speed; for example, an engine may have a volumetric efficiency of 75 percent at 1,000 rpm and increase to an efficiency of 85 percent at 2,000 rpm, then drop to 60 percent at 3,000 rpm. As engine speed increases, volumetric efficiency may drop to 50 percent.

If the airflow is drawn in at a low engine speed, the cylinders can be filled close to capacity. A required amount of time is needed for the airflow to pass through the intake manifold and pass the intake valve to fill the cylinder. As engine speed increases, the amount of time the intake valve is open is not long enough to allow the cylinder to be filled. Because of the effect of engine speed on volumetric efficiency, engine manufacturers use intake manifold design, valve port design, valve size, multiple valves, valve overlap, camshaft timing, exhaust tuning, variable length manifolds, and computer control to improve the engine's breathing. Current production models have fast computers that can calculate the volumetric efficiency by using computer data.

The VE of a typical modern spark ignition internal combustion normally aspirated engine is 75 to 95 percent. There are some examples of naturally aspirated engines having more than 100 percent VE. Older engines were limited in their designs and did not employ the same criteria of their modern counterparts. Today, many engines have four valves per cylinder, whereas an engine made in 1970 typically has only two valves per cylinder. A turbocharger or supercharger will force air into the engine (called forced induction). This process will also give the engine a VE of over 100 percent.

Other factors of VE are elevation, temperature, fuel type, and humidity. A higher VE is desired because it means less pumping and parasitic losses and greater power output of the engine. Volumetric efficiency should not be confused with the total engine efficiency, although it does have a great influence on it.

Thermal efficiency is a measurement comparing the amount of energy present in a fuel and the actual energy output of the engine. Thermal efficiency is the measurement of how much energy is actually used to turn the wheels as compared to the total energy available. Heat is a form of energy just like electricity. Heat energy can be measured using two terms: *intensity* and *quantity*. The measurement of the intensity of heat energy is measured in the temperature scale (Celsius or Fahrenheit). Temperature is the measurement of an object's internal energy. Something that is hot has more heat intensity than something that is cold; therefore, it has more heat intensity and more heat energy. For example, we may say that a liquid is hot because it is 200°F. This just means that it has a higher temperature than another familiar object.

Heat quantity (often referred to as heat value or heat's power) is the amount of heat energy a piece of matter has. The most commonly used units for measuring heat quantity are the Btu, the calorie (c), and the joule (J). One Btu is the amount of heat required to raise the temperature of 1 pound of pure liquid water by 1°F. The temperature at which water has its greatest density is 39°F. Testing for a Btu is done at this temperature. One Btu is equal to 1,055 J or 252 C.

Some of the different fuels used for internal combustion engines and their energy content are in the following list. Keep in mind that each of these values will vary with different mixtures, seasons, manufacturers, retailers, and geographic location.

1 gallon of midgrade gasoline $=$ 108,500 $-$ 117,000 Btus
1 gallon of ethanol fuel $=$ 76,100 Btus
1 gallon of no. 2 diesel $=$ 139,000 Btus
1 gallon of propane $=$ 91,600 Btus

The formula used to determine a gasoline engine's thermal efficiency has a couple of constants. First, 1 hp equals 42.4 Btus per minute. Second, gasoline has approximately 110,000 Btus per gallon. Knowing this, the formula is as follows:

$$\text{Thermal efficiency} = (\text{bhp} - 42.4 \text{ Bpmin})/(110{,}000 \text{ Bpg} - \text{gpmin})$$

where

bhp $=$ brake horsepower
Bpmin $=$ Btu per minute
Bpg $=$ Btu per gallon
gpmin $=$ gallons used per minute

As a whole, gasoline engines waste about two-thirds of the heat energy available in gasoline. Approximately one-third of the heat energy is carried away by the engine's cooling system. This is required to prevent the engine from overheating. Another third is lost in hot exhaust gases. The remaining third of heat energy is reduced by about 5 percent due to friction inside of the engine. Another 10 percent is lost due to friction in drivetrain components. Due to all of the losses of heat energy, only about 19 percent is actually applied to the driving wheels.

Thermal efficiency is the difference between potential and actual energy developed in a fuel measured in Btus per pound or gallon.

Of the 19 percent applied to the driving wheels, an additional amount is lost due to the rolling resistance of the tires against the road. Resistance to the vehicle moving through the air requires more of this energy. By the time the vehicle is actually moving, the overall vehicle efficiency is about 15 percent.

Manufacturers are continuing to increase the thermal efficiency of the engine in hopes that it will help increase fuel economy. As you begin to work with engines, you will be able to compare older engines to modern engines and identify different parts and computer-management strategies that have been employed to increase thermal efficiency.

OTHER ENGINE DESIGNS

Two-Stroke Engines

Throughout the history of the automobile, the gasoline four-stroke internal combustion engine has been the standard. As technology has improved, other engine designs have been used. This section of the chapter provides an overview of some of these other engine designs.

Two-stroke engines are capable of producing a power stroke every revolution. Valves are not used in most two-stroke engines; instead, the piston movement covers and uncovers intake and exhaust ports. The air-fuel mixture enters the crankcase through a **reed valve** or **rotary valve** (**Figure 3-26**). Upward movement of the piston causes pressure to increase in the cylinder above the piston and creates a slight vacuum below the piston. Atmospheric pressure pushes the air-fuel mixture into the low-pressure area below the piston. At the same time, a previously ingested air-fuel mixture is being compressed above the piston. This action combines the compression stroke and intake stroke.

When the compressed air-fuel mixture is ignited, the piston moves down the cylinder. This slightly compresses the air-fuel mixture in the crankcase. The air-fuel mixture in the crankcase is directed to the top of the piston through the transfer port to the intake port. The reed valve or rotary valve closes to prevent the mixture from escaping from the crankcase. The downward piston movement opens the exhaust port, and the pressure of the expanding gases pushes out the spent emissions. Continued downward movement of the piston uncovers the intake port and allows the air-fuel mixture to enter above the piston. This action combines the power and exhaust strokes.

Most gasoline two-stroke engines do not use the crankcase as a sump for oil, since it is used for air-fuel intake. To provide lubrication to the engine, the oil must be mixed with

A **reed value** is a one-way check valve. The reed opens to allow the air-fuel mixture to enter from one direction, while closing to prevent movement in the other direction.

A **rotary valve** is a valve that rotates to cover and uncover the intake port.

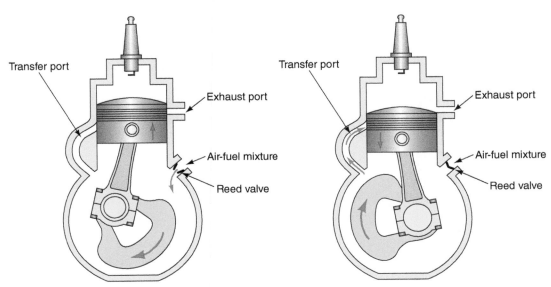

Figure 3-26 Two-stroke cycle engine design.

the fuel or injected into the crankcase. This results in excessive exhaust emissions. Diesel-design two-stroke engines may have a sump. Many automotive engine manufacturers are trying to find ways to reduce the emission levels of the two-stroke engine to allow for its use in automobiles. The advantage of the two-stroke engine is that it produces more power per cubic inch of displacement than the four-stroke engine.

Today, two-stroke engines are mostly used in small applications, such as lawn and garden equipment, and recreational and sports applications (snowmobiles, ATVs, water-craft, etc.). Although they are being quickly phased out by four-stroke diesel engines because of their greater efficiency and lower emissions. There have been a few automobiles, light trucks, and heavy trucks produced in the past that used two-stroke engines. One very famous example of this comes from the Detroit Diesel Company, which produced many two-stroke diesel engines for many military and on-highway commercial applications.

Diesel Engines

The first diesel engine was operated and invented by Rudolf Diesel in 1897. Since then, diesel engines have been mainly used in heavy-duty applications, including light-, medium-, and heavy-duty trucks, vans, tractors, and semis. Diesels used in marine and standby generator applications are desirable for their power output, low fuel consumption, and longevity. Diesel engines tend to be built heavier and often cost more to manufacture.

Domestic automobiles had a few diesel engine options for a few years, but the engine option in small cars has not sold very well over the past few decades with American consumers. Within recent years, the domestic pickup truck market has more widely accepted the diesel engine option. The more powerful and fuel-efficient diesel engine designs are suited well in pickups where the noise, vibration, and harshness level expectations are lower than they are for cars. By contrast, European automobiles are mostly powered by diesel engines and have been for many years. Within the last few years of the publish date of this book, the number of diesel engines in cars and light trucks has been increasing significantly. Examples of such vehicles with popular diesel engine options are Volkswagen cars and domestic light-duty trucks. The diesel engine in these vehicles is also popular in the North American market as well.

Diesel and gasoline engines have many similar components; however, the diesel engine does not use an ignition system consisting of spark plugs and coils. Instead of using a spark initiated and developed by the ignition system, the diesel engine uses the heat produced by compressing air in the combustion chamber to raise the temperature of the air (not the fuel) closer to (and sometimes above) the autoignition temperature of the fuel. Once the fuel is injected under high pressure, it instantly increases in temperature because of its contact with the high temperature air in the combustion chamber. This begins the combustion process of the diesel fuel in the power stroke (**Figure 3-27**). This explains why diesel engines have a higher compression ratio when compared to gasoline engines. Diesel engines are also categorized as CI (compression ignition) engines.

Because starting the diesel engine is dependent on heating the intake air to a high enough level to ignite the fuel, a method of preheating the intake air is required to start a cold engine. Some manufacturers use **glow plugs** to accomplish this. Another method includes using a heater grid in the air intake system.

In addition to ignition methods, there are other differences between gasoline and diesel engines (**Figure 3-28**). The combustion chambers of the diesel engine are designed to accommodate the different burning characteristics of diesel fuel. There are three common combustion chamber designs:

1. An open-type combustion chamber is located directly inside of the piston. The fuel is injected directly into the center of the chamber. Turbulence is produced by the shape of the chamber.

Glow plugs are threaded into the combustion chamber and use electrical current to heat the intake air on diesel engines.

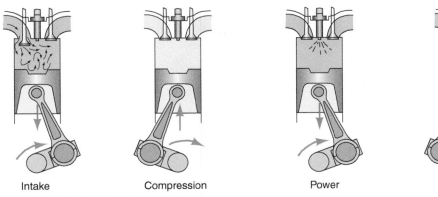

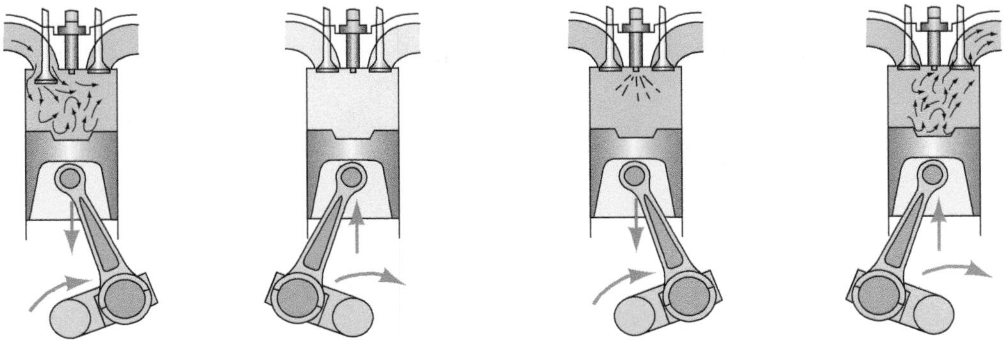

Intake　　　　Compression　　　　Power　　　　Exhaust

Figure 3-27 The four strokes of a diesel engine.

2. A precombustion chamber is a smaller second chamber connected to the main combustion chamber. Fuel is injected into the precombustion chamber, where the combustion process is started. Combustion then spreads to the main chamber. Since the fuel is not injected directly on top of the piston, this type of diesel injection may be called indirect injection.
3. A turbulence combustion chamber is a chamber designed to create turbulence as the piston compresses the air. When the fuel is injected into the turbulent air, a more efficient burn is achieved.

Diesel engines can be either four-stroke or two-stroke designs, although like gas engines, production automotive applications currently use four-stroke engines. Two-stroke engines complete all cycles in two strokes of the piston, much like a gasoline two-stroke engine.

Diesel engines typically have lower fuel consumption rates and a high-power output (compared to a same displacement gasoline engine). Many manufacturers are considering diesel engines as options for their line of cars. Diesel engines operate similarly to gasoline engines and have extra operating components to support clean and efficient operation.

It is very common to rebuild a diesel engine in heavy commercial applications because the cost of total engine replacement is very high and the cost of vehicle replacement is even higher.

	Gasoline	**Diesel**
Intake	Air-fuel	Air
Compression	8–10 to 1, 130 psi 545°F	13–25 to 1, 400–600 psi 1,000°F
Air-fuel mixing point	Carburetor or before intake valve with fuel injection	Near TDC by injection
Combustion	Spark ignition	Compression ignition
Power	464 psi	1,200 psi
Exhaust	1,300–1,800°F CO = 3%	700–900°F CO = 0.5%
Efficiency	22%–28%	32%–38%

Figure 3-28 Comparisons between gasoline and diesel engines.

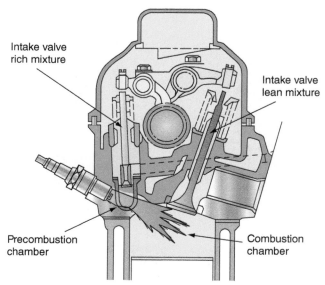

Intake valve
rich mixture

Intake valve
lean mixture

Precombustion
chamber

Combustion
chamber

Figure 3-29 Stratified engine design.

Stratified Charge Engine

The stratified charge engine uses a precombustion chamber to ignite the main combustion chamber (**Figure 3-29**). The advantage of this system is increased fuel economy and reduced emission levels.

In this engine type, the air-fuel mixture is stratified to produce a small, rich mixture at the spark plug while providing a lean mixture to the main chamber. During the intake stroke, a very lean air-fuel mixture enters the main combustion chamber. At the same time, a very rich mixture is pushed past the auxiliary intake valve and into the precombustion chamber (**Figure 3-30**). At the completion of the compression stroke, the spark plug fires to ignite the rich mixture in the precombustion chamber. The burning rich mixture then ignites the lean mixture in the main combustion chamber.

A stratified charge engine is in some ways similar to a diesel engine. The rich air-fuel mixture ignites easily in the precombustion chamber, which allows the excessively lean air-fuel mixture from the intake to ignite without detonation or preignition. This design has not been terribly successful in automobiles over the years, but manufacturers are still working toward a solution. One primary example of a stratified charge engine was the Honda CVCC engine produced in the 1970s.

Honda's CVCC engine lead to the development of other successful and efficient engines produced by other manufacturers, such as lean burn and direct injection designs.

ENGINE IDENTIFICATION

Proper engine identification is required for the technician or machinist to obtain correct specifications, service procedures, and parts. The vehicle identification number (VIN) is one source for information concerning the engine. The VIN is located on a metal tag that is visible through the vehicle's windshield (**Figure 3-31**). The VIN provides the technician with a variety of information concerning the vehicle (**Figure 3-32**). The eighth digit represents the engine code. The service information provides details for interpretation of the engine code. The VIN is also stamped in a number of other locations around the vehicle to make falsification more difficult. It is always important to make sure you get the correct engine code. Several vehicles have the same size engine but were available with different VIN codes. Not representing the eighth digit of the VIN correctly can cause issues with ordering parts, regular maintenance, specifications, and more.

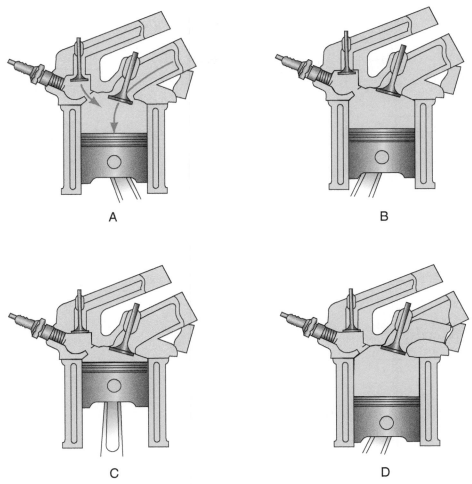

A   B

C   D

Figure 3-30 The stratified engine design uses a rich mixture in the precombustion chamber to ignite the lean mixture in the main chamber.

A vehicle code plate is also located on the vehicle body, usually inside the driver's side door or under the hood. This plate provides information concerning paint codes, transmission codes, and so forth (**Figure 3-33**). The engine sales code is also stamped on the vehicle code plate.

The engine itself may have different forms of identification tags or stamped numbers (**Figure 3-34**). In addition, color coding of engine labels may be used to indicate transitional low emission vehicles (TLEVs) and flex fuel vehicles (FFVs). These numbers may

Figure 3-31 The VIN number can be seen through the windshield.

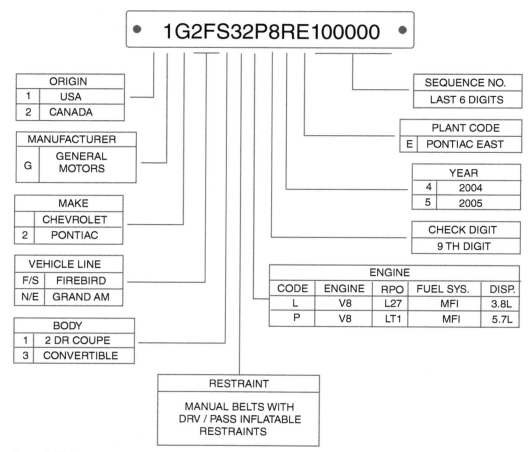

Figure 3-32 The VIN number is used to provide the required information about the vehicle and engine.

AUTHOR'S NOTE One day I was working on an older import vehicle, and a police officer walked up to the car and explained that he had reason to believe this was a stolen vehicle. To check it out, he lifted the rear carpet in the trunk and compared the VIN with the VIN located on the dashboard. Neither appeared to have been tampered with, and the number did not match the VIN of the stolen vehicle. He left me to my work and went on his way to search for the real stolen vehicle.

provide build date codes required to order proper parts. Use the service information to determine the location of any engine identification tags or stamps and the methods to interpret them; for example, a Jeep engine with the build date code of *401HX23* identifies a 2.5-liter engine with a multipoint fuel injection system, 9:1 compression ratio, built on January 23, 1994. This is determined by the first digit representing the year of manufacture (4 = 1994). The second and third digits represent the month of manufacture (01 = January). The fourth and fifth digits represent the engine type, fuel system, and compression ratio (HX = 2.5–liter engine with a multipoint fuel injection system, 9:1 compression ratio). The sixth and seventh digits represent the day of manufacture (23).

Some engines are manufactured with oversized or undersized components. This occurs due to mass production of the engines. Components that may be different sizes than nominal include:

■ Cylinder bores
■ Camshaft bearing bores

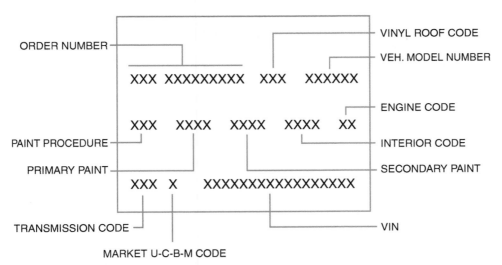

Figure 3-33 Vehicle code plates provide sales code information about the vehicle.

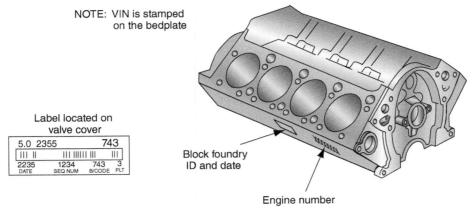

Figure 3-34 The engine may have identification numbers stamped on it.

- Crankshaft main bearing journals
- Connecting rod journals

Engines with oversize or undersize components (from the factory) are usually identified by a letter code stamped on the engine block (**Figure 3-35**). This information is helpful when ordering parts and measuring for wear.

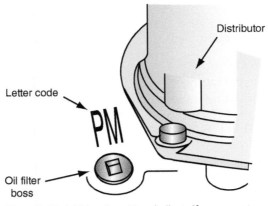

Figure 3-35 Additional markings indicate if components are undersize or oversize.

SUMMARY

- One of the many laws of physics used within the automotive engine is thermodynamics. The driving force of the engine is the expansion of gases. Heat is generated by compressing the air-fuel mixture within the engine. Igniting the compressed mixture causes the heat, pressure, and expansion to multiply.
- Engine classification is usually based on the number of cycles, the number of cylinders, cylinder arrangement, and valvetrain type. In addition, engine displacement is used to identify the size of the engine.
- Most automotive and truck engines are four-stroke cycle engines. A stroke is the movement of the piston from BDC to TDC of the cylinder and is usually measured in inches or millimeters.
- The first stroke of the cycle is the intake stroke. The compression stroke begins as the piston starts its travel back to TDC. When the spark is introduced to the compressed mixture, the rapid burning causes the molecules to expand, and this is the beginning of the power stroke. The exhaust stroke

of the cycle begins with the upward movement of the piston back toward TDC and pushes out the exhaust gases from the cylinder past the exhaust valve and into the vehicle's exhaust system.

- The three most commonly used valvetrains are the overhead valve (OHV), overhead cam (OHC), and dual overhead cam (DOHC).
- Other engine designs and features include two-stroke, diesel, stratified charge, gasoline direct injection.
- Displacement is the total volume of the cylinder between BDC and TDC measured in cubic inches, cubic centimeters, or liters. Engine displacement is usually an indicator of the engine's capability of producing power.
- Torque is a turning or twisting force. As the pistons are forced downward, this pressure is applied to a crankshaft that rotates. The crankshaft transmits this torque to the drivetrain and ultimately to the drive wheels.
- Horsepower is the rate at which an engine produces torque in a given amount of time.

REVIEW QUESTIONS

Short-Answer Essays

1. Describe how the basic laws of thermodynamics are used to operate the automotive engine.

2. List and describe the four strokes of the four-stroke cycle engine.

3. Describe the different cylinder arrangements and the advantages of each.

4. Describe the different valvetrains used in modern engines.

5. What is engine displacement and how is it determined?

6. Describe how the compression stroke of a gasoline four-stroke engine raises the temperature of the air-fuel mixture and why this step is so important.

7. Define the term *stroke*, as used as part of the operation of an engine.

8. Describe the differences of how the fuel and air are mixed and injected in a four-stroke diesel engine when compared to a four-stroke gasoline engine.

9. Explain the basic operation of the two-stroke engine.

10. Describe a situation where incorrectly identifying the eighth digit of the VIN may cause problems.

Fill-in-the-Blanks

1. Most automotive and truck engines are _____ cycle engines.

2. During the intake stroke, the intake valve is _____, while the exhaust valve is _____.

3. The _____ engine places all of the cylinders in a single row.

4. An _____ _____ engine has the valve mechanisms in the cylinder head, while the camshaft is located in the engine block.

5. _____ occurs when pressure and temperature build in gas chambers outside of the normal spark flame.

6. In a four-stroke engine, four strokes are required to complete one _____.

7. Instead of using a spark produced by the ignition system, the diesel engine uses the _____ produced by compressing air in the combustion chamber to ignite the fuel.

8. A _____ _____ engine has a power stroke every revolution.

9. In a two-cycle engine the air-fuel mixture enters the crankcase through a _____ _____ or _____ _____.

10. _____ is an explosion in the combustion chamber resulting from the air-fuel mixture igniting prior to the spark being delivered from the ignition system.

Multiple Choice

1. Increasing the compression ratio of an engine is being discussed.
   Technician A says increased compression ratios result in a reduction of power produced.
   Technician B says higher compression ratio engines require higher octane gasoline.
   Who is correct?
   A. A only   C. Both A and B
   B. B only   D. Neither A nor B

2. Engine measurement is being discussed.
   Technician A says the bore is the distance the piston moves within the cylinder.
   Technician B says the stroke is the diameter of the cylinder.
   Who is correct?
   A. A only   C. Both A and B
   B. B only   D. Neither A nor B

3. *Technician A* says to determine horsepower, torque must be known first.
   Technician B says horsepower will usually peak at a lower engine speed than torque.
   Who is correct?
   A. A only
   B. B only
   C. Both A and B
   D. Neither A nor B

4. *Technician A* says mechanical efficiency is a comparison between brake horsepower and indicated horsepower.
   Technician B says volumetric efficiency is a measurement of the amount of air-fuel mixture that actually enters the combustion chamber compared to the amount that could be drawn in.
   Who is correct?
   A. A only   C. Both A and B
   B. B only   D. Neither A nor B

5. During the compression stroke in a four-cycle engine
   A. the intake valve is open, and the exhaust valve is closed.
   B. the exhaust valve is open, and the intake valve is closed.
   C. both valves are open.
   D. both valves are closed.

6. All of these statements about diesel engines are true *except*:
   A. The air-fuel mixture is ignited by the heat of compression.
   B. Glow plugs may be used to heat the air-fuel mixture during cold starting.
   C. Spark plugs are located in each combustion chamber.
   D. Diesel engines may be two cycle or four cycle.

7. While discussing volumetric efficiency,
   Technician A says volumetric efficiency is not affected by intake manifold design.
   Technician B says volumetric efficiency decreases at high engine speeds.
   Who is correct?
   A. A only   C. Both A and B
   B. B only   D. Neither A nor B

8. In a two-stroke cycle engine,
   A. a power stroke is produced every crankshaft revolution.
   B. intake and exhaust valves are located in the cylinder head.
   C. the lubricating oil is usually contained in the crankcase.
   D. the crankcase is not pressurized.

9. A vehicle with sequential fuel injection will

   A. inject fuel only once per engine cycle.

   B. inject fuel on every stroke of a four-stroke cycle.

   C. have one fuel injector that is used for all of the cylinders in a multicylinder engine.

   D. inject fuel all of the time when an engine is running.

10. While discussing engine identification,

   Technician A says that the tenth character of the VIN defines the vehicle model year.

   Technician B says that an engine identification tag may identify the date on which the engine was built.

   Who is correct?

   A. A only

   B. B only

   C. Both A and B

   D. Neither A nor B

CHAPTER 4
ENGINE OPERATING SYSTEMS

Upon completion and review of this chapter, you should understand and be able to describe:

- The purpose of the starting system.
- The components of the starting system.
- The purpose of the battery.
- The purposes of the engine's lubrication system.
- The operation of the lubrication system and its major components.
- The basic types and purposes of additives formulated into engine oil.
- The purpose of the Society of Automotive Engineers' (SAE) classifications of oil.
- The purpose of the American Petroleum Institute's (API) classifications of oil.
- The purpose of the International Lubricant Standardization and Approval Committee's (ILSAC) classifications of oil.

- The two basic types of oil pumps: rotary and gear.
- The purpose of the cooling system.
- The operation of the cooling system and its major components.
- The advantages of a reverse-flow cooling system.
- The purpose of antifreeze and its characteristics.
- The purpose and basic operation of the fuel system.
- The purpose and basic operation of the ignition system.
- The purpose of the emission control systems.

Terms To Know

Armature	Gasoline	Oxidation inhibitors
Battery terminals	Head gasket	Piezoresistive sensor
Body control module (BCM)	Heater core	Pole shoes
Burn time	Ignition system	Positive displacement pump
C-class oil	Inlet check valve	Prove-out circuit
Cetane number	Jet valve	Radiator
Coolant recovery system	Knock sensor	Reverse-flow cooling system
Cooling system	Lubrication system	Ring gear
Detergents	Micron	Research Octane Number (RON)
Diesel	Motor Octane Number (MON)	Saddle
Distributor-ignition (DI)	Multipoint fuel injection	S-class oil
Electrolysis	Nucleate boiling	Tetraethyl (TEL) lead
Electronic ignition (EI)	Octane	Thermistor
Emission control system	Oil breakdown	Thermostat
Ethanol	Oil cooler	Viscosity
Flexplate	Oil pump	Viscosity index
Flywheel	Oil sludge	Warning light
Fuel system	Oxidation	
Full-filtration system		

INTRODUCTION

There are five basic requirements for the engine to run:

- The intake and exhaust systems must allow the engine to take air in and exhaust air out.
- The fuel system must supply the correct amount of fuel to mix with the air.
- The ignition system must deliver a spark to the cylinder at the correct time.
- The engine must develop adequate compression to allow ignition of the air and fuel.
- The electrical system must supply adequate electrical energy to start the engine and keep the auxiliary systems operating.

To maintain engine operation, the lubrication system must reduce friction in the engine and provide cooling. The cooling system must regulate engine temperature and remove excess heat. The emissions control systems reduce the environmentally damaging evaporative and exhaust emissions from the vehicle. The proper operation of these supporting systems helps ensure the engine's good performance, low emission output, efficiency, and longevity.

In this chapter, we will discuss the purpose and operation of these systems and their components. The engine cannot run properly without fully functional supporting systems. Failures in the lubrication system, as an example, can cause very serious engine problems.

As a professional technician, you will need to understand the operation of all the systems and components that keep the engine running properly. You will use this knowledge to help you provide excellent maintenance and perform expert diagnosis and repairs on the engine and its supporting systems.

THE STARTING SYSTEM

The **ring gear** around the circumference of a flywheel or flexplate allows the starter motor drive gear to turn the engine.

Some flexplates and flywheels are used as part of a sensor to determine the speed of the engine.

The **flywheel** is a heavy disk mounted at the end of the crankshaft to make the engine power pulses and vibrations more stable. It has a smooth surface to which the clutch assembly of a manual transmission is attached. This smooth surface is similar to the surface of a brake rotor.

A **flexplate** is a lighter version of a flywheel and is bolted to the torque converter of an automatic transmission.

The internal combustion engine must be rotated before it will run under its own power. It will need to build up speed and momentum before the ignition and fuel system can successfully take over and run the engine. The starting system is a combination of mechanical and electrical parts working together to start the engine. The starting system is designed to change the electrical energy that is being stored in the battery into mechanical energy. For this conversion to be accomplished, a starter or cranking motor is used. The conventional starting system includes the following components (**Figure 4-1**):

1. Battery
2. Cable and wires
3. Ignition switch
4. Starter solenoid or relay
5. Starter motor
6. Starter drive and flywheel ring gear
7. Starting safety switch

The battery delivers electrical power to the starter solenoid when the driver turns the ignition key to the crank position. The solenoid delivers battery power to the starter motor. The starter motor drive gear is engaged with the **ring gear** on the **flywheel** or **flexplate**; the flywheel (or flexplate) is bolted to the engine's crankshaft. When the starter motor spins, it rotates the engine at about 200–250 rpm (**Figure 4-2**). This is fast enough to allow the four-stroke cycle to produce sustainable combustion and begin running under its own power. Many newer vehicles have computer modules and antitheft systems installed on them. These computers are used to help control the engine's starting procedures.

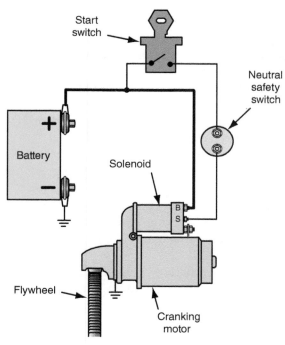

Start
switch

Neutral
safety
switch

Battery

Solenoid

B
S

Flywheel

Cranking
motor

Figure 4-1 The major components of a typical starting system.

Figure 4-2 A typical permanent magnet starter; it is more compact than a starter with
pole shoes and field windings.

The starting system of a hybrid electric vehicle (HEV) is different from the normal
starting system. The high-voltage battery pack is typically used to start the gasoline engine.
The same set of electric motors that are used for propulsion are used in conjunction with
a gear set to start the gasoline engine. Special care must be taken when servicing these
vehicles. This is explained in Chapter 14 of the *Shop Manual*.

 A BIT OF HISTORY

In the early days of the automobile, the vehicle did not have a starter motor. The operator had
to use a starting crank to turn the engine by hand. Charles F. Kettering invented the first electric
"self-starter," which was developed and built by Delco Electric. The self-starter first appeared on
the 1912 Cadillac and was actually a combination starter and generator.

The Battery

An automotive battery is an electrochemical device that provides for and stores potential electrical energy. When the battery is connected to an external load such as a starter motor, an energy conversion occurs, causing an electric current to flow through the circuit. Electrical energy is produced in the battery by the chemical reaction that occurs between two dissimilar plates that are immersed in an electrolyte solution.

When discharging the battery (current flowing away from the battery), the battery changes chemical energy into electrical energy. It is through this change that the battery releases stored energy. During charging (current flowing from the charging system to the battery), electrical energy is converted into chemical energy. Because of this, the battery can store energy until it is needed.

The largest demand is placed on the battery when it must supply current to operate the starter motor. The amperage requirements of a starter motor may be over 200 amperes. This requirement is also affected by temperature, engine size, and starter and engine condition.

A typical 12-volt automotive battery is made up of six cells connected in series (**Figure 4-3**). The six cells produce 2.1 volts each. Wiring the cells in series produces the 12.6 volts required by the automotive electrical system. A fully charged battery has 12.6 volts. The cell elements are submerged in a cell case that is filled with electrolyte solution.

All automotive batteries have two terminals. One terminal is a positive connection; the other is a negative connection. The **battery terminals** extend through the cover or the side of the battery case. Following are the most common types of battery terminals (**Figure 4-4**):

1. **Post or top terminals.** Used on most automotive batteries. The positive post is larger than the negative post to prevent connecting the battery in reverse polarity.
2. **Side terminals.** Positioned in the side of the container near the top. These terminals are threaded and require a special bolt to connect the cables. Polarity identification is by positive and negative symbols.
3. **L-terminals.** Used on specialty batteries and some imports.

Battery terminals provide a means of connecting the battery plates to the vehicle's electrical system.

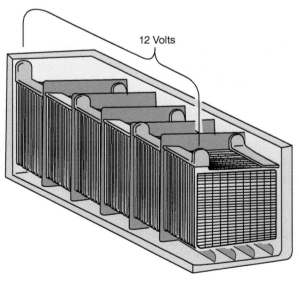

Figure 4-3 The 12-volt battery consists of six 2-volt cells connected together.

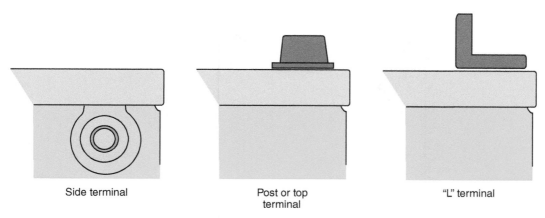

Side terminal   Post or top   "L" terminal
                        terminal

Figure 4-4 The most common types of automotive battery terminals.

Temperature	% of Cranking Power
80°F (26.7°C)	100
32°F (0°C)	65
0°F (−17.8°C)	40

Figure 4-5 As temperature decreases, so does the power available from the battery.

The condition of the battery is critical for proper operation of the starting system and all other vehicle electrical systems. The battery is often overlooked during diagnosis of poor starting or no-start problems. Many problems encountered within the starting system may be attributed to poor battery condition.

In addition, the proper battery selection for the vehicle is important. Some of the aspects that determine the battery rating required for a vehicle include engine size, engine type, climatic conditions, vehicle options, and so on. The requirements for electrical energy to crank the engine increase as the temperature decreases. Battery power drops drastically as temperatures drop below freezing (**Figure 4-5**). The engine also becomes harder to crank due to some oil's tendencies to thicken when cold, resulting in increased friction.

The battery's ability to deliver power to crank the engine is described by the cold cranking amperage (CCA). A battery with a CCA rating of 600 is able to deliver 600 amperes of electrical current for 30 seconds when the battery is at 0°F. A bigger engine will require a battery with a higher CCA than a small engine. The CCA requirement also increases in the northern regions of the continent, where temperatures regularly fall below freezing.

The battery that is selected must also fit the battery-holding fixture, and the hold down must be able to be installed. It is also important that the height of the battery not allow the terminals to short across the vehicle hood when it is shut. Battery Council International (BCI) group numbers are used to indicate the physical size and other features of the battery. This group number does not indicate the current capacity of the battery.

All batteries must be secured in the vehicle to prevent damage and the possibility of shorting across the terminals if the battery tips. Normal vibrations cause the plates to shed their active materials. Hold downs reduce the amount of vibration and help increase the life of the battery.

The battery's location may vary with different vehicles. Typically, the 12-volt battery is located under the hood and close to the front of the vehicle, in an area that is easy to get to for testing and replacement. Some vehicles may put the battery farther into the engine compartment, near the transaxle, in the trunk, or under the rear seat. These are known as remote location batteries (see **Figure 4-6**). There are many reasons for placing the battery

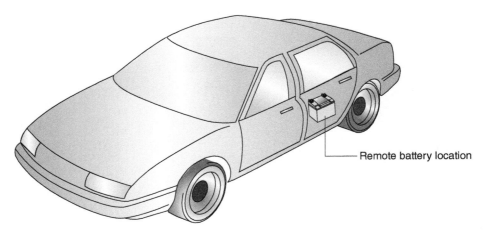

Figure 4-6 Some batteries are located remotely.

in these areas, which include weight distribution, temperature and insulation, and hood height and length. The batteries that are located inside the vehicle will have a vent tube that allows the battery to vent the gas during normal operation. It is important to refer to the manufacturer's service manual if you cannot locate the battery.

The Starter Motor

The **armature** is the movable component of the motor that consists of a conductor wound around a laminated iron core and is used to create a magnetic field.

Pole shoes are made of high-magnetic permeability material to help concentrate and direct the lines of force in the field assembly.

Starter motors use the interaction of magnetic fields to convert electrical energy into mechanical energy. **Figure 4-7** illustrates a simple electromagnet style of starter motor. The inside windings are called the **armature**. The armature rotates within the stationary outside windings, called the field, which has windings coiled around **pole shoes**. Some starters use stationary permanent magnets rather than the electromagnets created by the pole shoe windings. When current is applied to the field and the armature, both produce magnetic flux lines. The direction of the windings places the left pole at a south polarity and the right pole at a north polarity. The lines of force move from north to south in the field. In the armature, the flux lines circle in one direction on one side of the loop and in the opposite direction on the other side. Current now sets up a magnetic field around the loop of wire, interacting with the north and south fields and causing a turning force on the loop. This force causes the loop to turn in the direction of the weaker field (**Figure 4-8**).

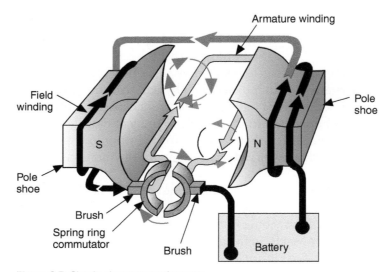

Figure 4-7 Simple electromagnetic motor.

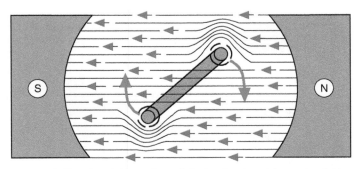

Figure 4-8 Rotation of the conductor is in the direction of the weaker field.

Problems within the starter or in the wiring to the starter can prevent it from turning the engine over fast enough to start. Engine mechanical problems can also cause this symptom. You must evaluate the starting system before condemning the engine.

LUBRICATION SYSTEMS

When the engine is operating, the moving parts generate heat due to friction. If this heat and friction were not controlled, the components would suffer severe damage and even weld together. In addition, heat is created from the combustion process. It is the function of the engine's **lubrication system** to supply oil to the high-friction and high-wear locations and to dissipate heat away from them (**Figure 4-9**).

It is impossible to eliminate all of the friction within an engine, but a properly operating lubrication system will work to reduce it. Lubrication systems provide an oil film to prevent moving parts from coming in direct contact with each other (**Figure 4-10**). Oil molecules work as small bearings rolling over each other to eliminate friction (**Figure 4-11**). Another function of the lubrication system is to act as a shock absorber between the connecting rod and the crankshaft.

Besides reducing friction, engine oil absorbs heat and transfers it to another area for cooling. As the oil flows through the engine, it conducts heat from the parts it comes in contact with. When the oil is returned to the oil pan, it is cooled by the air passing over the pan. Other purposes of the oil include improving the seal between the piston rings and

The engine's **lubrication system** supplies oil to high friction and high wear locations.

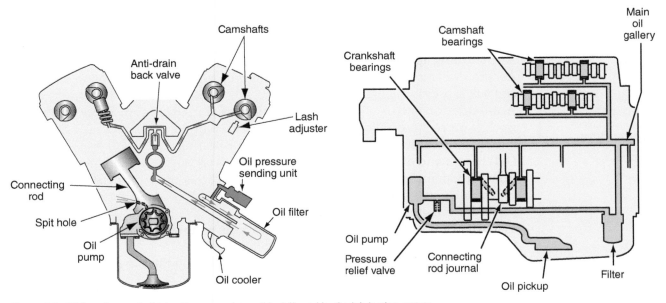

Figure 4-9 Oil flow diagram indicates the areas where oil is delivered by the lubrication system.

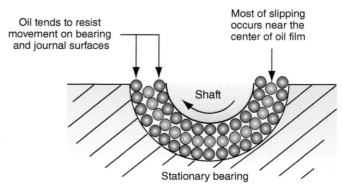

Figure 4-10 The oil develops a film to prevent metal-to-metal contact.

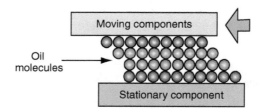

Figure 4-11 Oil molecules work as small bearings to reduce friction. An oil cooler is a heat exchanger that looks much like a small radiator.

the cylinder walls, and washing away abrasive metal and dirt. To perform these functions, the engine lubrication system includes the following components:

- Engine oil
- Oil pan or sump
- Oil filter
- Oil pump
- Oil galleries
- Oil coolers (heavy-duty applications)

Engine Oil

Engine oil must provide a variety of functions under all of the extreme conditions of engine operation. To perform these tasks, additives are mixed with natural oil. A brief description of the types of additives used is as follows:

1. **Antifoaming agents**. These additives are included to prevent aeration of the oil. Aeration will result in the oil pump not providing sufficient lubrication to the parts and will cause low oil pressure.
2. **Antioxidation agents.** Heat and oil agitation result in **oxidation**. These additives work to prevent the buildup of varnish and to prevent the oil from breaking down into harmful substances that can damage engine bearings.
3. **Detergents and dispersants.** These are added to prevent deposit buildup resulting from carbon, metal particles, and dirt. Detergents break up larger deposits and prevent smaller ones from grouping together. A dispersant is added to prevent carbon particles from grouping together.
4. **Viscosity index improver.** As the oil increases in temperature, it has a tendency to thin out. These additives prevent oil thinning.
5. **Pour point depressants.** Prevent oil from becoming too thick to pour at low temperatures.
6. **Corrosion and rust inhibitors.** Displace water from the metal surfaces and neutralize acid.

Oxidation occurs when hot engine oil combines with oxygen and forms carbon.

7. **Cohesion agents.** Maintain a film of oil in high-pressure points to prevent wear. Cohesion agents are deposited on the parts as the oil flows over them and remain on the parts as the oil is pressed out.

Oil has traditionally been rated by two organizations: the Society of Automotive Engineers (SAE) and the American Petroleum Institute (API). The SAE has standardized oil viscosity ratings, while the API rates oil to classify its service or quality limits. Engine oil containers have markings on them indicating the qualities defined by these two organizations (**Figure 4-12**). The API and SAE ratings are written in the "doughnut" on oil containers (**Figure 4-13**). These rating systems were developed to allow proper selection of engine oil.

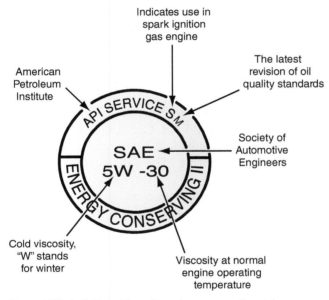

Figure 4-12 Definition of the ratings shown on an oil container.

Figure 4-13 The engine oil "doughnut" displays oil rating information such as the quality and viscosity of an oil.

S-class oil is used for automotive gasoline engines.

C-class oil is designed for commercial or diesel engines. The S stands for *spark ignition*, and the C stands for *compression ignition*. In some areas, the S stands for *service engines*, and the C stands for *commercial engines*.

The API classifies engine oils by a two-letter system. The prefix letter is either listed as S-class or C-class to classify the oil's intended usage. **S-class oil** is designated for use in spark ignition engines; **C-class oil** should be used in compression ignition diesel engines. The second letter denotes various grades of oil within each classification and denotes the oil's ability to meet the engine manufacturers' warranty requirements. Each category of oil that meets the qualifications has been tested and approved by the API under set conditions. As new oil standards are developed, the next letter in the alphabet is used to designate the latest revision. As vehicles, engines, and driving habits continue to change, the oil formula must change along with it. This system began many years ago and has developed since then. The sequence started with an SA-approved oil. The SA oil is a plain mineral oil that doesn't contain any additives. It was primarily used in the 1920s. Currently, the SN rating is used for all automotive engines in use. It provides improved oxidation resistance, improved deposit protection, better low-temperature performance, and better wear protection over the life of the oil. All newer API oils have a limitation on the amount of phosphorus that can be used. Phosphorus can be harmful to a catalytic converter. Newer engines are developed using the new oil standards. This means that you can use an SN grade oil in any vehicle, but you cannot use an SM or older oil in a newer vehicle that requires an SN grade oil.

Diesel engine oils are much more complicated. The most current oil ratings for compression ignition engines are CJ-4, CI-4, CH-4, CG-4, CF-2, and CF. Some smaller stationary compression ignition engines (e.g., generators, pumps, and other commercial applications) use an older CC rating. The exhaust on these engines are regulated differently.

Figure 4-14 charts the current production oil classifications for gasoline and diesel engines. The API intentionally omits the SI and SK from the sequence of categories. SA through SH are now obsolete.

At present, many manufacturers of gasoline engines demand an SM or SN oil classification for warranty requirements. Most gasoline engine manufacturers also demand an engine oil with the API Starburst symbol displayed on the oil container (**Figure 4-15**). In addition to the SAE and API ratings, other ratings are common. The International Lubricant Standardization and Approval Committee (ILSAC) was formed to approve

Figure 4-14 SAE 0W-16, designed for greater fuel economy, came from the SAE 16 oil classification for oils having lower than 20 viscosity.

API Categories	
Category	**Service Recommendations**
SN	Introduced in October 2010, the SN oil will work for all automotive gasoline engines currently in use.
SM	For use in 2010 and older automotive gasoline engines currently in use.
SL	For use in 2004 and older automotive gasoline engines currently in use.
SJ	For use in 2001 and older automotive gasoline engines currently in use.
CJ-4	Introduced in 2006, CJ-4 was designed to meet the requirements of 2007 model year on-highway exhaust emission standards.
CI-4	For use in high-speed, four-stroke diesel engines designed to meet the 2004 exhaust emission standards.
CH-4	For use in high-speed, four-stroke diesel engines designed to meet the 1998 exhaust emission standards.
CG-4	For use in severe-duty, high-speed, four-stroke diesel engines designed to meet the 1994 exhaust emission standards.
CF-4	For use in high-speed, four-stroke diesel engines designed to meet the 1990 exhaust emission standards.
CF-2	For use in severe-duty, two-stroke diesel engines.
CF	For use in off-road indirect-injected diesel engines, including those that use a 0.5% or greater sulfur content fuel.

motor oils for use across domestic and import lines and to help speed up the certifying process. SAE, API, the Japanese Automobile Standards Organization (JASO) and Daimler-Chrysler, General Motors, and Ford have all been involved in the development of the latest ILSAC standard oil GF-5 introduced late 2010. GF-4 oils are no longer available. Only oils with an SAE viscosity rating of 0W, 5W, or 10W may carry the GF rating.

Other oil ratings that you may encounter include ACEA (common with European manufacturers) and JASO (Japanese Automotive Standards Organization). Some manufacturers have their own specific oil rating systems. One example of this is GM's Dexos system. GM recommends the Dexos oil classification for its vehicles. According to GM it is designed to reduce emissions, provide for better fuel economy, last longer, and prolong the life of the emissions system. Many oil brands meet the Dexos specification. Dexos 1 is designed for gasoline engines, and Dexos 2 is designed for compression ignition engines. Different oils are being developed and tested for GDI systems because of soot accumulation in the oil.

Figure 4-15 The API Starburst symbol is displayed on oil containers.

SAE ratings provide a numeric system of determining the oil's **viscosity**. The higher the viscosity number, the thicker or heavier the weight of the oil. For example, oil classified as SAE 50 is thicker than SAE 20 oil. The thicker the oil, the slower it will pour. Thicker oils should be used when engine temperatures are high, but they can cause lubrication problems in cooler climates. Using an engine oil with a thicker viscosity than is recommended can cause premature engine wear. The thicker oil will not reach high-friction areas fast enough during the critical start-up and warm-up periods. Viscosity recommendations have become lighter over the years, as engine machining improvements produce engines with closer clearances. The thinner oils are needed to circulate between the close-fitting components. Thinner oils will flow through the engine faster and easier at colder temperatures, but **oil breakdown** may occur under higher temperatures or heavy engine loads. Oil also has to be thick enough to seal piston rings, yet if it is too thin, the engine will not develop enough oil pressure. The oil viscosity will also change with temperature. The higher the temperature, the lower the viscosity, and vice versa.

A compromise is needed. Multigrade oils were developed for this reason. Multigrade oils contain two viscosities and are designated by their label. For example, an SAE 5W30 oil is a multigrade oil. The *W* stands for *winter*. SAE has established a standard for testing these oils. The lower number (5 in this case) is the viscosity of the oil at 0°C, and the higher number (30 in this case) is the viscosity of the oil at 100°C (32°F and 212°F, respectively).

The manufacturing of multigrade oil starts out with a base oil. This base oil is typically the lower number (5 in our case). **Viscosity index** modifiers are then added to change the viscosity of the oil by changing its behavior at higher temperatures. A single-grade oil is different than a multigrade oil because it cannot pass the low temperature SAE standard.

Oils have been developed that may be labeled "energy conserving." These oils use friction modifiers to reduce friction and increase fuel economy. The energy-conserving designation is displayed in the lower portion of the "doughnut."

Selection of oil is based upon the type of engine and the conditions it will be running in. The API ratings must meet the requirements of the engine and provide protection under the normal expected running conditions. One consideration in selecting SAE ratings is ambient temperatures. Some manufacturers will recommend an oil viscosity rating for normal driving and suggest thicker oil while driving at high sustained speeds in a hot climate. A motorist may request that you refill the engine with higher-viscosity oil before a summer vacation while pulling a camper, for example. Many consumers run the standard oil in the vehicle year-round. Always select oil that meets or exceeds the manufacturer's recommendations.

To a great degree it is the improvement in the quality of oils that has allowed the tighter clearances and improvements in piston fit and ring selection. The incredible gains in oil technology are a significant part of the reason that engines now often last 150,000 miles or longer before needing major service. Use a high-quality oil with the viscosity rating specified by the manufacturer, and change it regularly to get the most life out of an engine. SL, GF-4 oil is able to meet the extreme demands of late-model, precisely machined engines.

Synthetic Oils

Synthetic oils are oils primarily composed of chemical compounds that are synthetically made in a laboratory. Synthetic oils are manufactured as a substitute for and generally provide superior properties when compared to conventional petroleum oils. Most synthetic oil ratings not only meet but exceed the API and ILSAC ratings for their given SAE viscosity.

Synthetic engine oils have been in use for a long time, but it hasn't been until recently that their use as an automotive engine oil has been widely accepted. Synthetic oils are chemically more stable, have extended oil change intervals, have better viscosity performance at extreme temperatures, increase fuel mileage, provide better lubrication on cold starts, and are very resistive to oil breakdown, sludge, and oxidation. In some cases, using synthetic engine oils are a part of the initial engine designs to help lower emissions.

Viscosity is the measure of the resistance of the oil when placed under shear and extensional stress (the same stresses placed on oil by the crankshaft and connecting rod bearings). It is used to describe an oil's internal resistance to flow. Technicians often use viscosity to describe an oil's thickness or weight.

All oils will break down their additives over time. **Oil breakdown** usually refers to the results from prolonged high temperatures and pressures that quickly degrade the oil and do not allow it to perform its duties.

The **viscosity index** of an oil is a measure of the change in viscosity an oil will have with changes in temperature.

Today's modern engines are built to provide more power from a smaller engine. Engine compartments are typically smaller and have less wind movement and heat dissipation than the previous older, larger vehicles. Combustion chamber temperatures are typically higher as well. For these reasons, all modern engines are under more stress to perform. Overall, engine temperatures are warmer, and if the cooling system is neglected, they are more susceptible to overheating and damage. In a modern engine, one of the major problems with oil is sludge buildup.

Oil sludge buildup is when the oil gels or solidifies after it has been broken down due to high stresses in the engine. Oil sludge is a preventable disaster. Regular maintenance and inspections can prevent major buildup. When oil additives break down the oil, it oxidizes (mixes with oxygen) and begins to gel. When the oil gels, it flows slower and sticks to the walls of the oil galleries. Often, this gel will stick to oil return galleries and cause low oil levels in the oil sump. This, of course, leads to oil starvation and engine damage. Oil sludge problems, emissions standards, fuel mileage standards, and customer demand have led many manufacturers to install synthetic oils in brand new vehicles. Manufacturers that choose to add synthetic oils to their vehicles may even suggest that synthetic oil continue to be used during regular oil changes.

The manufacturers of these vehicles are also using synthetic engine oils because of their superior qualities to regular oil. If a customer already has synthetic engine oil in his or her engine from the factory, it is recommended to continue to use synthetic oil.

Synthetic oils do not gel as much as traditional oils do in colder temperatures. An engine may start easier and fully lubricate faster in extreme cold temperatures with synthetic oils. Synthetic oils are mainly used for their superior engine protection, especially during cold starts when most engine damage occurs. Synthetic oils also have more stability over time, better lubricity, and better resistance to thermal breakdown. Some customers claim increased fuel mileage, but the increase in mileage by just changing the type of oil alone is very small and difficult to measure.

If customers want to switch to synthetic oil during a routine oil change, try to advise them with the knowledge you have and the written material the synthetic oil manufacturer provides. Also check with the manufacturer's service literature and technical service bulletins for information about the warranty. Most warranties are not voided when using synthetic oils at recommended intervals, but you should still check. Synthetic oils can be used in gas or diesel engines.

Some engine rebuilders do not recommend using synthetic oils during the engine break-in period (typically the first 500 miles after a rebuild). Other sources may point to the fact that some new vehicles are using synthetic oils as a factory fill. New engines from the manufacturer are machined to a lower tolerance than most rebuilt engines and are manufactured in very controlled environments. The decision to use a synthetic oil during engine break-in should be made after considering the extent of the rebuild, the type, and the quality of any machining performed, and the type and quality of the parts that were installed.

> **Oil sludge** buildup occurs when the oil gels and thickens due to oxidizing.

Oil Pumps

The **oil pump** is the heart of the lubrication system. It develops oil pressure to pump oil throughout the engine. There are two basic types of oil pumps: rotor and gear. Both types are **positive displacement pumps**. The rotary pump generally has a four-lobe inner rotor with a five-lobe outer rotor (**Figure 4-16**). The outer rotor is driven by the inner rotor. As the lobes come out of mesh, a vacuum is created, and atmospheric pressure on the oil pushes it into the pickup tube. The oil is trapped between the lobes as it is directed to the outlet. As the lobes come back into mesh, the oil is expelled under pressure from the pump.

Gear pumps can use two gears riding in mesh with each other or use two gears and a crescent design (**Figure 4-17**). Both types operate in the same manner as the rotor-type pump. The advantage of the rotor-type pump is its capability to deliver a greater volume of oil since the cavities are larger.

> The **oil pump** creates a vacuum so atmospheric pressures can push oil from the sump. The pump then pressurizes the oil and delivers it throughout the engine by use of galleries.

> **Positive displacement pumps** deliver the same amount of oil with every revolution, regardless of speed.

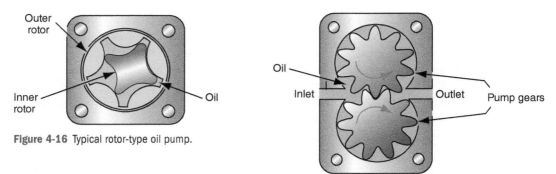

Figure 4-16 Typical rotor-type oil pump.

Figure 4-17 Typical gear-type oil pump.

Figure 4-18 This crankshaft-driven oil pump sits under the timing cover.

Many of today's engines drive the oil pump by the front of the crankshaft (**Figure 4-18**). In the past, most oil pumps were driven off of the camshaft by a drive shaft fitting into the bottom of the distributor shaft, but not many engines still use an ignition distributor (**Figure 4-19**).

Most lubrication systems use oil pressures between 10 and 80 psi, depending on engine speed and temperature. A general rule of thumb is that an engine should develop a minimum of 10 psi of oil pressure for each 1,000 rpm of engine speed once the engine is at operating temperature. Since oil pumps are positive displacement types, output pressures must be regulated to prevent excessive pressure buildup. A pressure relief valve opens to return oil to the sump or pump inlet if the specified pressure is exceeded (**Figure 4-20**). A calibrated spring holds the valve closed, allowing pressure to increase. Once the oil pressure is great enough to overcome the spring pressure, the valve opens and returns the oil to the sump or pump inlet.

Oil Filter

As the oil flows through the engine, it works to clean dirt and deposits from the internal parts. These contaminants are deposited into the oil pan or sump with the oil. Since the oil pump pickup tube syphons the oil from the sump, it can also pick up these contaminants. The pickup tube has a screen mesh to prevent larger contaminants from being picked up

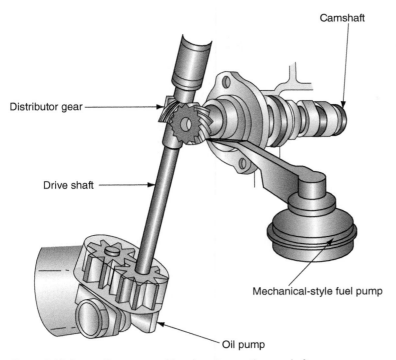

Camshaft

Distributor gear

Drive shaft

Mechanical-style fuel pump

Oil pump

Figure 4-19 Some oil pumps are driven by a gear on the camshaft.

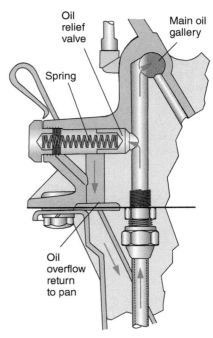

Oil relief valve

Spring

Main oil gallery

Oil overflow return to pan

Figure 4-20 The oil pressure relief valve opens to return the oil to the sump if oil pressures are excessive.

and sent back into the engine. The finer contaminants must be filtered from the oil to prevent them from being sent with the oil through the engine. Oil filter elements have a pleated paper or fibrous material designed to filter out particles between 20 and 30 **microns** (**Figure 4-21**). Most oil filters use a **full-filtration system**. Under pressure from the pump, oil enters the filter on the outer areas of the element and works its way toward the center. In the event the filter becomes plugged, a bypass valve will open and allow the oil to enter the oil galleries. If this occurs, the oil will no longer be filtered (**Figure 4-22**).

> A **micron** is a thousandth of a millimeter or about 0.0008 inch.

> A **full-filtration system** means all of the oil is filtered before it enters the oil galleries.

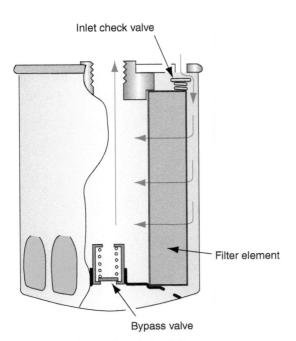

Inlet check valve

Filter element

Bypass valve

Figure 4-21 Oil flow through the oil filter.

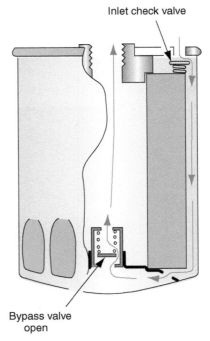

Inlet check valve

Bypass valve open

Figure 4-22 If the oil filter becomes plugged, the bypass valve opens to protect the engine.

The **inlet check valve** keeps the oil filter filled at all times, so when the engine is started, an instantaneous supply of oil is available.

An **oil cooler** is a heat exchanger that looks much like a small radiator.

The **saddle** is the portion of the crankcase bore that holds the bearing half in place.

Jet valves are used to help cool pistons.

The connecting rod journal is also referred to as the crank pin.

An **inlet check valve** is used to prevent oil draining back from the oil pump when the engine is shut off. The check valve is a rubber flap covering the inside of the inlet holes. When the engine is started and the pump begins to operate, the valve opens and allows oil to flow to the filter.

Oil Coolers

Turbocharged, supercharged, and other high-performance engines or engines in vehicles designed to haul or tow frequently use an **oil cooler** to help keep oil temperatures at a safe level. Oil temperature should generally run between 200°F and 250°F. Excess heat will prematurely break down the oil and allow premature engine wear. Oil coolers look like small radiators and are usually mounted near the radiator. Some designs exchange their heat with cooler incoming air, similar to how a radiator works. Others use engine coolant to remove the heat. Some manufacturers have the oil filter located remotely (not directly mounted on the engine). This also acts like a small oil cooler because cooler air will flow over hoses and the filter.

Oil Flow

The oil pump sends pressurized oil to the oil filter. After the oil leaves the filter, it is directed through galleries to various parts of the engine. A main oil gallery is usually drilled through the length of the block. From the main gallery, pressurized oil branches off to upper and lower portions of the engine (**Figure 4-23**). Oil is directed to the crankshaft main bearings through the main bearing **saddles**. Passages drilled in the crankshaft then direct the oil to the rod bearings (**Figure 4-24**). Some manufacturers drill a small spit or squirt hole in the connecting rod to spray pressurized oil delivered to the bearings out and onto the cylinder walls (**Figure 4-25**). When the spit hole aligns with the oil passage in the rod journal, it squirts the oil onto the cylinder wall. Some engines use **jet valves** to spray oil into the underside of the piston to cool it (**Figure 4-26**).

Each camshaft bearing (on OHV engines) also receives oil from passages drilled in the block (Figure 4-26). The valvetrain can receive oil through passages drilled in the

 A BIT OF HISTORY

Ernest Sweetland and George H. Greenhalgh patented the first oil filter in 1923. It was a device that was designed to trap particles in the oil and allow clean oil to pass through. Their patent was named "Purolator," which was short for "pure oil later." You can find Purolator oil filters still sold today.

Figure 4-23 Some oil filters do not have a housing around the filter media.

block and then through a hollow rocker-arm shaft. Engines using pushrods usually pump oil from the lifters through the pushrods to the valvetrain (**Figure 4-27**). Oil sent to the cylinder head is returned to the sump through drain passages cast into the head and block.

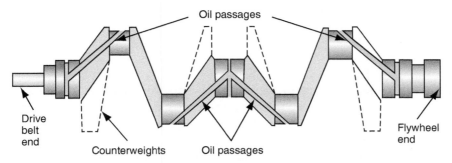

Figure 4-24 Oil passages drilled into the crankshaft provide lubrication to the main bearings.

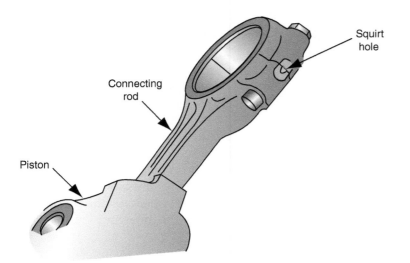

Figure 4-25 A squirt hole is used to lubricate the cylinder walls and remove excess heat from the piston head.

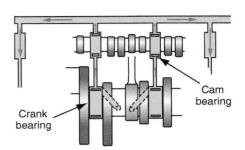

Figure 4-26 OHV engines have branches from the main oil gallery to direct pressurized oil to the camshaft journals.

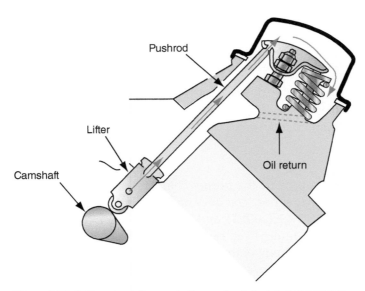

Figure 4-27 OHV engines often use hollow pushrods to deliver oil pressure to the cylinder head and valvetrain.

Not all areas of the engine are lubricated by pressurized oil. The crankshaft rotates in the oil sump, creating some splash-and-throw effects. Oil that is picked up in this manner is thrown throughout the crankcase. This oil splashes onto the cylinder walls below the pistons. This provides lubrication and cooling to the piston pin and piston rings. In addition, oil that is forced out of the connecting rod bearings is thrown to lubricate parts that are not fed pressurized oil. Valve timing chains are often lubricated by oil draining from the cylinder head being dumped onto the chain and gears.

AUTHOR'S NOTE During my experience as an automotive technician and instructor, I encountered a significant number of vehicle owners who attempted to save money on car maintenance by extending engine oil and filter changes beyond the manufacturer's recommended interval. Unfortunately, they spent a lot more on premature and expensive engine repairs than they saved on maintenance costs. Proper lubrication system maintenance is extremely important to provide long engine life; explain this to your customers. In some cases, the engine damage will cause excessive exhaust emissions. A vehicle with this situation will likely not pass the required emissions testing found in some states and regions of the United States.

COOLING SYSTEMS

> A **cooling system** controls the temperature of an engine by allowing it to heat up quickly, operate at the correct temperature, and remove excess heat.

The heat created during the combustion process could quickly increase to a point where engine damage would occur. The **cooling system** is used to disperse the heat to the atmosphere. The cooling system also allows the engine to warm up quickly and provides an even operating temperature. This is important for minimizing emissions and providing efficient engine operation. The cooling system also provides heat for the passenger compartment. The components of the cooling system include the following:

- Coolant
- Radiator
- Radiator cap
- Recovery system
- Coolant jackets
- Hoses
- Water pump
- Thermostat
- Cooling fan
- Heater system
- Transmission cooler
- Temperature warning system

Coolant

> **Electrolysis** is the chemical and electrical process of decomposition that occurs when two dissimilar metals are joined with the presence of moisture. The lesser of the two metals is eaten away.

Water by itself does not provide the proper characteristics needed to protect an engine. It works quite well to transfer heat, but does not protect the engine in colder climates. The boiling point of water is 212°F (100°C), and it freezes at 32°F (0°C). Temperatures around the cylinders and in the cylinder head can reach above 500°F (250°C), and the engine will often be exposed to ambient temperatures below 32°F (0°C). In addition, water reacts with metals to produce rust, corrosion, and **electrolysis**. Coolant must protect against freezing, reduce rust and corrosion in the cooling system, and lubricate the water pump and seals. Most manufacturers use a glycol-based engine coolant. In older engines, there was only one type of coolant. Today there are several types of coolants and many different applications for each.

Traditional green ethylene glycol (EG) coolant has been used for many years. It uses inorganic borate salts, silicate, and phosphate to minimize rust and corrosion. These additives break down over time, and the coolant becomes acidic. The acidity accelerates corrosion. The newer extended-life, EG-based coolants have a different additive package designed to lengthen the effective life of the coolant. These coolants use a new type of corrosion inhibitor called Organic Additive Technology (OAT) corrosion inhibitors. Many manufacturers using extended-life coolant now specify flushing the cooling system every 5–10 years or 100,000–150,000 miles rather than the traditional coolant flush recommendation of two years or 30,000 miles. Many European manufacturers specify coolants with no phosphates to reduce sediment and scale buildup when mixed with hard water. Asian manufacturers may specify a coolant with little or no silicates. These different coolants may be blue, pink, yellow, orange, or red.

Propylene glycol (PG) coolant is manufactured because it is less toxic than EG coolant. Less than half a cup of EG coolant is potentially fatal to pets and humans. Animals are attracted to coolant because it is sweet. Some consumers may choose to fill their cooling systems with these environmentally friendly coolants. If EG and PG are mixed, it is more difficult to detect the antifreeze protection level. Propylene glycol is not recommended by any vehicle manufacturers and, if used, could possibly void the vehicle's warranty.

Each manufacturer installs its own type of coolant from the factory for specific reasons. Different types and brands of coolant protect engine parts in different ways. Since each manufacturer uses different types and blends of metals, runs different temperatures, and uses different computer programs to control the engine and the emissions, it is important to recognize that you should be using the type of coolant that is recommended by the manufacturer.

Not surprisingly, cooling system failures are the leading cause of other mechanical breakdowns. One of the most common problems related to coolant degradation is electrolysis. When two or more dissimilar metals are used in the parts of the engine that touch the coolant, a small electrical charge will occur. This electrical charge is similar to the electrical charge produced by the vehicle's lead acid battery. A battery uses two different metal plates to create and hold its charge. An engine is similar in that some parts are aluminum, iron, copper, or brass. Electrolysis can damage the cooling system walls and make them thinner. Having the proper coolant in the engine protects the metals in the engine and cooling-heating system from electrolysis damage. Excessive electrolysis can lead to overheating, leaking, seal damage, hose damage, and gasket damage. In some cases, it will make component removal difficult.

Air should not be trapped in the coolant. If air is trapped in the coolant, the engine may not cool properly, and the heater system may not heat properly. Air may get into the system for several reasons: component replacement, coolant leakage, or improper refill. Air pockets can explode in the coolant and cause metal cavitation. Trapped air can also cause the coolant to become acidic. It is important to keep the system bled properly. Some vehicles have a coolant bleed screw to remove the air from the system, while others may need special tools or procedures to remove the air. Removing the air from the cooling system is covered in the *Shop Manual*.

Use the proper coolant and water mixture to protect the engine under most conditions. Water expands as it freezes. If the coolant in the engine block freezes, the expansion could cause the block or cylinder head to crack. If the coolant boils, liquid is no longer in contact with the cylinder walls and combustion chamber. The vapors are not capable of removing the heat, and the pistons may collapse, or the head may warp or crack from the excessive heat.

Nucleate boiling is the process of maintaining the overall temperature of a coolant to a level below its boiling point, but allowing the portions of the coolant actually contacting the surfaces (the nuclei) to boil into a gas.

The **radiator** is a heat exchanger used to transfer heat from the engine to the air passing through it.

For every pound of pressure increase, boiling point of water is raised about 3 ¼ °F.

Many vehicles that experience difficulty with air entering the cooling system during service add a bleeder valve at the high point of the cooling system to allow the technician to bleed the system of air.

Ethylene glycol by itself has a boiling point of 330°F (165°C), but it does not transfer heat very well. The freezing point of EG is 8°F (−13°C). To improve the transfer of heat and lower the freezing point, water is added in a mix of about 50/50. At a 50/50 mix, the boiling point is 226°F (108°C), and the freezing point is 34°F (1°C). In some very cold climates, coolant is mixed with water in a 60/40 ratio to prevent freezing to at least 40°F. These characteristics can be altered by changing the mixture (**Figure 4-28**).

It may appear that lowering the boiling point by adding water is not desirable. In fact, this is required for proper engine cooling. Under pressure, the boiling point of the coolant mix is about 263°F (128°C). As was stated earlier, the temperature next to the cylinder or cylinder head may be in excess of 500°F (260°C). As coolant droplets touch the metal walls, they boil and are turned into a gas. The superheated gas bubbles are quickly carried away into the middle of the coolant flow, where they cool and condense back into a liquid. If this **nucleate boiling** did not occur, the coolant would not be capable of removing heat fast enough to protect the engine.

Over time, all coolant can become slightly acidic because of the metals and minerals in the cooling system. All vehicle manufacturers provide a maintenance schedule recommending cooling system flushing and refill based on time and mileage.

Radiator

The **radiator** contains a series of tubes and fins that transfer the heat in the coolant to the air. As the coolant is circulated throughout the engine, it attracts and absorbs the heat within the engine, and then flows into the radiator intake tank. The coolant then flows

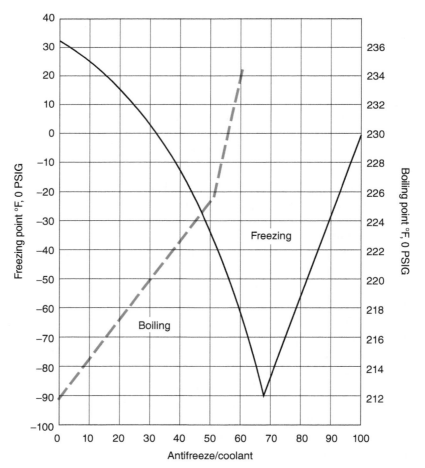

Figure 4-28 Changing the strength of the coolant mixture affects its boiling and freezing points.

through the tubes to the outlet tank. As the heated coolant flows through the radiator tubes, the heat is dissipated into the air through the fins.

There are two basic radiator designs: downflow and crossflow. Downflow radiators have vertical fins that direct coolant flow from the top inlet tank to the bottom outlet tank (**Figure 4-29**). Crossflow radiators use horizontal fins to direct coolant flow across the core (**Figure 4-30**). The core of the radiator can be constructed using a tube-and-fin or a cellular-type design (**Figure 4-31**). The material used for the core is usually copper, brass, or aluminum. Aluminum-core radiators generally use nylon-constructed tanks.

For heat to be transferred to the air effectively, there must be a significant difference in temperature between the coolant and the air. The greater the temperature difference, the more effectively heat is transferred. Manufacturers increase the temperature and boiling

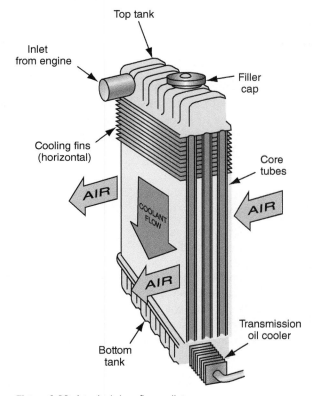

Figure 4-29 A typical downflow radiator.

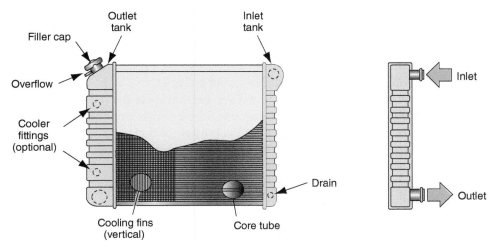

Figure 4-30 A typical crossflow radiator.

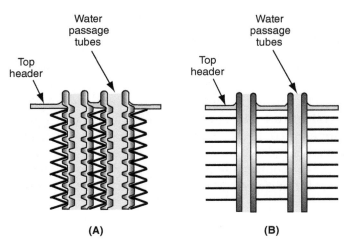

Figure 4-31 (A) Construction of a cellular radiator core. (B) Construction of a tube and fin radiator.

point of the coolant by pressurizing the radiator. The radiator cap allows for an increase of pressure within the cooling system, increasing the boiling point of the coolant; for example, if the pressure is increased to 15 psi (103 kPa), the boiling point of the 50/50 mix would be increased to 266°F (130°C). Because the boiling point is increased, the operating temperature of the engine can also be increased.

A BIT OF HISTORY

In the early years of the automobile, water was used as engine coolant. This is where some of the engine parts got their name, like the water pump. The water worked well to remove heat from the engines, but it would freeze, resulting in cracked blocks. Motorists used several different substances to prevent the water from freezing. Most of these substances prevented freezing but had other adverse effects to the engine. Some of these early substances included salt, calcium chloride, soda, sugar, honey, engine oil, and coal oil. The first successful antifreeze was made from wood or grain alcohol. However, these substances lowered the boiling point of the water and evaporated at higher engine temperatures. The first permanent antifreeze (meaning it would not evaporate) was made with a glycerin base.

The radiator cap uses a pressure valve to pressurize the radiator to between 14 and 18 pounds per square inch (**Figure 4-32**). If the pressure increases over the setting of the cap, the cap's seal will lift against a spring and release the pressure into a recovery tank (**Figure 4-33**). The cap also has a vacuum valve that will release any vacuum from the system as the coolant temperature decreases after the engine is turned off. If a vacuum is developed in the cooling system, the radiator could collapse.

The **coolant recovery system** contains a reservoir that is connected to the radiator by a small hose. The coolant expelled from the radiator during a high-pressure condition is sent to this reservoir. When the engine cools, a void is created in the radiator, and the vacuum valve in the radiator cap opens (**Figure 4-34**). Under this condition, atmospheric pressure on the coolant in the overflow reservoir pushes it back into the radiator. Due to lower hood lines, many vehicles now use a remote pressure tank. There is no cap on the radiator; the system is filled and checked at the remote pressure tank, sometimes called the reservoir. These are used because the top of the radiator is lower than the highest point on the cooling system. During filling, the system would collect air pockets in the higher spots of the engine cooling systems. These reservoirs are not the same as coolant recovery tanks; they contain coolant under pressure, and the top should never be removed while the coolant is hot.

The **coolant recovery system** prevents loss of coolant if the engine overheats, and it keeps air from entering the system.

Figure 4-32 The radiator cap seals the radiator or reservoir and regulates system pressure.

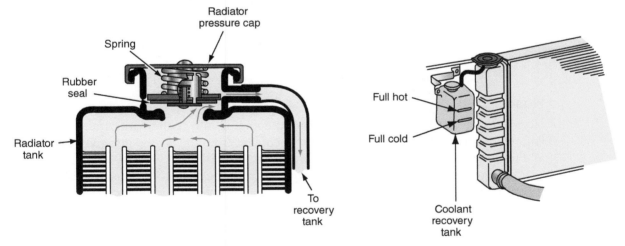

Figure 4-33 A typical coolant recovery system holds coolant released from the radiator.

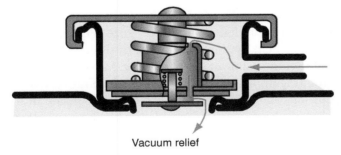

Vacuum relief

Figure 4-34 When the radiator cap vacuum valve opens, coolant from the recovery reservoir enters the radiator.

Whenever an engine is rebuilt, the radiator should be thoroughly cleaned and then inspected by pressure testing it. The radiator cap should be replaced with a new one.

Heater Core

The **heater core** is similar to a small version of the radiator. The heater core is located in a housing, usually in the passenger compartment of the vehicle (**Figure 4-35**). Some of the hot engine coolant is routed to the heater core by hoses. The heat is dispersed to the air inside the vehicle, thus warming the passenger compartment. To aid in quicker heating of the compartment, a heater fan blows the radiated heat into the compartment.

The **heater core** dissipates heat like a radiator. The radiated heat is used to warm the passenger compartment.

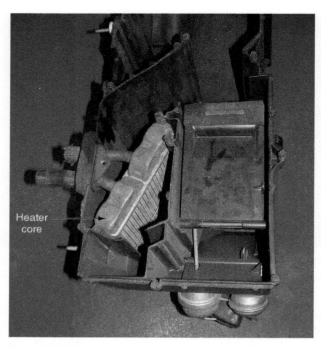

Figure 4-35 The heater core is shown in this cutaway of a heater distribution box.

Hoses

Hoses are used to direct the coolant from the engine into the radiator and back to the engine. In addition, hoses are used to direct coolant from the engine into the heater core and (on some engines) to bypass the thermostat.

Since radiator and heater hoses deteriorate from the inside out, most manufacturers recommend periodic replacement of the hoses as preventive maintenance. Whenever the engine is rebuilt, the hoses should always be replaced. It is true that newer and higher-quality hoses are built to last longer and stand up better to electrolysis degradation.

Water Pump

The water pump is the heart of the engine's cooling system. It forces the coolant through the engine block and into the radiator and heater core (**Figure 4-36**). The water pump is driven by an accessory belt, by the timing belt, or directly from the camshaft. Some hybrids drive the water pump using an electric motor. Most water pumps have a centrifugal design, using a rotating impeller to move the coolant (**Figure 4-37**). When the engine is running, the impeller rotates, forcing coolant from the inside of the cavity outward toward the tips by centrifugal force. Once inside the block, the coolant flows around the cylinders and into the cylinder heads, absorbing the heat from these components. If the thermostat is open, the coolant will then enter the radiator. The void created by the empty impeller cavity allows the pressurized coolant to be pushed from the radiator to fill the cavity and repeat the cycle. When the thermostat is closed because coolant temperatures are too cold, the coolant is circulated through a bypass. This keeps the coolant circling through the engine block until it gets warm enough to open the thermostat.

Thermostat

The **thermostat** allows the engine to quickly reach normal operating temperatures and maintains the desired temperature.

Control of engine temperature is the function of the **thermostat** (**Figure 4-38**). The thermostat is usually located at the outlet passage from the engine block to the radiator. When the coolant is below normal operating temperatures, the thermostat is closed, preventing coolant from entering the radiator. In this case, the coolant flows through a bypass passage and returns directly to the water pump (**Figure 4-39**).

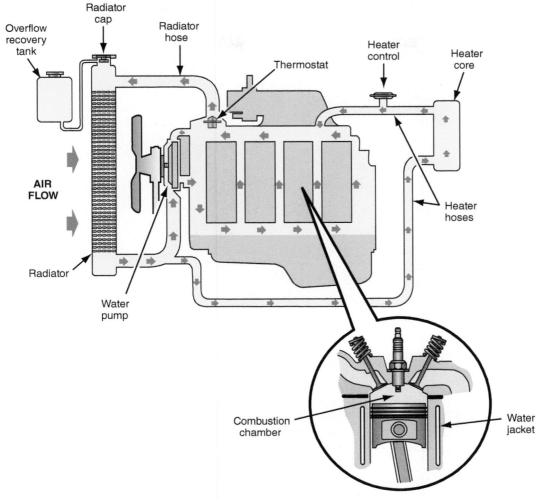

Figure 4-36 Coolant flow through the engine.

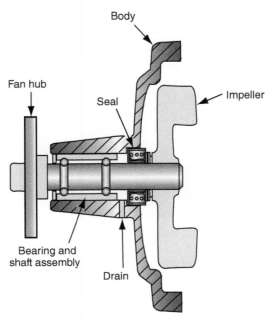

Figure 4-37 Most water pumps use an impeller to move coolant through the system.

Figure 4-38 The thermostat controls the temperature of the coolant.

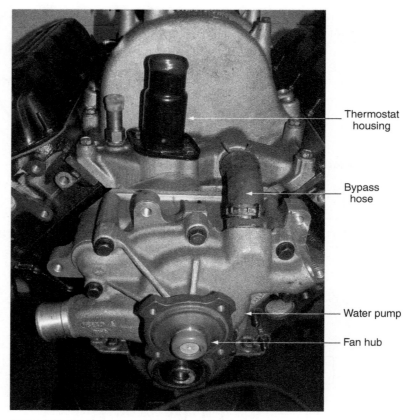

Thermostat housing

Bypass hose

Water pump

Fan hub

Figure 4-39 The bypass hose lets coolant return to the water pump.

The thermostat is rated at the temperature it opens in degrees Fahrenheit. If the rating is 195°F, this is the temperature at which the thermostat *begins* to open. When the temperature exceeds the rating of the thermostat by 20°F, the thermostat will be totally open. Once the thermostat is open, it allows the coolant to enter the radiator to be cooled. The thermostat cycles open and closed to maintain proper engine temperature.

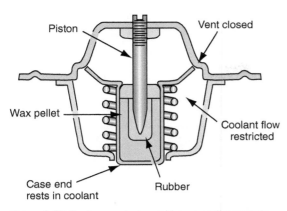

Figure 4-40 The temperature-sensitive wax pellet controls the valve in the thermostat.

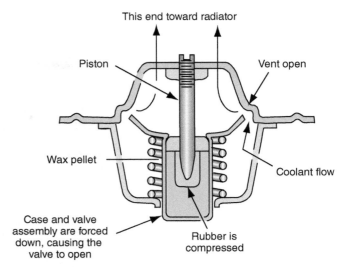

Figure 4-41 When the pellet expands, the thermostat opens and allows the coolant to flow through the radiator.

Operation of the thermostat is performed by a specially formulated wax and powdered metal pellet located in a heat-conducting copper cup (**Figure 4-40**). When the wax pellet is exposed to heat, it begins to expand. This causes the piston to move outward, opening the valve (**Figure 4-41**).

Thermostats have a high failure rate and are a very common cause of cooling system concerns. Do not install a thermostat with a lower opening temperature or run the engine without a thermostat. Allowing the engine to run cooler than designed accelerates engine wear and increases emissions and fuel consumption.

Coolant Flow

Coolant flow through the engine can be one of two ways: parallel or series. In a parallel flow system, the coolant flows into the engine block and into the cylinder head through passages beside each cylinder. In a series flow system, the coolant enters the engine block and flows around each cylinder (**Figure 4-42**). It then flows to the rear of the block and through passages to the cylinder head (**Figure 4-43**). The coolant then flows through the head to the front of the engine (usually the highest point of the cooling system) and out to the radiator.

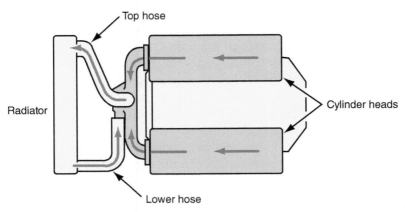

Figure 4-42 The parallel flow sends the coolant around each cylinder and up to the cylinder head. There is more than one path for coolant to flow to the head.

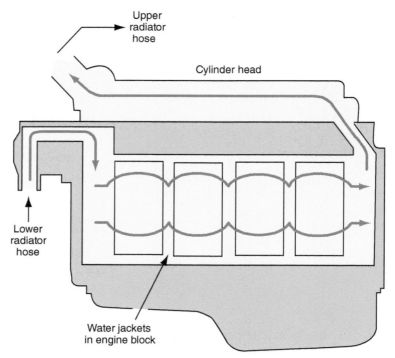

Figure 4-43 A series flow directs the coolant around all of the cylinders before sending it up through the cylinder head.

The head gasket is used to prevent compression pressures, gases, and fluids from leaking. It is located on the connection between the cylinder head and engine block.

The cooling system passages are more than hollow portions in a casting. They are designed and located so no steam pockets can develop. Any steam will go to the top of the radiator. This is accomplished by bleed holes among the cylinder block, **head gasket**, and cylinder head.

Corrosion or contaminants can plug the coolant passages. This results in reduced heat transfer from the cylinder to the coolant. Deposits can build up in small corners around the valve seats, causing hot spots in the cylinder head. This can allow the cylinder head to crack or the valves to burn.

Reverse-Flow Cooling Systems

In a **reverse-flow cooling system**, coolant flows from the water pump through the cylinder heads and then through the engine block.

A **reverse-flow cooling system** was used in some engines, such as the General Motors LT1 5.7 L V8 engine in some model years of Camaro, Firebird, and Corvette. In these engines, the water pump forces coolant through the cylinder heads first and then through the block passages (**Figure 4-44**). The water pump is driven from a gear and shaft mounted on the front of the engine. The water pump drive gear is rotated by gear teeth on the back of the camshaft sprocket (**Figure 4-45**). Coolant flows from the block through other water pump passages and through the upper radiator hose to the radiator. Coolant returns through the lower radiator hose to the inlet side of the water pump. On a conventional flow cooling system, the coolant flows through the block first and then through the cylinder heads. When the engine is running, the hottest part is the exhaust valve seat. Circulating coolant through the cylinder head first means that the exhaust valve seats (and the entire cylinder head) will operate at cooler temperatures because they have cooler coolant flowing through them versus warmer coolant that picked up the heat from the cylinder block. This allows the combustion chamber to operate at a higher temperature. The higher temperature is a result of a higher compression ratio. The compression ratio in the LT1 engine is 10.25:1. Improved combustion chamber cooling also allows the engine to operate with increased spark advance without encountering detonation problems. This action improves fuel economy and engine performance.

Many manufacturers now have systems in which the coolant leaves the radiator through the top radiator hose and then enters the engine through the intake manifold or head. It then passes through the head or heads and engine block to the thermostat located in or near the lower radiator hose. As the coolant heats up, the thermostat opens, allowing the hot coolant back into the lower end of the radiator to release its heat.

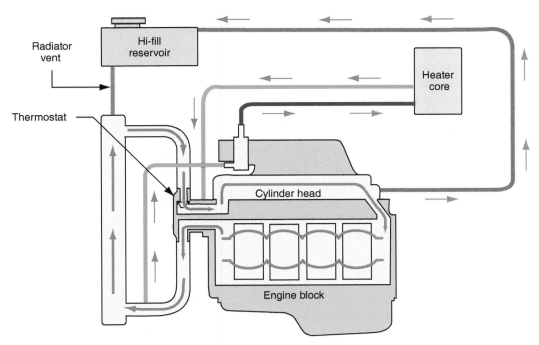

Figure 4-44 Reverse-flow cooling system.

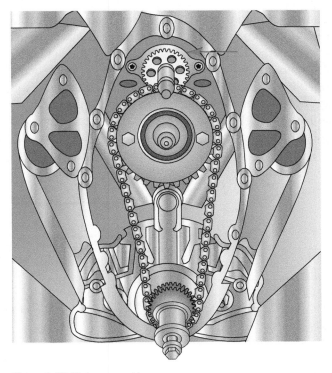

Figure 4-45 Water pump drive.

COOLING FANS

To increase the efficiency of the cooling system, a fan is mounted to direct flow of air through the radiator core or past the radiator tubes. In the past, at high vehicle speeds airflow through the grill and past the radiator core was sufficient to remove the heat. The cooling fan was really only needed for low-speed conditions when airflow was reduced. As modern cars have become more aerodynamic, airflow through the grill has declined; thus proper operation of the cooling fan has become more critical.

The cooling fan can be driven either mechanically by the engine or by an electric motor. Electric drive fans are common on today's vehicles because they operate only when needed, thus reducing engine loads. Some of the earlier designs of electric cooling fans use a temperature switch in the radiator or engine block that closes when the temperature of the coolant reaches a predetermined value. With the switch closed, the electrical circuit is completed for the fan motor relay control circuit and the fan turns on (**Figure 4-46**).

The power feed to the relay is supplied through the ignition switch or directly from the battery. If it is supplied directly from the battery, the cooling fans can come on, even though the engine is not running.

Now, most cooling fans are controlled by the powertrain control module (PCM). The PCM receives engine coolant temperature input signals from the thermistor-type engine coolant temperature (ECT) sensor. When the ECT sensor indicates that the temperature is hot enough to turn on the fans, usually around 200°F–230°F, the PCM activates the fan control relay (**Figure 4-47**). In this diagram, the relay has power from the fuse waiting at pins number 2 and number 3. When the PCM decides to turn the fan on, it provides a ground to pin number 1 of the relay. This allows current to flow from the 10-amp fuse, through pin number 2 and the relay coil, and to ground through the PCM. Current flowing through the coil inside the relay creates a magnetic field that pulls the contacts between pins number 3 and number 4 closed. Now power can flow from the 40-amp fuse, to pin number 3 and through the closed relay contact points, out through pin number 4 and to the cooling fan. The cooling fan already has a ground supply, so it begins to spin. On some vehicles, the PCM also turns on the fans whenever air-conditioning is turned on, regardless of the engine temperature. In some cases, the electric cooling fan may turn on because the engine is hot, even if the ignition is off and the key is out.

Belt-driven fans usually attach to the water pump pulley (**Figure 4-48**). Some belt-driven fans use a viscous-drive fan clutch to enhance performance (**Figure 4-49**). The clutch operates the fan in relation to the engine temperature using silicone oil (**Figure 4-50**). If the engine is hot, the silicone oil in the clutch expands and locks the fan to the pump hub. The fan now rotates at the same speed as the water pump. When the engine is cold, the silicone oil contracts and the fan rotates at a reduced speed. Some fan clutches

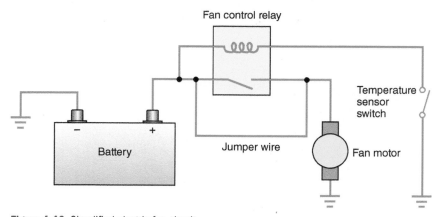

Figure 4-46 Simplified electric fan circuit.

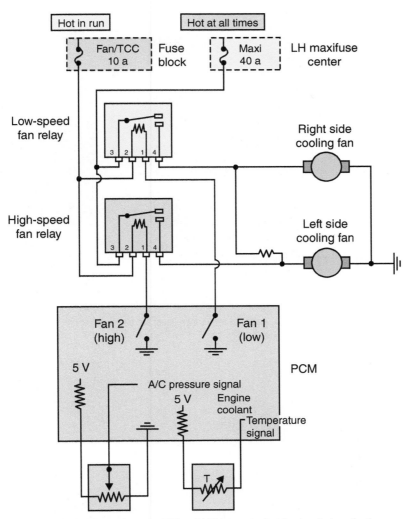

Figure 4-47 The PCM looks at the ECT and A/C inputs to decide when to turn the fans on.

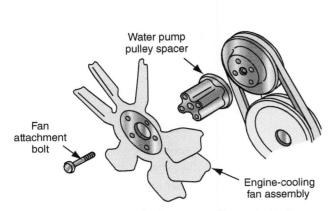

Figure 4-48 Belt-driven cooling fans are usually mounted to the water pump pulley.

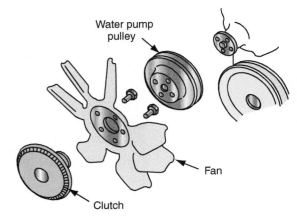

Figure 4-49 Most mechanical fans use a clutch to minimize noise and load on the engine when the engine is cooler.

use a thermostatic coil that winds and unwinds in response to the engine temperatures (**Figure 4-51**). The coil operates a valve that prevents or allows silicone to leave the reservoir and flow into the working chamber. Some performance engines use a clutch fan that is electronically operated (**Figure 4-52**).

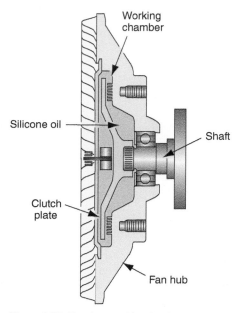

Figure 4-50 The viscous-drive clutch uses silicone oil to lock the fan to the hub.

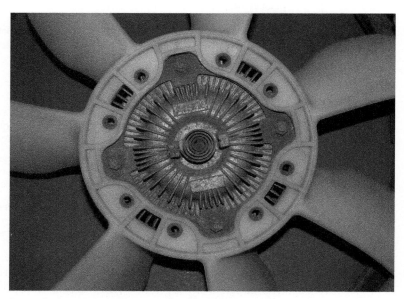

Figure 4-51 As temperature increases, the bimetallic spring expands to operate the clutch and lock the fan on.

Figure 4-52 This clutch fan is electronically controlled by the PCM.

LUBRICATION AND COOLING WARNING SYSTEMS AND INDICATORS

Vehicle manufacturers provide some method of indicating problems with the lubrication or cooling system to the driver. This is done by the use of an indicating gauge or by the illumination of a light. Some manufacturers control the operation of the gauge or light by computers. Regardless of the type of control, sensors or switches provide the needed input.

Gauge Sending Units

The most common sensor type used for monitoring the cooling system is a **thermistor**. The lubrication system uses piezoresistive sensors to measure oil pressure.

In a simple coolant temperature-sensing circuit, current is sent from the gauge unit into the top terminal of the sending unit, through the variable resistor (thermistor), and to the engine block (ground). The resistance value of the thermistor changes in proportion

The **thermistor** is a resistor whose resistance changes in relation to changes in temperature. It is often used as a coolant temperature sensor.

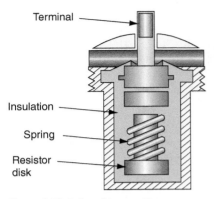

Figure 4-53 A thermistor used to sense engine temperature.

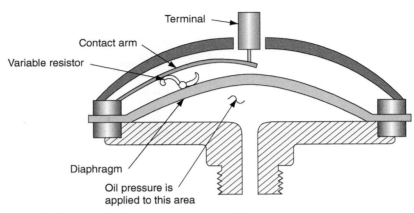

Figure 4-54 A piezoresistive sensor used to sense oil pressure.

to the coolant temperature (**Figure 4-53**). As the temperature rises, the resistance decreases, and the current flow through the gauge increases. As the coolant temperature lowers, the resistance value increases, and the current flow decreases. Modern instrument panels are operated by an internal computer, a **body control module (BCM)**, or an instrument panel (IP) module. The coolant temperature sensor is still a thermistor, but its signal serves as an input to the IP or to the BCM. The computer then drives an electronic gauge assembly to display the coolant temperature.

The **piezoresistive sensor** sending unit is threaded into the oil delivery passage of the engine, and the pressure that is exerted by the oil causes the flexible diaphragm to move (**Figure 4-54**). The diaphragm movement is transferred to a contact arm that slides along the resistor. The position of the sliding contacts on the arm in relation to the resistance coil determines the resistance value and the amount of current flow through the gauge to ground. Again, the oil pressure sending unit may not be wired directly to the gauge; it may deliver its signal to the instrument panel cluster or the BCM. See the diagram showing this circuit (**Figure 4-55**).

> A **body control module (BCM)** is a computer used to control many systems such as the instrument panel, radio, interior lighting, climate control, and driver information.

> A **piezoresistive sensor** is sensitive to pressure changes. The most common use of this type of sensor is to measure the engine oil pressure.

Warning Lamps

A **warning light** may be used to warn of low oil pressure or high coolant temperature. Unlike gauge sending units, the sending unit for a warning light is nothing more than a simple switch. The style of switch can be either normally open or normally closed, depending on the monitored system.

Most oil pressure warning circuits use a normally closed switch (**Figure 4-56**). A diaphragm in the sending unit is exposed to the oil pressure. The switch contacts are controlled by the movement of the diaphragm. When the ignition switch is turned to the on position with the engine not running, the oil warning light turns on. Since there is no pressure to the diaphragm, the contacts remain closed, and the circuit is complete to ground. When the engine is started, oil pressure builds and the diaphragm moves the contacts apart. This opens the circuit, and the warning light goes off. The amount of oil pressure required to move the diaphragm is about 3 psi. If the oil warning light comes on while the engine is running, it indicates that the oil pressure has dropped below the 3-psi limit.

Most coolant temperature warning light circuits use a normally open switch (**Figure 4-57**). The temperature sending unit consists of a fixed contact and a contact on a bimetallic strip. As the coolant temperature increases, the bimetallic strip bends. As the strip bends, the contacts move closer to each other. Once a predetermined temperature level has been exceeded, the contacts are closed, and the circuit to ground is closed. When this happens, the warning light is turned on.

> A **warning light** is a lamp that is illuminated to warn the driver of a possible problem or hazardous condition.

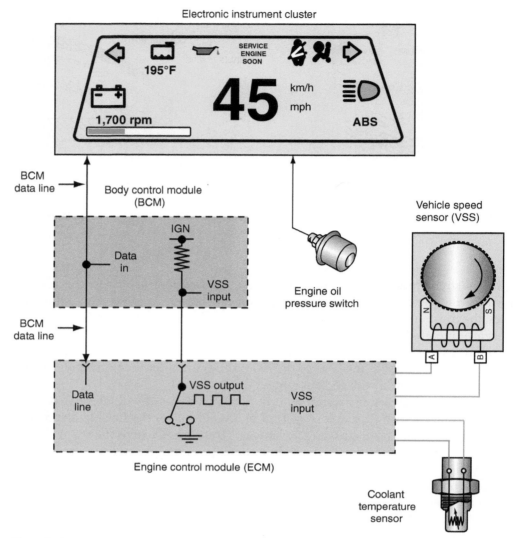

Figure 4-55 An electronic instrument panel with an internal integrated circuit.

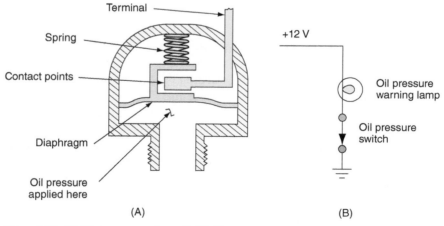

Figure 4-56 (A) Oil pressure sending unit. (B) Oil pressure warning light circuit.

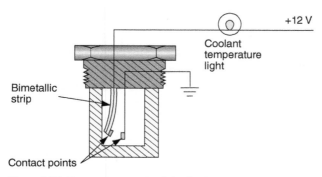

Figure 4-57 Temperature warning light circuit.

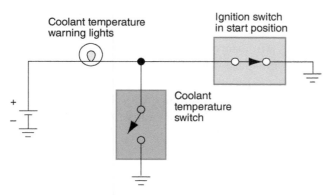

Figure 4-58 A prove-out circuit included in the normally open (NO) coolant warning light circuit.

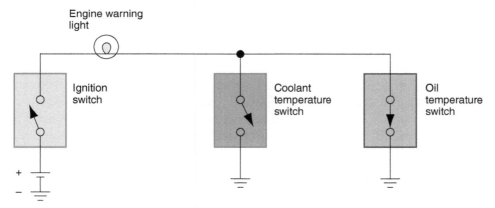

Figure 4-59 One warning light used with two sensors.

With normally open switches, the contacts are not closed when the ignition switch is turned to the on position. To perform a bulb check on normally open switches, a **prove-out circuit** is included (**Figure 4-58**).

It is possible to have more than one sender unit connected to a single bulb (**Figure 4-59**). The light will come on whenever oil pressure is low or coolant temperature is too high.

Message Center

Some vehicles have a message center that displays a number of warning messages to alert the driver regarding dangerous vehicle-operating conditions. Warning messages related to the lubrication and cooling system include LOW COOLANT, CHECK COOLANT TEMPERATURE, ENGINE OVERHEATED, CHECK ENGINE OIL PRESSURE, CHECK ENGINE OIL LEVEL, and CHANGE ENGINE OIL. The messages are displayed in the message center by

A **prove-out circuit** completes the warning light circuit to ground through the ignition switch when it is in the start position. The warning light will be on during engine cranking to indicate to the driver that the bulb is working properly.

Refer to *Today's Technician: Automotive Engine Performance* for a detailed explanation of fuel system operation.

the PCM. The PCM receives input signals from the ECT sensor. If the engine coolant reaches a specific temperature above the normal operating temperature, the PCM illuminates the CHECK COOLANT TEMPERATURE. If the engine coolant temperature increases a specific number of degrees, the PCM turns on the ENGINE OVERHEATED message. Fluid level sensors are required to send a low coolant signal or a low engine oil signal to the PCM. The PCM looks at several different engine operating parameters to illuminate the CHANGE ENGINE OIL message. If the engine operating parameters such as engine temperature, engine on time, and rpm are continually in the normal operating range, the PCM will probably turn on the CHANGE ENGINE OIL message at the approximate interval recommended by the vehicle manufacturer for oil change intervals. However, under severe vehicle-operating conditions such as trailer towing, the engine temperature and load may be above normal, and the PCM may illuminate the CHANGE ENGINE OIL message at a lower mileage interval. As an example, on some vehicles, the CHANGE ENGINE OIL message comes on for 10 to 25 seconds each time the engine is started. After this time period, the message is turned off. After the oil is changed, the CHANGE ENGINE OIL message must be reset. As an example, on some vehicles, this is done by pushing the accelerator pedal wide open three times in a 5-second interval with the ignition switch on and the engine not running. Some vehicles may have a small button that can be pushed only by a pen tip, and others may require a scan tool to reset. Always consult the vehicle manufacturer's service manual for the proper procedure.

FUEL SYSTEM

The fuel system includes the intake system that is used to bring air into the engine and the components used to deliver the fuel to the engine.

Multipoint fuel injection systems uses one injector for each cylinder to deliver fuel.

The **fuel system** is designed to deliver the correct amount of fuel to each of the cylinders for any engine operating condition. In this way, the engine can perform well while minimizing toxic emissions and increasing fuel economy. The PCM precisely controls the fuel delivery through its operation of fuel injectors. Air is brought into the engine through the intake system. A fuel injector is placed in the intake manifold near the opening to the cylinder. Most modern vehicles use one fuel injector for each cylinder; this is called **multipoint fuel injection (MPFI)**. A few newer vehicles inject the fuel directly into the cylinder; this is called gasoline direct injection (GDI). Another system called the central port fuel injection (CPFI) uses poppet valves underneath the intake plenum. The fuel delivery system includes the fuel tank to store fuel, an electric fuel pump mounted in the tank to deliver fuel under pressure to the injectors, a fuel filter, a fuel-pressure regulator, and fuel injectors (**Figure 4-60**). The fuel control system includes the PCM and several inputs used to provide the PCM with enough information to determine how much fuel is required. The primary input for fuel delivery calculations is either a mass airflow (MAF) sensor or a manifold absolute pressure (MAP) sensor. The MAF sensor measures the mass of air flowing into the engine so the computer can calculate the proper amount of fuel to mix with it (**Figure 4-61**). A vehicle may use a MAP sensor instead to provide input for the PCM. The MAP sensor measures the pressure in the intake system to infer the amount of load on the engine. When the driver has the accelerator pedal all the way to the floor, the throttle opens the passage into the intake manifold and the pressure increases to atmospheric pressure. The PCM interprets this as a very heavy engine load and demand and delivers more fuel. The PCM also receives information about engine temperature, intake air temperature, air pressure, engine speed, throttle position, the level of oxygen in the exhaust, and more depending on the vehicle. When the PCM looks at all the data, it decides how much fuel should be injected and it opens the injector for exactly the right amount of time to deliver that quantity. The theoretically perfect air-to-fuel ratio is 14.7 parts of air to 1 part of fuel. During acceleration, the mixture must be "richer" (have more fuel). During deceleration, the PCM will "lean" out the mixture (give the engine less fuel to improve fuel economy). On many hybrids the engine will completely shut off when decelerating to save fuel. The injector has power applied all the time, and the PCM provides a ground pulse that times the injector opening; while the engine is idling, the injector on time may be 1 to 2 milliseconds. The injector has very small openings at its base to atomize the fuel. This means that it breaks the fuel into extremely small particles that can easily mix with the air to help support complete burning and powerful combustion (**Figure 4-62**).

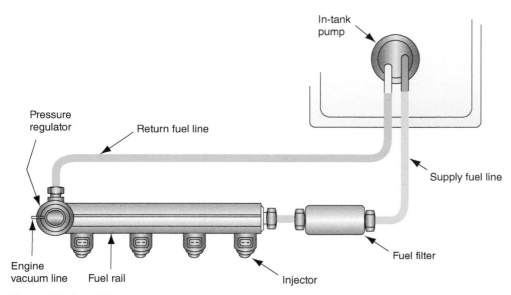

Figure 4-60 Fuel delivery system components.

Figure 4-61 The MAF sensor sits in the incoming airstream and senses the mass of airflow.

Figure 4-62 The fuel injector has a small pinhole-sized opening for fuel to be sprayed out under pressure.

Refer to *Today's Technician: Automotive Engine Performance* for a detailed explanation of ignition system operation.

The delivery of the spark used to ignite the compressed air-fuel mixture is the function of the **ignition system**.

Burn time is the amount of time from the instant the mixture is ignited until the combustion is complete.

An **electronic-ignition** system uses an electronic module to turn the coils on and off.

A **distributor-ignition (DI)** system has a distributor that distributes spark to each spark plug.

Ignition System

The compressed air-fuel mixture must be ignited at the correct time. The delivery of the spark is the function of the **ignition system**. If the spark is not delivered at the correct time, poor engine performance and fuel economy will result. Remember that fuel will burn inside the cylinder (called combustion), and so it does not explode. **Burn time** of the fuel plays into the calculation of spark delivery. Combustion should be completed by the time the piston is at 10 degrees after top dead center (ATDC) during the power stroke. If the spark occurs too early, when the piston is moving up during the compression stroke, the piston will have to overcome the combustion pressures. If the spark occurs too late, the combustion pressures are not as effective in pushing the piston down during the power stroke. In either case, power output is reduced.

Engine speed is another consideration in spark delivery. As engine speed increases, the amount of piston movement also increases for the same amount of time. Burn time of the gasoline remains constant, but the piston will travel more as engine speed increases. To compensate for this, the spark must be delivered at an earlier time in the piston movement.

The PCM controls the spark timing. It uses a crankshaft position sensor to determine the precise position of each piston. It may also use a camshaft position sensor to help identify each cylinder. The PCM also looks at many of the same inputs it uses for fuel delivery and then sends an output that allows the ignition coils to fire. The majority of modern ignition systems use multiple coils to fire the spark plugs. These systems are called **electronic-ignition (EI)** systems. Some EI systems use one coil to provide spark for two cylinders; these are called waste spark systems (**Figure 4-63**). Each coil fires two spark plugs at the same instant. One of the spark plugs is in the cylinder that is near TDC on the compression stroke; this spark will begin combustion on the power stroke. The other spark plug's cylinder is on the exhaust stroke. The spark has no effect on engine operation, hence the term *waste spark*. Other EI systems place a coil on top of each spark plug; this eliminates sensitive high-voltage wiring. These are called coil-on-plug (COP) systems (**Figure 4-64**). Other ignition systems use one coil to fire all the spark plugs. These systems use a mechanical distributor to deliver the spark to the correct cylinder; these are termed **distributor-ignition (DI)** systems.

EI systems use coils to deliver high voltage to the spark plug. The high voltage forms a spark as it jumps across the gap of the spark plug electrodes. The coils have power available to them whenever the engine is running. The PCM provides a ground to turn

Figure 4-63 This coil pack is for a four-cylinder engine; it has two coils, each with two outputs.

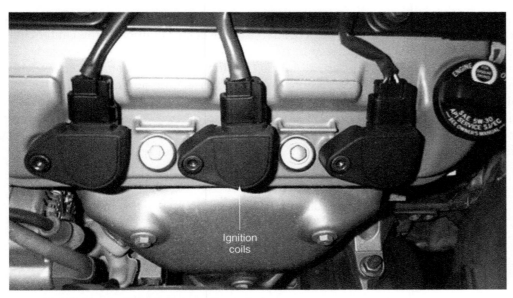

Figure 4-64 These coils sit directly on top of the spark plugs.

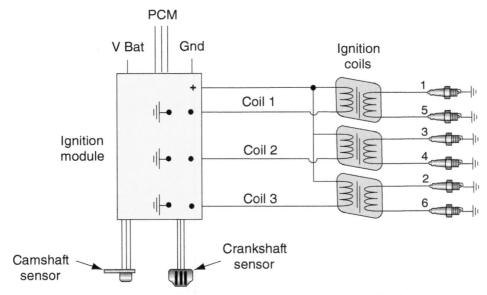

Figure 4-65 This EI system uses an ignition module to provide a ground for the coils.

the coils on and off. In some cases, the PCM provides the ground directly to the coils; in other cases, the PCM provides inputs to an ignition module and it provides the ground to the coils (**Figure 4-65**).

Emission Control System

The **emission control system** helps reduce the harmful emissions resulting from the combustion process. The PCM and its engine management system are integral parts of controlling emissions. The PCM is part of a complex system that controls many aspects of how the vehicle operates. It is often paired up and coordinated with other computers in the vehicle to make the vehicle operate smoothly. Spark timing, fuel injection, and several other systems are controlled by the PCM to minimize the output of toxic emissions. Modern engine management systems do an excellent job of cleaning up tailpipe and evaporative emissions. One study showed that driving a two-stroke snowmobile for 7 hours produced more emissions than a 2000 model year vehicle driven 100,000 miles.

The **emission control system** includes various devices connected to the engine or exhaust system to reduce harmful emissions of hydrocarbons (HC), carbon monoxide (CO), and oxides of nitrogen (NO_x).

Refer to Today's Technician: Automotive Engine Performance for a detailed explanation of emission system operation.

While modern engines release less than 10 percent of the toxic pollutants that they did in the 1970s, the automobile is still a primary source of air pollution in the United States, Canada, and other industrialized countries. Hydrocarbons (HC), carbon monoxide (CO), and oxides of nitrogen (NO_x) are all formed during the combustion process. When an engine is running poorly, the emissions greatly increase. Emission control devices minimize the pollutants by affecting combustion, cleaning up the post-combustion emissions, and controlling the evaporative emissions that occur when a vehicle is not being driven or when it is being refueled. The actual controls on an engine depend upon the design of the engine. Many vehicles use the following emission control systems:

- Catalytic converter
- Secondary air injection system
- Positive crankcase ventilation valve (PCV) system
- Exhaust gas recirculation (EGR) system
- Evaporative emissions system
- Variable valve timing (VVT) system

AUTOMOTIVE FUELS

Gasoline

Gasoline is a mixture of many ingredients, such as hydrocarbons and various other chemicals and additives. Hydrocarbon compounds are refined from crude oil that is removed from the earth. Crude oil is refined into many products, including gasoline, diesel fuel, plastics, soaps, fertilizers, and asphalts. The refining process separates the light liquids from the heavy liquids. The lighter liquids are typically used for producing gasoline, and the heavier liquids are used for producing waxes and asphalts.

Octane

The antiknock quality of gasoline is rated with the term octane. The advertised octane number on the gas station pump is the average of two methods of measuring a fuel's octane. Both RON (research octane number) and MON (motor octane number) are laboratory test methods of measuring a fuel's octane (Figure 4-66). The octane rating of a fuel is used by manufacturers to determine which fuel is appropriate to use for a specific engine.

Figure 4-66 These are the most common octane ratings available. Note the (R 1 M/2) method.

Gasoline has many ingredients and compounds. It is refined from crude oil and is the most common fuel used in today's engines.

The octane number is the ratio of two test fuels that match the knock characteristic of the test fuel. The advertised octane number at a gas station is an average of RON and MON and is an indicator of the fuel's antiknock quality. Lower octane fuels will burn faster. Higher octane fuels will burn slower.

RON (research octane number) and MON (motor octane number) are octane testing numbers that are averaged together to get the advertised octane number. Each uses a different test procedure that represents different real-world driving situations.

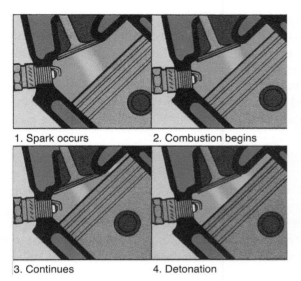

1. Spark occurs   2. Combustion begins

3. Continues   4. Detonation

Figure 4-67 Detonation can cause severe damage.

Figure 4-68 Knock sensor

Higher-performance engines with higher compression ratios typically use a premium gas that has a higher octane number. More normal engines with normal compression ratios will most likely use a regular or mid-grade fuel. The compression ratio is the main determining factor in the octane choice of a fuel, but other factors such as fuel delivery method, intake manifold design, and combustion chamber design are also considered.

The minimum octane rating must always be observed and used for several reasons. The main reason is to protect the engine from detonation (also known as spark knock). Detonation occurs when two flame fronts collide in the combustion chamber (**Figure 4-67**). The first flame front is normal and created by the spark plug. The second flame front is created because of high temperatures. The collision could cause serious damage to the piston if it is not corrected immediately. Fuels with a lower octane number will burn faster than those with a higher octane number. A higher octane fuel will slow down the flame speed and give less of a chance for detonation to occur. Modern engines and computer systems use a **knock sensor** to detect if there is an abnormal explosion in the combustion chamber (**Figure 4-68**). If an abnormal explosion is detected, the PCM will usually retard the ignition timing, which lowers the temperature of the combustion chamber and usually stops the detonation.

The octane number itself does not have any effect on fuel efficiency or economy; however, if a lower-than-recommended octane number is used, the PCM may change the spark timing to compensate and the fuel economy may suffer. Using a higher-than-recommended octane will not directly affect the engine's performance or the fuel economy. After an engine rebuild, the octane requirement may change due to the excessive machining and possible exchange of parts, such as pistons, crankshaft, or cylinder heads, especially if you rebuild an engine to have more performance or a higher compression ratio. During an engine rebuild, you should always remain aware of how machining and upgrading components will affect the compression ratio. The octane requirement can also be affected by the air-fuel mixture, valve timing, coolant temperature, carbon deposits, and turbo-/supercharging.

A **knock sensor** is used to detect abnormal combustion. Its data is used by the PCM to change the ignition timing.

Volatility

Gasoline has to be very volatile so it can easily evaporate and readily mix with air to support combustion. Vaporized fuel and air will support combustion, and volatility describes how easily the gasoline evaporates.

Gasoline blended in the summertime (or year-round in very warm climates) is less volatile than winter blends. The more volatile a fuel is, the more vapor pressure it will create in the fuel line and system components at a given temperature. Different pressures are required because engines need to operate in a variety of conditions.

During hot temperatures, a fuel that vaporizes too quickly can develop vapor bubbles in the fuel line and block the flow of liquid to the engine. This is called vapor lock. During cold temperatures, the fuel needs to vaporize quickly so that it can mix with the cold incoming air in the combustion chamber.

Ethanol

Ethanol is an alcohol
that can be used as an
automotive fuel.

Ethanol is an alcohol that can be used as an automotive fuel. Ethanol (ethyl alcohol) behaves similarly to gasoline and is commonly blended with gasoline as an additive. Ethanol is a noncorrosive, low-toxic-level alcohol that is made from renewable biological sources. In the United States, ethanol is typically made from corn. The starch and sugars from the corn are processed in a fermenting tank at an ethanol-producing plant. Distillation towers are then used to separate water from the alcohol.

Ethanol is typically blended in with gasoline and is used as an additive and octane improver. Ethanol is an oxygenated fuel that helps reduce the toxicity of burning regular gasoline. Ethanol also helps keep the fuel system cleaner and less subject to corrosion. Ethanol, like other alcohols, absorbs water readily. Water is formed in the fuel system by condensation. If it is not absorbed, it will rust out system components and possibly "freeze" the fuel line in the winter. Currently, ethanol blends of up to 10 percent are approved by all manufacturers and do not void their warranty. Ethanol's volatility is significantly lower to gasoline, the only condition that makes it unsuitable for cold weather use.

Ethanol burns cooler than gasoline does. This reduces the risk of valve burning and detonation, two common side effects of excessive heat. Ethanol also has an octane rating of over 100, which makes it a suitable fuel for advancing the ignition timing and running a higher compression ratio.

Ethanol has a lower energy rating than gasoline does. One hundred percent ethanol has about 76,000 Btus/gallon, and 100 percent gasoline has about 125,000 Btus/gallon. The traditional gasoline that you would find at a gas station is mixed with ethanol, which reduces the amount of energy in the fuel because of the mixture rate.

Ethanol is typically found blended at a rate of 85 percent ethanol and 15 percent gasoline. This blend is called E85. Flex fuel vehicles are specifically set up to operate on E85, gasoline, or any blend in between. Vehicles that are not designed by the manufacturer to operate on E85 should not be operated with E85 without extensive modifications, as it could cause engine or fuel system damage. More about ethanol and flex fuel vehicles is found in Chapter 14.

Additives

Gasoline by itself does not properly serve the needs of the modern engine. Gasoline additives have changed over the years to meet the demands of different engine types and driving patterns.

TEL is not used as an
additive in gasoline
anymore. Unleaded
gasoline is readily
available and is less
toxic.

Tetraethyl (TEL) lead was added to gasoline in the 1920s. Originally, lead was added to gasoline to improve the antiknock quality. After lead was added to gasoline, engines started to have higher compression ratios, and vehicles went faster and farther on a tank of fuel. On vehicles that were driven hard, the added lead also helped to lubricate hot exhaust valves to lower the chances of valve recession. Years later, it was found that the lead was contaminating the drinking water and causing a few cases of lead poisoning. Lead also poisons the catalytic converter. The practice of using lead as an additive was stopped. Unleaded fuels were commonly used as early as the mid-1970s. Leaded fuel is no longer used or available for on-road automotive use. Lead also poisons catalytic convertors by coating its active precious metals.

Oxidation inhibitors prevent the formation of gum and varnish on components. When gasoline is not used for a prolonged period of time, its evaporates can mix with oxygen and form a residue on system components.

Detergents are designed to clean fuel systems and components. Gasoline can sometimes contain microscopic particles of dirt that can, with time, cause fuel filter or fuel injector clogging. The amount of detergents varies greatly between different manufacturers. The amount of detergent necessary may also be dependent on the design of the fuel system.

Diesel Fuel

Diesel fuel acts very different than gasoline or ethanol and has very different characteristics. A diesel engine does not use spark plugs for the ignition of the fuel-air mixture (**Figure 4-69**). It uses the heat developed by the compression stroke. It is very important that diesel fuel is clean and able to flow at low temperatures for good engine longevity and performance.

Diesel is typically a product of crude oil at a distillation station, similar to gasoline. But there are other ways to produce diesel fuel. Some examples are biodiesel, which can be produced from crop and renewable products. Another example is biomass-to-liquid (BTL), which uses waste, garbage, sewage, and used cooking oils. Recently, gas-to-liquid (GTL) technology has proven successful. GTL is a diesel fuel that is produced from natural gas reserves found in the earth. Diesel fuel is not rated by the octane scale, but rather it uses a different scale called the cetane scale. The **cetane number** is a measure of how easily the fuel can be ignited. The cetane number is an indicator of the fuel's ability to start the engine in low temperatures and smoothly warm up the engine. Typically, the cetane rating for automotive and light truck diesel fuel is between 40 and 50.

Diesel fuel is denser than regular gasoline. Diesel fuel weighs about 7.09 lbs/gallon, while gasoline weighs about 6.01 lbs/gallon.

Diesel fuel contains significantly more energy than gasoline or ethanol. At approximately 138,000 BTU's per gallon, it gives diesel engines that extra energy to make more power and get better fuel economy. Diesel engines are also more efficient because of the

The formation of gum and varnish is prevented by **oxidation inhibitors**.

Detergents are gasoline additives that clean the fuel system.

Diesel fuel is any fuel that will operate in a diesel (compression ignition) engine. It is named after its inventor, Rudolf Diesel. Diesel fuel has a lower volatility rating than gasoline. It can be derived from petroleum or alternative sources such as biomass, natural gas, or soy beans.

The **cetane number** of a fuel is the measurement of how easily a diesel fuel can be ignited.

Figure 4-69 Diesel engines use high-pressure injectors to inject the fuel directly into the cylinder.

higher compression ratio, average lower operating speeds, and a slower, more complete burn inside the cylinder. Diesel fuel inherently has high sulfur content. The sulfur was traditionally used for lubrication in the cylinder. Recent ultra-low sulfur diesel (ULSD) laws have forced the sulfur level to decrease, and additives are now being used. Biodiesel is becoming a common additive because of its good lubricity properties. Sulfur in the exhaust and the atmosphere are major components of acid rain. Ultra-low sulfur diesel standards have been in place in the United States since 2007. These new fuels allow for the use of new technology emissions devices that significantly clean up the exhaust.

Diesel fuel quickly increases its viscosity as the temperature decreases. Two grades of diesel fuel are commonly available. Diesel number 1 is a less dense fuel with lower heat content. Diesel number 2 is more popular and is more common during summer months and for on-road vehicles. It is common to blend a little bit of number 1 into number 2 in colder temperatures to keep the fuel from gelling. Off-road diesel (non-highway commercial use) is also available in the United States. This fuel is untaxed and does not meet the emissions levels for on-road use. It is primarily used for agriculture applications and is dyed red in color.

Diesel fuel is also immiscible with water—meaning it does not mix with water. Small microbes can grow where the water and fuel come in contact. These microbes feed off the fuel and cause filter clogging, injection problems, and engine problems. Diesel engines have filters and water separators that need to be maintained.

Biodiesel fuels are blends of organically made fuel and regular fuel. Biodiesel itself burns cleaner and does not contain any petroleum. Biodiesel blends with regular fuel are very common. Common blends are B20 and B5. Many manufacturers have approved their engines to operate on a small mixture of biodiesel.

SUMMARY

- The starting system is a combination of mechanical and electrical parts that work together to start the engine.
- The starting system components include the battery, cable and wires, the ignition switch, the starter solenoid or relay, the starter motor, the starter drive and flywheel ring gear, and the starting safety switch.
- An automotive battery is an electrochemical device that provides for and stores electrical energy.
- Electrical energy is produced in the battery by the chemical reaction that occurs between two dissimilar plates that are immersed in an electrolyte solution.
- The lubrication system provides an oil film to prevent moving parts from coming in direct contact with each other. Oil molecules work as small bearings rolling over each other to eliminate friction. Another function is to act as a shock absorber between the connecting rod and the crankshaft.
- Many different types of additives are used in engine oils to formulate a lubricant that will meet all of the demands of today's engines.
- Synthetic oils are superior to conventional oils because they provide better protection from engine wear.
- Oil is rated by two organizations: the Society of Automotive Engineers (SAE) and the American Petroleum Institute (API). The SAE has standardized oil viscosity ratings, while the API rates oil to classify its service or quality limits.
- Oil filter elements have a pleated paper or fibrous material designed to filter out particles between 20 and 30 microns.
- There are two basic types of oil pumps: rotor and gear. Both types are positive displacement pumps.
- The radiator is a series of tubes and fins that transfer heat from the coolant to the air.
- The radiator cap allows for an increase of pressure within the cooling system, increasing the boiling point of the coolant.
- The water pump is the heart of the engine's cooling system. It forces the coolant through the engine block and into the radiator and heater core.
- Control of engine temperatures is the function of the thermostat. When the coolant is below normal operating temperature, the thermostat is closed, preventing coolant from entering the radiator. When normal operating temperature is obtained, the thermostat opens, allowing the coolant to enter the radiator to be cooled.
- In a reverse-flow cooling system, coolant flows from the water pump through the cylinder heads and then through the engine block.

- In some instrument panels, a message center displays specific warning messages to alert the driver regarding dangerous vehicle operating conditions.
- Modern vehicles use multipoint fuel injection systems that provide precise control of fuel delivery.
- Some engines have electronic distributor-ignition (DI) systems, while most engines manufactured at present have electronic-ignition (EI) systems, which do not have a distributor.
- Emission control systems minimize emission of toxic pollutants NO, CO, and NO_x.
- Gasoline is a mixture of many ingredients, such as hydrocarbons and various other chemicals and additives.

- Gasoline has to be very volatile so that it can easily evaporate and readily mix with air to support combustion.
- Ethanol (ethyl alcohol) is a noncorrosive, low-toxic-level alcohol that is made from renewable biological sources. It behaves similarly to gasoline and can be mixed with 15 percent gasoline to form E85, which is used in flex fuel vehicles.
- Detergents are designed to clean fuel systems and components.
- Diesel fuel contains significantly more energy than gasoline or ethanol and is rated by its cetane number.

REVIEW QUESTIONS

Short-Answer Essays

1. What are the purposes of the engine's lubrication system?

2. Explain the purpose of the Society of Automotive Engineers' (SAE) classifications of oil.

3. List the three most recent API's classifications of oil.

4. Describe the two basic types of oil pumps: rotary and gear.

5. What is the purpose of the starting system?

6. Explain the operation of the thermostat.

7. Describe the function of the radiator.

8. Explain the function of the water pump.

9. Explain the purpose of the pressure and vacuum valves used in the radiator cap.

10. Describe the purpose of antifreeze, and explain its characteristics.

Fill-in-the-Blanks

1. It is the function of the engine's lubrication system to supply oil to the _____ and _____ locations and to remove heat from them.

2. The Society of Automotive Engineers has standardized oil _____ ratings, while the American Petroleum Institute rates oil to classify its service or quality limits.

3. Oil pumps are _____ displacement pumps.

4. A pressure relief valve opens to return oil to the sump if the pressure is _____.

5. If the oil filter becomes plugged, a _____, _____ will open and allow the oil to enter the galleries.

6. Coolant can become slightly acidic because of the minerals and metals in the cooling system. A small _____ may flow between metals through the acid and have a corrosive effect on the metals.

7. For heat to be effectively transferred to the air, there must be a _____ in temperature between the coolant and the air.

8. E85 is composed of _____ percent ethanol and _____ percent gasoline.

9. Diesel contains significantly _____ energy than gasoline or ethanol.

10. The fuel system uses a _____ sensor or a _____ sensor as its primary input for fuel calculations.

Multiple Choice

1. Each of the following is a function of the lubrication system, *except*:

   A. Provides oil pressure to the crankshaft journals

   B. Cushions the shock between the connecting rod and the crankshaft

   C. Splashes lubricate the cylinder walls

   D. Provides oil pressure to the piston pins

2. *Technician A* says that the oil pressure increases as the temperature increases.
   Technician B says that the oil pressure decreases as the engine speed increases.
   Who is correct?
   A. A only
   B. B only
   C. Both A and B
   D. Neither A nor B

3. A likely cause of low oil pressure is
   A. the oil pressure relief valve sticking open.
   B. a defective inlet check valve.
   C. the oil filter bypass valve opening.
   D. dirty oil galleries.

4. Coolant and water should be mixed in the ratio of
   A. 70/30.
   B. 50/50.
   C. 40/60.
   D. 100/1.

5. Each of the following is a function of the radiator cap, *except*:
   A. Holds pressure on the system to raise the boiling point
   B. Releases excess pressure to prevent damage
   C. Seals the radiator to prevent toxic leaks
   D. Holds a vacuum on the system to maintain a proper seal

6. *Technician A* says that when the thermostat opens the coolant flows through the radiator.
   Technician B says that thermostats usually last the life of the engine.
   Who is correct?
   A. A only
   B. B only
   C. Both A and B
   D. Neither A nor B

7. *Technician A* says that electric cooling fans reduce the load on the engine, saving horsepower.
   Technician B says that a clutch-type fan slips on the pump hub when the engine is hot.
   Who is correct?
   A. A only
   B. B only
   C. Both A and B
   D. Neither A nor B

8. Each of the following is an input to the PCM for fuel delivery, *except*:
   A. Engine coolant temperature
   B. Engine speed
   C. Exhaust oxygen content
   D. Thermostat position

9. The type of ignition system that uses one coil to fire two cylinders.
   A. DI
   B. Coil-on-plug
   C. Distributor
   D. Waste spark

10. Each of the following is a toxic pollutant created by the automobile, *except*:
    A. NO_x
    B. CO
    C. N_2
    D. HC

CHAPTER 5
FACTORS AFFECTING ENGINE PERFORMANCE

Upon completion and review of this chapter, you should understand and be able to describe:

- The common engine mechanical causes of improper engine performance.
- The components that seal the combustion chamber.
- The effects of combustion chamber sealing problems on engine performance.
- Fuel volatility and its effect on engine performance.
- The octane rating and explain its effects on combustion.
- The different forms of abnormal combustion.
- The potential causes and effects of abnormal combustion.

Terms To Know

Autoignite	Misfire	Reid vapor pressure (RVP)
Detonation	Octane rating	Volatility
Knock sensor (KS)	Preignition	

INTRODUCTION

Proper engine performance requires that the engine is mechanically sound and that its support systems are functioning as designed. One of the most important tasks in engine repair is determining the precise cause of the concern. Before you make engine mechanical repairs, you must gather as much information as possible. There are many different reasons why an engine can perform poorly. You will need to understand the requirements for proper engine performance before you can successfully diagnose the reasons for performance failures. In this chapter, we will look at the areas of the engine that can affect performance and create a need for mechanical repairs.

Components as common as spark plugs can cause an engine to perform as though it has a "dead" cylinder(s). Problems that occur related to the combustion chamber may require special test procedures.

Smoke coming from the tailpipe is never a good sign (**Figure 5-1**). Some causes may be relatively simple and inexpensive to rectify, while others may require a complete engine overhaul or replacement. You will need to understand the possible causes of engine smoking to correctly diagnose the cause of the unwanted emission of smoke.

As engine components wear, clearances become greater, parts break, and noises develop in the engine. These can be warning signs that a component is wearing, or they can predict imminent engine failure. Training, experience, and careful listening can help you narrow down the possible causes of unusual noises and recommend effective repairs (**Figure 5-2**).

Figure 5-1 A lot of thick blue smoke can require serious engine mechanical repairs or relatively inexpensive and simple cures. Your task will be to properly diagnose the cause to make an effective repair.

Figure 5-2 You will listen in different engine areas to identify unusual noises.

Figure 5-3 The spark plugs must be functioning well for the engine to provide good performance.

SPARK PLUGS

Spark plugs are the workhorses of the ignition system. All the buildup of electrical energy of the coil(s) comes to a head at the spark plug gap (**Figure 5-3**). The electrodes of the spark plug must provide an adequate electrical path. If the spark is well timed and strong enough, it jumps the gap with a powerful burst. Then the conditions within the combustion chamber determine whether the spark will ignite the air-fuel mixture well enough to foster good combustion.

Inspecting the spark plugs is a method you will use to gather information about how the engine is operating. You may use this during a routine tune-up, when there is a drivability concern, and as part of gathering information about an engine mechanical failure. Spark plugs can give you tremendous insight into what has been occurring within the combustion chambers. When you have a drivability concern, one of the first areas to begin investigating is the spark plugs. This can help identify if one or more cylinders are acting up or if all cylinders are affected. Spark plugs can also help guide your diagnosis toward a fuel, ignition, or mechanical issue, for example.

AUTHOR'S NOTE Many technicians will use nothing but OEM (original equipment from the manufacturer) spark plugs on late-model vehicles. Today's PCMs and their monitoring systems are very sensitive; using aftermarket spark plugs is a very common cause of drivability problems and customer complaints. Vehicle manufacturers work closely with spark plug manufacturers, and both spend a lot of time and money designing the spark plug that will deliver the perfect spark for the individual combustion chamber. Many aftermarket companies produce excellent spark plugs. Unfortunately, they are usually designed to fit several applications. This does not guarantee that they will operate exactly like original equipment in each vehicle. In some cases you will find that aftermarket spark plugs will not cause major issues. If you do find quality aftermarket spark plugs that perform properly in an engine, it is appropriate to use them. Many technicians have a spark plug brand they have come to trust and use it on many, if not all, applications successfully. You will have to make this judgment call as a professional technician.

COMBUSTION CHAMBER SEALING

Proper combustion is the key to the engine's power production. Fuel must be delivered in the correct quantity; a strong spark must be ignited at precisely the right instant, and the combustion chamber must be full of a fresh charge of air. Then the combustion chamber must be properly sealed for the engine to be able to harness the power of combustion. The spark plug, the engine valves, the head gasket, as well as the piston, rings, and cylinder walls seal the combustion chamber (**Figure 5-4**). Each of those components must be functioning properly, or the piston strokes cannot achieve their optimum result and engine performance will deteriorate.

On the intake stroke, the engine creates a vacuum so a charge of fresh air can enter into the cylinder. If any of the components that form a seal for the combustion chamber are leaking, adequate vacuum level will not develop. Reduced vacuum levels can affect the ratio of air and fuel, causing the cylinder not to contribute its share of power (**Figure 5-5**). The more air an engine can take in, the more power it can put out. If an exhaust valve is leaking, residual pressure will be present in the cylinder when the piston moves down on the intake stroke. Exhaust from the exhaust manifold will join with the fresh air from the intake manifold and dilute the charge. These effects will create significantly rough engine operation and a loss of engine power. Any other combustion chamber leaks would have a similar effect on the intake stroke.

The compression stroke should develop somewhere between 125 and 200 psi of pressure while the engine is cranking. This is a widely generalized estimate; always refer to the

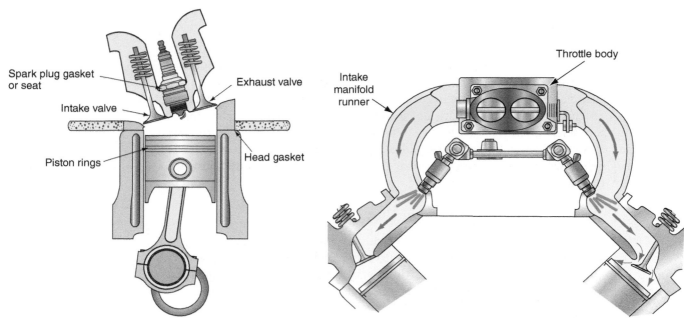

Figure 5-4 A leak through any of these combustion chamber sealing points will lead to engine performance problems.

Figure 5-5 The engine needs a strong vacuum from the cylinder to pull air through the intake runners and into the cylinder.

manufacturer's specifications when evaluating a specific engine. Compression of the air-fuel mixture adds heat and pressure to the mix to make it more combustible. If compression is low, the spark may not be able to develop a strong flame front. Or the flame may start but sputter out as it hits loosely held areas of cooler air and fuel. This would cause incomplete combustion or even a total **misfire**. A misfire is when combustion does not take place. Any leaks past the valves, head gasket, piston, rings, and cylinder walls would lower the amount of compression a cylinder can develop.

Leaks in the combustion chamber reduce the power of combustion. All the force of the expanding gases should be exerted on the top of the piston. This pushes the piston down and turns the crankshaft through the connecting rod. A cylinder has the potential to develop roughly 2,000 lbs. of force on top of the piston. If one quarter of that force leaks past worn piston rings and cylinder walls, the driver will definitely notice that lack of power. This is a typical scenario of a worn, high-mileage engine. The engine could also leak those forces past a burned intake or exhaust valve (**Figure 5-6**). The power loss would be evident, but the consumer would also likely notice a popping noise in the intake or exhaust, depending on which valve(s) had failed.

> A **misfire** means that combustion is either incomplete or does not occur at all.

> Variable valve timing (VVT) can be used to lower the emission of NO_x. It does this by closing the exhaust valve later and opening the intake valve earlier, thereby increasing a valve overlap, and causing exhaust gas to dilute the incoming intake charge.

Ring and Cylinder Wall Wear

Piston rings will normally develop wear as the engine accumulates miles. Out of the box they have very sharp edges; these wear down over time and compromise their ability to seal against the cylinder walls (**Figure 5-7**). When piston rings wear, they may allow combustion gases to leak down into the crankcase. They may also let oil track up the cylinder walls into the combustion chamber, where they are burned. This is one of the primary causes of bluish smoke exiting the tailpipe. It is called oil consumption when oil is burning in the combustion chamber. It is important to remember that some oil consumption is considered normal. The piston's rings will not completely seal all of the oil from the combustion chamber. The amount of oil consumption (oil burning) varies with engine load, driving style, type of oil, and amount of wear of the engine. One of the primary processes of an engine overhaul is replacing the piston rings and refurbishing the cylinder walls.

Figure 5-6 A burned valve will significantly reduce the power output of an engine and cause rough running.

Figure 5-7 The cylinder walls must be in excellent condition to allow the rings to seal the combustion chamber properly.

Valve Wear

Intake and exhaust valves can wear in a few different ways. The most dramatic problem they encounter is when they burn. Valves generally burn when they are open during combustion and exposed to the extreme combustion temperatures. This can happen when the valve springs become weak and cannot close the valves at the proper time, particularly when the engine is spinning at high speeds. When the engine is revving high, the momentum of the valve opening is greater and a slightly weak spring may not be able to close the valve at the proper time. With the valve face and margin exposed to combustion temperatures, the valve will rapidly burn.

The valve faces and seats can also become pitted from the corrosive by-products of combustion or can develop carbon buildup on the seats (**Figure 5-8**). In either case, leakage past the valve face and the seat can occur. Again, once combustion temperatures can leak past parts of a valve, the parts tend to burn rapidly.

Figure 5-8 A pitted valve face allows leakage that will eventually cause the valve to burn.

Misadjusted valves can also cause valve leakage and burning. If a valve is adjusted too tightly, it will be held open longer than it was designed to. This can cause a reduction in performance through poor sealing of the combustion chamber during the appropriate strokes. In extreme cases, it can also cause the valves to burn if they are exposed to excessive temperatures.

 A BIT OF HISTORY

As recently as the mid-1980s, it was not uncommon for engines to require valve reconditioning as early as 60,000 or 75,000 miles. Now due to advances in materials and machining, most of today's engines can run at least 150,000 miles before requiring valve service.

Head Gasket Damage

When a head gasket leaks, it can present a whole host of different symptoms. In the context of combustion chamber sealing, a failure usually results in a leak between two adjoining cylinders. This will lower the compression and combustion of both cylinders significantly. This will cause rough running and a lack of power. Combustion gases can also leak out into the cooling system. This can cause the cooling system pressure cap to release pressure and coolant. The most common symptom of a blown head gasket is coolant leaking into the combustion chamber. Another common symptom of head gasket failure is the presence of coolant in the oil. The oil dipstick will show signs of coolant mixing with the oil, and it will look foamy and brownish, like a coffee milkshake. In these situations, a complete engine rebuild may be required. This burning coolant causes clouds of white, sweet-smelling exhaust to exit the tailpipe (**Figure 5-9**).

There is a significant difference in cost, labor, and technique, depending on what problems exist with combustion chamber sealing. It will be your job to recognize the possible causes of low performance, in order to offer the customer responsible repair options with a realistic estimate.

Figure 5-9 A blown head gasket often results in a cloud of white smoke.

FUEL AND COMBUSTION

When combustion does not occur normally, severe engine damage can result. It is important that you are aware of the causes of abnormal combustion. When you repair or replace an engine with a catastrophic failure, you need to find the source of the problem so it doesn't happen again (**Figure 5-10**). The fuel that customers use can have a significant impact on their engine performance and durability. Your customers may ask your advice about what type of fuel to use and why. You will also see drivability problems caused by fuel issues affecting combustion.

Octane Rating

The **octane rating** of a fuel describes its ability to resist spontaneous ignition or engine knock. Engine knocking (detonation) or pinging (preignition) results when combustion occurs at the wrong time or at the wrong speed. Low octane fuel can be a cause of

A **octane rating** of a fuel describes its ability to resist detonation (also known as spark knock).

Figure 5-10 This diesel head gasket has multiple steel layers; yet it did not hold up to the pressures and stresses of the combustion chamber and engine when it overheated due to a cooling system malfunction.

Figure 5-11 Make sure your customer is using the correct octane fuel to prevent abnormal combustion.

preignition and detonation. If combustion begins before the spark, for example, combustion pressures may try to push the piston backward at the end of the compression stroke. This results in a rattling noise from the piston. This is called preignition, and the sound is often described as pinging. Fuel is generally available for automobiles as regular, 87 octane; mid-range, 89 octane; or as premium, 92 octane or 93 octane (**Figure 5-11**). The higher the octane number, the greater its resistance to knock. Octane is tested in two ways, by the research method and by the motor method. The advertised octane rating is the average of the two ratings. You will often see this described on the pumps as:

$$\text{Octane} = \frac{\text{RON} + \text{MON}}{2}$$

Volatility

Fuel **volatility** is the ability of the fuel to vaporize (evaporate). The **Reid vapor pressure (RVP)** defines the volatility of the fuel. The RVP is the pressure of the vapor above the fuel in a sealed container heated to 100°F. The higher the pressure of the vapor, the greater the volatility of the fuel. This means that the fuel will more readily vaporize.

Fuel volatility is adjusted seasonally in many parts of North America. A higher volatility fuel is allowed in the winter to help the engine start when it is cold. The fuel vaporizes more easily during compression rather than puddling along the cool walls of the combustion chamber. A fuel with very low volatility will cause hard starting and rough running at start-up. A lower volatility fuel is used in the summer to reduce the amount of evaporative emissions and to prevent vapor lock in the fuel lines. A fuel with very high volatility used in the summer months can cause extended cranking times. This occurs because the fuel actually boils in the fuel lines, allowing air in the fuel stream after the vehicle is shut off. When the vehicle is restarted, it takes a significant time for the pump to develop fuel pressure.

Fuel **volatility** is the ability of the fuel to vaporize. The higher the volatility, the easier it is for the fuel to evaporate; this allows easier cold starts.

Reid vapor pressure (RVP) is a method of describing the volatility of the fuel. You can test the RVP to rate the fuel's volatility.

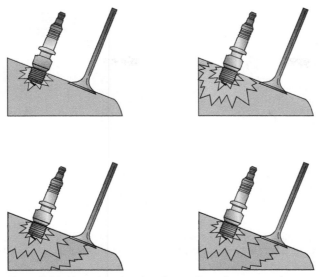

Figure 5-12 Normal combustion allows a flame front to travel rapidly across the cylinder and develop the peak force of combustion at about 10 degrees after top dead center.

Combustion Issues

Combustion is the chemical reaction between fuel and oxygen that creates heat. It is a closely controlled burning of the air and fuel. Spark ignition occurs before top dead center on the compression stroke. The hot, compressed, air-fuel mixture is ignited, and a flame front develops. For normal combustion to occur, the air-fuel mixture must be delivered in the proper proportions and mixed well, the spark must be timed precisely, and the temperatures inside the combustion chamber must be controlled. The flame can then move quickly and evenly (propagate) across the combustion chamber, harnessing the power of the fuel as heat. Pressure builds steadily as the gases expand from heat. The peak of this pressure develops around 10° ATDC to push the piston down on the power stroke (**Figure 5-12**).

Abnormal Combustion

Normal combustion takes about 3 milliseconds (3/1,000 of a second). One form of abnormal combustion, detonation, is more like an explosion, occurring as fast as 2 millionths of a second (2/100,000 of a second). The explosive nature of detonation can potentially cause engine damage. Preignition is a form of abnormal combustion when part of the compressed air-fuel mixture ignites before the spark. Engine misfire is another form of abnormal combustion; it causes the engine to run poorly and lack power.

Preignition

Preignition means that a flame starts before intended spark time. This premature ignition can happen when a hot spot in the combustion chamber autoignites the fuel. A flame front develops and starts moving across the chamber (**Figure 5-13**). Then the spark occurs, and the normal flame front develops. When these two flame fronts collide, a pinging or knocking is heard. Preignition causing pinging can lead to the more damaging detonation by overheating the cylinder. Hot spots in the combustion chamber, the wrong spark plug, cooling system problems, lean air-fuel ratio, low octane fuel, over-advanced timing or metal edges sticking out from the head gasket are some of the reasons preignition can occur. Preignition can lead to engine damage such as melting the center of pistons. But preignition is less likely to cause major engine damage than detonation. It is also less likely than detonation to cause loud spark knock before it causes engine damage.

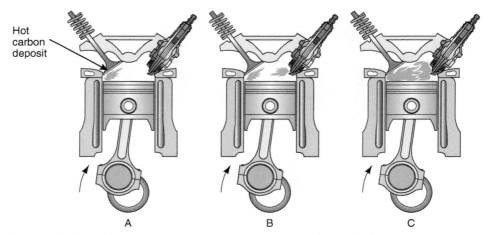

Hot
carbon
deposit

A   B   C

Figure 5-13 The preignition process is started when an uncontrolled flame front is started too soon. (A) An uncontrolled flame front is started before the spark by a hot carbon deposit. (B) The spark plug is fired. (C) The flame fronts collide.

Detonation

Detonation occurs when combustion pressures develop so fast that the heat and pressure will "explode" the unburned fuel in the rest of the combustion chamber. Before the primary flame front (ignited by the spark plug) can sweep across the cylinder, the end gases ignite in an uncontrolled burst (**Figure 5-14**). The dangerous knocking results from the violent explosion that rapidly increases the pressure and temperature in a cylinder. This can be caused by engine overheating, over-advanced timing, low EGR system flow, low octane fuel, and other factors affecting the combustion chamber. In a short period of time detonation can cause serious engine damage, such as bent connecting rods and badly damaged pistons and ring lands (**Figure 5-15**).

Knock Sensor

A **knock sensor (KS)** screws into the engine and creates an electrical input for the PCM when abnormal knocking is detected.

Most modern engines are equipped with a **knock sensor (KS)**. The knock sensor creates an electrical signal when it senses a particular frequency of knocking or detonation. This signal serves as an input to the engine computer, the powertrain control module (PCM). When knocking is detected, the PCM modifies the spark timing to reduce the potentially dangerous knocking (**Figure 5-16**).

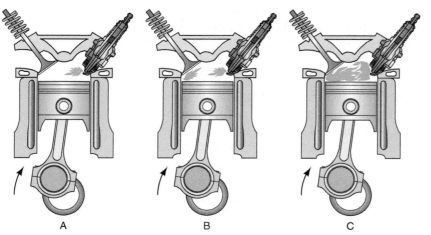

A   B   C

Figure 5-14 Detonation results from an uncontrolled flame front that is started after the spark plug is fired. (A) Combustion begins. (B) Detonation or postspark begins a second flame front. (C) The two flame fronts collide to create a knocking sound.

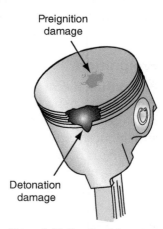

Preignition
damage

Detonation
damage

Figure 5-15 Results of the extra stress placed on the piston due to detonation or preignition.

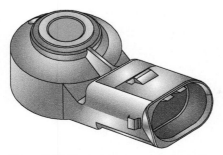

Figure 5-16 The knock sensor screws into the engine to detect abnormal engine knocking.

Misfire

Misfire is another type of abnormal combustion. When an engine misfires, it means that one or more of the cylinders are not producing their normal amount of power. The cylinder(s) is (are) unable to burn the air-fuel mixture properly and extract adequate energy from the fuel. The misfire may be total, meaning that a flame never develops and the air and fuel are exhausted out of the cylinder unburned. Hydrocarbon emissions increase dramatically. Misfire may also be partial when a flame starts but sputters out before producing adequate power (due to a lack of fuel, compression, or good spark). When an engine is misfiring, the engine bucks and hesitates; it is often more pronounced under acceleration. Technicians normally call this a miss or a skip. On modern vehicles (1996 and newer) equipped with the OBD II system, a misfire DTC will be set when the PCM detects a single- or multiple-cylinder misfire. This can help guide your diagnosis.

A P0300 DTC means that the PCM has detected a multiple-cylinder misfire. A DTC P0301 means the PCM has detected a misfire on cylinder number 1, and right up to DTC P0312, a misfire on cylinder number 12.

Support Systems' Contribution to Abnormal Combustion

When the cooling, lubrication, intake, exhaust, or fuel system is not functioning properly, it can cause or contribute to improper combustion. If the engine is running too hot due to a malfunctioning cooling system, detonation and preignition are much more likely to occur. A buildup of deposits in a corner of a cooling passage near the combustion chamber can have devastating effects on the engine due to detonation and preignition. The lubrication system must also reduce engine friction and heat to keep the cylinders cool enough to allow normal combustion.

A restriction in the intake or exhaust system can cause the engine to misfire from a lack of air in the cylinders. If an engine cannot breathe, it cannot produce normal combustion. When the fuel system is failing to deliver adequate fuel, the engine can suffer misfire and preignition from running too hot. A lean air-fuel mixture causes hot engine temperatures; the fuel actually helps cool the mixture. If the compressed air-fuel mixture is too hot, a portion of it can **autoignite** either before or after the spark begins.

ENGINE NOISES

The engine can create a wide variety of noises due to worn or damaged parts. It will be your job to diagnose the causes of these noises and to correct them. Most noises occur because of worn components causing excessive clearances. A typical wear item is crankshaft bearings, which cause expensive damage heard as a deep lower-end knock (**Figure 5-17**). Valve lifters also commonly wear with higher mileage and cause a higher pitch clatter. These are just a couple of examples of possible noises; in the *Shop Manual*, you will learn to isolate noises and their causes.

Figure 5-17 The bearings on the left are worn down to the copper underlayer; these created engine knocking from the bottom end.

SUMMARY

- Proper engine performance requires that the engine is mechanically sound and that its support systems are functioning as designed.
- The combustion chamber must be properly sealed to provide good engine performance.
- The valves, spark plug, rings, and head gasket seal the combustion chamber.
- The octane rating of gasoline describes its ability to resist knocking; the higher the number, the greater the resistance to knocking.

- Higher volatility fuel should be used in the winter to assist cold starts; lower volatility fuel should be used in the summer to prevent excessive HC emissions and vapor lock.
- Misfire, preignition, and detonation are three types of abnormal combustion that can cause serious engine damage.
- Failures in the cooling or lubrication systems can cause abnormal combustion or serious engine defects.
- Normal engine wear will eventually lead to abnormal noises and reduced performance.

REVIEW QUESTIONS

Short-Answer Essays

1. What components seal the combustion chamber?

2. What problems can occur from improper combustion chamber sealing?

3. What effect will a burned valve have on engine performance?

4. What does a gasoline's octane rating describe?

5. What problems can occur when fuel with an inappropriate volatility is used?

6. Define preignition.

7. Define detonation.

8. Describe some causes of abnormal combustion.

9. What engine problems can detonation lead to?

10. What causes abnormal engine noises?

Fill-in-the-Blanks

1. A leak past the _____ gasket can cause compromised engine performance.

2. The _____ _____, _____ _____, and _____ in addition to the spark plug and cylinder wall seal the combustion chamber.

3. Malfunctions in the _____ system and the _____ system can cause engine mechanical problems.

4. Three types of abnormal combustion are _____, _____, and _____.

5. Preignition occurs when a flame front starts _____ the spark.

6. _____ is likely to damage the piston.

7. _____ describes a fuel's ability to vaporize.

8. Over _____ timing, low _____, and _____ octane fuel can lead to detonation.

9. A common cause of clatter from the top end of a high-mileage engine is worn _____.

10. Causes of engine misfire are lack of _____ and good _____, _____ and _____.

Multiple Choice

1. Fuel is generally available in octane ratings of all of the following *except*:
   - A. 87 octane
   - B. 89 octane
   - C. 92 octane
   - D. 99 octane

2. Normal combustion takes approximately:
   - A. 3 milliseconds
   - B. 30 milliseconds
   - C. 2 nanoseconds
   - D. 20 nanoseconds

3. Misfire means.
   - A. the cylinder fires at the wrong time.
   - B. the cylinder does not fire.
   - C. two sparks occur.
   - D. the wrong spark plug is installed.

4. Each of the following is a likely cause of detonation *except*:
   - A. Carbon buildup in the combustion chamber
   - B. Engine overheating
   - C. A lean air-fuel mixture
   - D. Fuel with too high of an octane rating

5. A _____ informs the PCM about unusual frequency vibrations in the engine.
   - A. coolant temperature sensor
   - B. vibrator actuator
   - C. knock sensor
   - D. PCM mounting sensor

6. *Technician A* says that worn piston rings can cause a lack of power.
   Technician B says that a burned valve can cause a popping noise out of the intake or exhaust.
   Who is correct?
   - A. A only
   - B. B only
   - C. Both A and B
   - D. Neither A nor B

7. Spark ignition occurs:
   - A. ATDC on the power stroke
   - B. ATDC on the compression stroke
   - C. ATDC on the intake stroke
   - D. BTDC on the compression stroke

8. *Technician A* says that hydrocarbon emissions typically increase when an engine misfires.
   Technician B says that low compression can cause misfire.
   Who is correct?
   - A. A only
   - B. B only
   - C. Both A and B
   - D. Neither A nor B

9. Too low a fuel volatility used in the winter will cause.
   - A. excessive evaporative emissions.
   - B. hard starting and rough running when cold.
   - C. extended cranking times when the engine is hot.
   - D. detonation.

10. A burned intake valve will cause.
    - A. weak engine vacuum.
    - B. lower compression pressures.
    - C. reduced engine power.
    - D. all of the above.

CHAPTER 6
ENGINE MATERIALS, FASTENERS, GASKETS, AND SEALS

Upon completion and review of this chapter, you should understand and be able to describe:

- The various types of materials used in engine construction, including iron, steel, aluminum, plastics, ceramics, and composites.
- The common usages of aluminum alloys in engine construction. The different manufacturing processes, including casting, forging, machining, stamping, and powdered metal.
- The various treatment methods, including heat treating, tempering, annealing, case hardening, and shot peening.
- The methods used to locate cracks in castings.

- How to properly identify, inspect, and select the correct fasteners required to assemble engine components to the block.
- The purposes of engine gaskets.
- The purpose of engine seals.
- The requirements and construction of a head gasket.
- The different gasket materials.
- The proper application of room-temperature vulcanizing (RTV) sealant.
- The proper application of anaerobic sealant.

Terms To Know

Aerobic sealant	Gaskets	Sealants
Alloys	Grade	Seals
Anaerobic sealant	Grade marks	Service sleeve
Annealing	Gray cast iron	Shot peening
Bolt diameter	Intake manifold gasket	Sintering
Bolt length	Lip seal	Stainless steel
Carbon steel	Metallurgy	Steel
Case hardening	Nodular iron	Strip seal
Ceramics	Nonferrous metals	Tempering
Coke	Oil pan gaskets	Tensile strength
Composites	Pitch	Thread depth
Exhaust manifold gasket	Plastic	Torque-to-yield
Ferrous metals	Rear main seal	Valley pan
Fire ring	Room-temperature vulcanizing (RTV)	Valve cover gasket
Front main seal		

INTRODUCTION

This chapter covers the materials and machining processes used to construct the components of the engine. Today's technician should be able to identify these characteristics properly in order to perform repairs or machining operations on the components.

The mating surfaces of the engine require proper sealing to prevent leakage and for proper engine performance. **Gaskets** are used to prevent gas, coolant, oil, or pressure from escaping between two stationary parts. In addition, gaskets are used as spacers, shims, wear indicators, and vibration dampers. In recent years, many gaskets have been replaced with the use of sealants. The sealant is less expensive and easy to apply. **Seals** are used to prevent leakage of fluids around a rotating part.

ENGINE MATERIALS

Today's engines are constructed with the use of several different types of materials. The same engine can have iron, steel, aluminum, plastic, ceramic, and composite components. Although a complete understanding of **metallurgy** is not a prerequisite for performing engine repairs, today's technician is challenged to perform repairs requiring the proper handling, machining, and service of these materials.

Metals are divided into two basic groups: **ferrous metals** and **nonferrous metals**. All metals have a grain structure that can be seen if the metal is fractured (**Figure 6-1**). Grain size and position determine the strength and other characteristics of the metal.

Iron

Iron is a very common metal used to produce engine components. It comes from iron ore, which is retrieved from the earth. The ore is heated in blast furnaces to burn off impurities. **Coke** is used to fuel the furnaces, resulting in some of the carbon from the coke being deposited into the iron. When the liquid iron is poured from the coke furnace, the resultant billet (called pig iron) will have about a 5 percent carbon content. This makes the iron brittle.

Cast Iron

Once the iron cools, it can be shipped for several uses. When shipped to an engine manufacturing plant, the pig iron is remelted and poured into a cast or mold. The resultant form, when the iron cools, is referred to as cast iron. During the process of remelting the iron, the amount of carbon content can be controlled, **alloys** can be added, or special heat treatments can be performed to achieve the desired strength and characteristics of the cast iron. The most common type of cast iron used in automotive engine construction is **gray cast iron**. Gray cast is easy to cast and machine. It also absorbs vibrations and resists corrosion.

Figure 6-1 The properties and strength of a metal can be seen in its grain.

Gaskets are used to prevent engine fluids or pressure from leaking between two stationary parts.

Seals are used to prevent leakage of fluids around a rotating part.

Metallurgy is the science of extracting metals from their ores and refining them for various uses.

Ferrous metals contain iron. Cast iron and steel are examples of ferrous metals. These metals are easy to identify because they will naturally attract a magnet.

Nonferrous metals contain no iron. Aluminum, magnesium, and titanium are examples of nonferrous metals. These metals will not attract a magnet.

Coke is a very hot-burning fuel formed when coal is heated in the absence of air. Coal becomes coke at temperatures above 1,022°F (550°C). It is produced in special coke furnaces.

Alloys are mixtures of two or more metals. For example, brass is an alloy of copper and zinc.

Gray cast iron is a type of cast iron that has a graphitic microstructure. It is a widely used material because of its tensile strength to weight ratio. A common consideration for engines because of their high-strength and low-weight requirements.

Nodular iron, also known as ductile iron, is a version of gray cast iron but is more flexible and elastic.

Some engine applications require additional strength above that provided by gray cast. **Nodular iron** is used in some engines for the construction of crankshafts, camshafts, and flywheels. Nodular iron contains between 2 and 2.65 percent carbon, with small amounts of magnesium and other additives. It is also heat treated, causing the carbon to form as small balls or nodules. The result is a cast iron with reduced brittleness.

Steel

Iron containing very low carbon (between 0.05 and 1.7 percent) is called **steel**.

Steel is produced by heating the iron at a controlled temperature to burn off most of the carbon, phosphorus, sulfur, silicon, and manganese. As the process continues, the correct amount of carbon is then readded. This type of **steel** is referred to as **carbon steel**.

Carbon steel has a specific carbon content.

Low-carbon steel (carbon content between 0.05 and 0.30 percent) can be easily bent and formed. As the carbon content of the steel increases, so does its resistance to denting and penetration. However, the higher the carbon content, the more brittle the steel. Medium-carbon steel (carbon content between 0.30 and 0.60 percent) is used for connecting rods, crankshafts, and camshafts in some engines. It is commonly used in high-performance and race engines. High-carbon steel (carbon content between 0.70 and 1.70 percent) has limited use in engine applications but is used to make drill bits, files, hammers, and other tools.

Stainless steel is an alloy that is highly resistant to rust and corrosion.

The quality of steel can be improved through the addition of alloys. Some of the most common alloys include nickel, molybdenum, tungsten, vanadium, silicon, manganese, and chromium. Steel with 11 to 26 percent content of chromium is referred to as **stainless steel**. Nickel is often added to make the steel able to withstand sudden shock loads. If increased **tensile strength** is required, vanadium can be added.

Tensile strength is the metal's resistance to being pulled apart.

The addition of chromium and molybdenum (sometimes called chrome-moly) produces a very strong steel. This alloy is often used to make crankshafts in high-performance engines. Chrome-moly is also used in the construction of tube frames used in racing cars.

Magnesium

Magnesium is a very lightweight metal, about two-thirds the weight of aluminum. Pure magnesium is expensive and has a low tensile strength; however, it can be alloyed with other metals (such as aluminum). Magnesium alloy is used in some engine valve, covers, or in the case of the Porsche 911, the engine block.

Aluminum

Aluminum is a silver-white ductile metallic element. It is the most abundant metal in the earth's crust. There is no pure aluminum found in the earth, though. It is always mixed with other elements. The aluminum we use comes from an ore called bauxite, containing 50 percent alumina (aluminum oxide). The bauxite is crushed and then ground into a fine powder. Then the impurities are separated from the powdered ore by mixing it with a hot caustic soda solution. It is then pumped into large pressure tanks called digesters. The temperature in the tanks, which is maintained at 300°F (149°C), dissolves the alumina, but the impurities remain solid. The impurities can then be filtered out. The remaining liquid is pumped into precipitator tanks, where it is allowed to cool slowly. The alumina comes out of the liquid as crystals when it cools. The crystals are then heated until white hot to remove any water. The result is a dry, white powder. To complete the process of turning alumina into aluminum, it is dissolved in a substance called cryolite. An electric current is passed through the pot containing the mixture from carbon anodes that hang in the liquid from overhead bars. The current flows through the liquid and causes the liquid to break up. Pure aluminum then falls to the bottom of the pot.

Although aluminum is not as strong as steel, it weighs about one-third less than steel. Pure aluminum is not strong enough to be used for most engine applications, but the

Figure 6-2 This is a lightweight piston used in a racing engine.

attractiveness of weight reduction has led to the development of many types of aluminum alloys that are proving acceptable. Today, most pistons used in automotive engines are constructed of aluminum alloys.

The light weight of aluminum pistons allows for higher engine speed and increased engine responsiveness. The lighter pistons also reduce the amount of load on the connecting rods and crankshaft (**Figure 6-2**). A few racing applications use aluminum connecting rods to reduce the amount of reciprocating weight. This allows the engine to accelerate faster, operate smoother, last longer, and have an increased maximum speed (also known as the red line on the tachometer).

Many manufacturers are now using aluminum alloys for engine block construction. Aluminum is too soft to be used by itself and is unable to withstand the wear caused by the piston rings traveling in the cylinder. Most aluminum blocks of today's engines are either fitted with special cylinder liners (made of iron, steel, or composite) or special casting. A few engines have used plated aluminum cylinders in the past.

Most manufacturers use an aluminum alloy to construct cylinder heads, engine mounts, water pump housings, air-conditioning generator housings, intake manifolds, and valve covers.

The Audi A8 was the first production vehicle to use an all-aluminum body. The higher cost of making components with aluminum has increased the basic cost of a vehicle, but has made it more fuel efficient.

Titanium

Titanium alloys offer the benefits of light weight and high strength. Titanium alloys are almost as strong as steel, at about half the weight. The most common use for titanium is in racing applications for the construction of connecting rods and valves. The high cost of titanium alloys, coupled with their difficulty in being welded and machined, have limited titanium use in production vehicles. However, the NSX from Acura was introduced in 1990 with titanium connecting rods.

Plastics

Plastic is a substance made from petroleum by a special method that joins atoms together to make long chains of atoms. The word **plastic** means something that can be pressed into a new shape. After pure plastic is made, it can be improved by adding other materials to give it additional strength, stiffness, or density.

Plastics are made from petroleum and are used in several engine components because of their light weight and heat transfer properties.

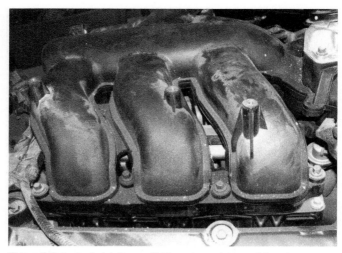

Figure 6-3 A plastic intake manifold used on a late-model engine.

Plastics were first introduced in the automotive engine when American Motors used plastic to make valve covers on their 258 CID 6-cylinder engine. In subsequent years, other manufacturers have used plastics to help reduce the overall weight of the engine. Today some manufacturers are using forms of plastics to make intake manifolds, pulleys, oil pans, and valve timing train covers. Plastic intake manifolds transfer less heat to the intake air compared to cast iron or aluminum intake manifolds (**Figure 6-3**). This action reduces intake air temperature and engine detonation. In many cases, plastic components are easier to make production modifications in the middle of a production year. Plastics also do not transfer noise and vibration as much as metals making them quieter.

Composites

Composites are human-made materials using two or more different components tightly bound together. The result is a material that has characteristics that neither component possesses on its own.

Manufacturers are experimenting with increased use of **composite** materials. These human-made materials are showing great promise in engine applications. Graphite-reinforced fiber or nylon materials have been successfully used as connecting rods, pushrods, rocker arms, intake manifolds, and cylinder liners.

Ceramics

Ceramics is a combination of nonmetallic powdered materials fired in special kilns. The end product is a new product.

Ceramics also show good promise in automotive engine uses. They are lightweight, provide good frictional reduction, are heat resistant, isolate sound and vibrations, and are brittle. Engine components constructed from ceramics require special handling due to being so brittle. Ceramic components that are in use today include the compressor and turbine wheels of some General Motors turbochargers and the rocker arm pads on some Mitsubishi engines. In addition, the aftermarket parts suppliers are making ceramic valves, valve seats, valve spring retainers, and wrist pins available.

Typical uses of ceramics today also include the rotor of some turbochargers, the liner for exhaust ports, intake and exhaust valves, valve seats, and piston pins.

MANUFACTURING PROCESSES

Not only do different types of irons, steels, and aluminum have different qualities, the manufacturing process also determines their strength and properties. The most common processes are casting, forging, machining, and stamping. In addition to the manufacturing process, the material may be specially treated to increase strength.

Casting

As discussed earlier, casting requires heating the metal to a liquid and then pouring it into a mold. When the metal cools, it returns to a solid state. The molds can be made of foundry sand. After the metal cools, the sand is broken away to expose the part.

The mold is made by packing the foundry sand around a wooden pattern. The pattern is removed prior to the liquid metal being poured in. This type of casting usually leaves a rough, grainy appearance.

The lost foam casting method incorporates the use of a polystyrene foam pattern. This pattern is left in the mold when the molten metal is poured in. The heat of the molten metal vaporizes the foam. This method leaves the casting with a surface texture similar to Styrofoam (**Figure 6-4**).

Cast iron is commonly used for engine block, camshaft, connecting rod, and crankshaft construction.

Aluminum is usually cast into permanent, reusable molds. The molten aluminum alloys are forced into the mold under pressure or through the use of centrifugal force. The pressure is used to help eliminate any air pockets that may affect the machining process.

Cast aluminum is commonly used for cylinder head, bell housing, piston, and intake manifold construction.

Forging

Forging heats the metal to a state in which it can be worked and reshaped. It is not heated until it becomes a liquid, as it is when casting. When the metal is hot enough to be worked, a forging die is forced onto the metal under great pressure. The metal then assumes the shape of the forging die. If the metal must be worked into complex shapes, several forging dies may be used. Each die will alter the shape until the desired results are obtained.

When the forging die is closed around the hot metal, some of the metal is forced into the parting lines of the die. This excess metal is called flash and is usually removed after the part is removed from the die.

Figure 6-4 This engine is formed by lost foam casting and has a surface appearance similar to Styrofoam.

Forged parts are very strong because the high pressure used to force the die onto the component causes the grain structure to follow the shape of the part. The pressure also causes the molecules of the metal to become very compact and tightly bound.

Most engines equipped with turbochargers use forged pistons, and some use forged connecting rods and crankshafts. High-performance engines often use a forged crankshaft and camshaft to increase the endurance of the engine.

Machining

To construct a component from a piece of steel or iron billet is very time consuming and very expensive. Despite this, many engine components are constructed in this manner. Machined components are stronger than cast, but not as strong as forged. Machined components used in today's engines include rocker arms, piston pins, lifters, and followers. High-performance engines use machined steel crankshafts and camshafts.

Stamping

Some engine components of simple design, and not requiring much thickness, can be stamped out of a sheet of metal. The metal is not heated during this process. Depending on the final design of the component, it can be stamped on a press punch. This process is used if the piece is to remain flat. The punch is in the shape of the piece to be cut. The press forces the punch through the sheet of metal, much like a cookie cutter through dough. If the finished product must have some bends to it, it is bent by the stamping process and then trimmed to final specifications. Common components made from stamped steel include oil pans, valve covers, timing train covers, and heat shields.

Powder Metallurgy

Metal powder may be derived by cooling melted metal very quickly. Other methods include reducing the metal oxide, electrolysis, and crushing. The powder can then be blended with other metals to produce an alloy. The powder is then poured into a die and compacted by a cold press. Next a process of **sintering** is used to make the powder bond together.

> Powder metallurgy is the manufacturing of metal parts by compacting and sintering metal powders. **Sintering** is done by heating the metal to a temperature below its melting point in a controlled atmosphere. The metal is then pressed to increase its density.

This process is currently being used to construct some connecting rods and valve seats. After the connecting rod is removed from the die, it is forged into final shape and then shot peened to increase its strength. This process provides a strong component at a light weight. It also eliminates many of the machining processes forged or machined connecting rods must undergo.

Treatment Methods

To obtain the desired result of the component, engine manufacturers may have the component treated by heat, chemical, or shot-peening processes. Heating the metal will change its grain structure. By heat treating a metal component, its properties can be altered. To heat treat a metal properly, it is heated to a desired temperature, depending on the metal used and the desired results. Once the temperature is reached, it is maintained for a specific amount of time. The last step is to cool the metal at a controlled rate.

To harden carbon steel, it is heated to $1,400°F$ ($760°C$) and then quickly cooled. Heating the steel causes its grain structure to become finer. When it is cooled very fast, the grain structure does not have time to change and it remains very fine. The harder the steel becomes, the more brittle it becomes.

> In **tempering**, metal is heated to a specific temperature to reduce the brittleness of hardened carbon steel.

To counteract the effects of hardening steel, **tempering** of the metal may be the next step. The steel is heated to a temperature between $300°F$ and $1,100°F$ ($150°C$ and $600°C$) and then cooled slowly. The higher the temperature used to temper the steel, the more hardening is lost, yet the toughness of the metal is increased.

Hard metals must be softened to be machined. This process is called **annealing**. Annealing is much like tempering except the temperatures are increased to above 1,000°F (550°C). The metal is then allowed to cool at a slower rate than used for tempering. This makes the grain structure of the metal coarse. If desired, the metal can be hardened and tempered again after the machining process is completed.

To help protect the shell or outer surfaces of a component, the manufacturer may require it to be case hardened. This process does not alter the core structure of the metal. **Case hardening** is performed after all machining processes are completed. Case hardening of crankshafts protects the journals from wear and fractures. Most manufacturers now use a process called ion nitriding to case harden their components. The component is placed into a pressurized chamber filled with hydrogen and nitrogen gases. An electrical current is then applied through the component. This changes the molecular structure and allows the induction of gases onto the surface area of the component.

Another form of metal treatment is **shot peening**. Shot peening is performed by blasting the component's surface with steel or glass shot. When the balls hit the component, a small dent is made. Thousands of these dents are applied to the component's surface. The dents overlap each other and compress the component's surface. This process of prestressing means any tension forces applied to the part must overcome the compression forces before the part will crack.

Cracks

Cracks are the result of stress in the casting. A crack can cause fluid or compression leakage. When performing engine service requiring the removal of the intake manifold, exhaust manifold, cylinder heads, and so on, check the component for cracks. The following is a list of some of the most common causes for this stress:

1. Fatigue
2. Excessive flexing
3. Impact damage
4. Extreme temperature changes in a very short time
5. Freezing of the engine coolant
6. Excessive overheating
7. Detonation
8. Defects during the casting process

> **AUTHOR'S NOTE** It has been my experience that one of the common causes of cracked cylinder heads is adding cold water to the cooling system on an overheated engine. This action subjects the cylinder head to a very sudden change in temperature, resulting in a cracked head. If the engine overheats, do not loosen the radiator cap, and do not add cold water or antifreeze to the cooling system until the engine cools down.

Detecting Cracks

Many cracks are detected by a thorough visual inspection. However, very small stress cracks may not be detected in this manner. There are several different methods of crack detection. Following is a sample of the four most common methods:

1. *Magnetic particle inspection (MPI).* Uses an electromagnet to create a magnetic field in the casting (**Figure 6-5**). Because all magnets have north and south poles, the magnetic field runs from one pole to the other through the casting. A crack in

Annealing is a heat-treatment process to reduce metal hardness or brittleness, relieve stresses, improve machinability, or facilitate cold working of the metal.

Case hardening is a heating and cooling process for hardening metal. Because the case hardening is only on the surface of the component, it is usually removed during machining procedures performed at the time the engine is being rebuilt. It is not necessary to replace the case hardening in most applications.

Shot peening is a cold-working process used to make the outer layers of a metal more compressive.

Figure 6-5 This damaged racing crankshaft is being magnafluxed to check for cracks.

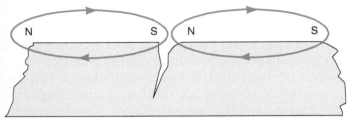

Figure 6-6 A crack causes a break in the magnetic field, creating opposite magnetic poles.

Magnaflux is another name for magnetic particle inspection.

the casting causes a break in the field, creating opposite poles (**Figure 6-6**). The magnetic powder is attracted to this area. Magnetic particle inspection cannot be used to detect cracks in aluminum parts because they cannot be magnetized.

2. *Magnetic fluorescent.* This method uses a fluorescent paste dripped onto the casting. The casting is then placed in a magnetic field and observed under a black light. Cracks are visible by white, gray, or yellow streaks.

3. *Penetrant dye.* Using dye penetrant is a three-step process. First, the special dye is sprayed onto the casting surface and allowed to dry. The excess dye is then wiped away. Next, a special remover is sprayed over the surface and the casting is rinsed with water. Third, the developer is sprayed onto the casting. As it dries, any dye left in the cracks seeps through the developer. The crack shows as a red line against a white background.

4. *Pressurizing the block or head.* The casting can be pressurized with air or water. When air is used, the casting is attached to a special fixture and 40–60 psi of air pressure is applied. A soapy solution is then sprayed onto the casting so the technician can look for bubbles, indicating a leak. Water pressure testing uses hot water to cause casting expansion. If the casting is leaking, water will be found on the outside of it.

FASTENERS

There are many different types of fasteners used throughout the engine. The most common are threaded fasteners, including bolts, studs, screws, and nuts. These fasteners must be inspected for thread damage, fillet damage, and stretch before they can be reused (**Figure 6-7**). Some bolts stretch in different areas than that shown in Figure 6-7. Some manufacturers require that you measure the diameter of the bolt's shank to determine if it can be reused. Other manufacturers may call for the measurement of bolt length to inspect for excessive bolt stretch. In addition, many threaded fasteners are not designed for reuse; the service manual should be referenced for the manufacturer's recommendations. If a threaded fastener requires replacement, there are some concerns the technician must be aware of:

■ Select a fastener of the same diameter, thread pitch, strength, and shank length as well as overall length as the original.

■ Note that all bolts in the same connection must be of the same grade.

■ Use nut grades that match their respective bolts.

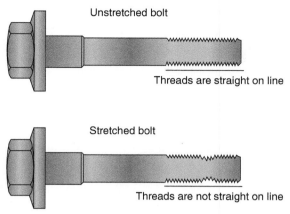

Figure 6-7 Check all bolts for stretch and other damage before reusing them.

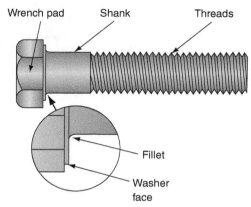

Figure 6-8 A typical bolt.

- Use the correct washers and pins as originally equipped.
- Torque the fasteners to the specified value.
- Use torque-to-yield bolts where specified.

Standard threaded fasteners used in automotive applications are classified by the Unified National Series using four basic categories:

1. Unified National Coarse (UNC or NC)
2. Unified National Fine (UNF or NF)
3. Unified National Extrafine (UNEF or NEF)
4. Unified National Pipe Thread (UNPT or NPT)

In recent years, the automotive industry has switched to the use of metric fasteners. Metric threads are classified as coarse or fine, as denoted by an SI or ISO lettering.

The most common type of threaded fastener used on the engine is the bolt (**Figure 6-8**). To understand proper selection of a fastener, terminology must be defined (**Figure 6-9**). The head of the bolt is used to torque the fastener. Several head designs are used, including hex, torx, slot, and spline. **Bolt diameter** is the measure across the threaded area or shank. The **pitch** (used in the English system) is the number of threads per inch. In the metric system, thread pitch is a measure of the distance (in millimeters) between two adjacent threads. **Bolt length** is the distance from the bottom of the head to the end of the bolt. The grade of the bolt denotes its strength and is used to designate the amount of stress the bolt can withstand. The grade of the bolt depends upon the material it is constructed from, bolt diameter, and **thread depth**. **Grade marks** are placed on the top of the head (in the English system) to identify the bolt's strength (**Figure 6-10**). In the metric system, the strength of the bolt is identified by a property class number on the head (**Figure 6-11**): the larger the number, the greater the tensile strength.

Like bolts, nuts are graded according to their tensile strength (**Figure 6-12**). As discussed earlier, the nut grade must be matched to the bolt grade. The strength of the connection is only as strong as the lowest grade used; for example, if a grade 8 bolt is used with a grade 5 nut, the connection is only a grade 5.

Proper fastener torque is important to prevent thread damage and to provide the correct clamping forces. The service manual provides the manufacturer's recommended torque value and tightening sequence for most fasteners used in the engine. The amount of torque a fastener can withstand is based on its tensile strength (**Figure 6-13**). To obtain proper torque, the fastener's threads must be cleaned and *may* require light lubrication. *Always clean out bolt holes before installing the bolts.* Parts can crack from hydrostatic pressure if a bolt is installed in a threaded hole with fluid in it.

Unified National Fine-thread or Unified National Course thread bolts may be referred to as fine-thread or coarse-thread bolts.

Bolt diameter indicates the diameter of an imaginary line running through the center of the bolt threaded area.

Pitch is the number of threads per inch on a bolt in the USC system.

Bolt length indicates the distance from the underside of the bolt head to the tip of the bolt threaded end.

The **grade** of a bolt is the classification of its strength.

Thread depth is the height of the thread from its base to the top of its peak.

Grade marks are radial lines on the bolt head.

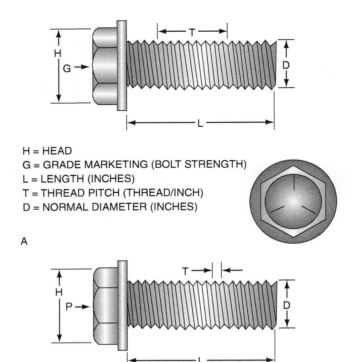

H = HEAD
G = GRADE MARKETING (BOLT STRENGTH)
L = LENGTH (INCHES)
T = THREAD PITCH (THREAD/INCH)
D = NORMAL DIAMETER (INCHES)

A

H = HEAD
G = PROPERTY CLASS (BOLT STRENGTH)
L = LENGTH (MILLIMETERS)
T = THREAD PITCH (THREAD/MILLIMETER)
D = NORMAL DIAMETER (MILLIMETER)

B

Figure 6-9 Bolt terminology.

SAE grade markings					
Definition	No lines: unmarked indeterminate quality SAE grades 0-1-2	3 lines: common commercial quality Automotive and AN bolts SAE grade 5	4 lines: medium commercial quality Automotive and AN bolts SAE grade 6	5 lines: rarely used SAE grade 7	6 lines: best commercial quality NAS and aircraft screws SAE grade 8
Material	Low-carbon steel	Medium-carbon steel tempered	Medium-carbon steel quenched and tempered	Medium-carbon alloy steel	Medium-carbon alloy steel quenched and tempered
Tensile strength	65,000 psi	120,000 psi	140,000 psi	140,000 psi	150,000 psi

Figure 6-10 Bolt-grade identification marks.

Torque-to-Yield

Modern engines are designed with very close tolerances. These tolerances require an equal amount of clamping forces at mating surfaces. Normal head bolt torque values have a calculated 25 percent safety factor; that is, they are torqued to only 75 percent of the bolt's

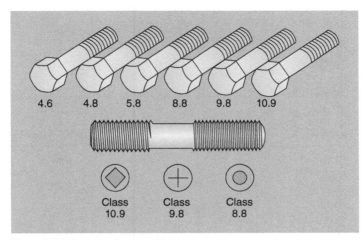

Figure 6-11 Property class numbers.

Inch system		Metric system	
Grade	Identification	Class	Identification
Hex nut grade 5	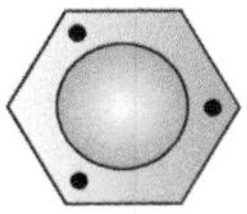 3 dots	Hex nut property grade 9	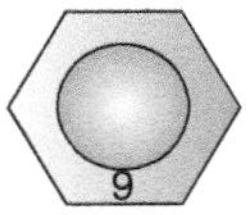 Arabic 9
Hex nut grade 8	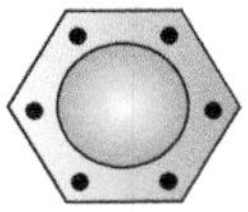 6 dots	Hex nut property grade 10	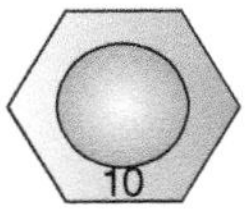 Arabic 10
Increasing dots represent increasing strength.		Can also have blue finish or paint dab on hex flat. Increasing numbers represent increasing strength.	

Figure 6-12 Nut grade markings.

STANDARD BOLT AND NUT TORQUE SPECIFICATIONS					
Size Nut or Bolt	Torque ft-lb	Size Nut or Bolt	Torque ft-lb	Size Nut or Bolt	Torque ft-lb
1/4–20	7–9	7/16–20	57–61	3/4–10	240–250
1/4–28	8–10	1/2–13	71–75	3/4–16	290–300
5/16–18	13–17	1/2–20	83–93	7/8–9	410–420
5/16–24	15–19	9/16–12	90–100	7/8–14	475–485
3/8–16	30–35	9/16–18	107–117	1–8	580–590
3/8–24	35–39	5/8–11	137–147	1–14	685–695
7/16–14	46–50	5/8–18	168–178		

Figure 6-13 Standard bolt and nut torque specifications.

maximum proof load (**Figure 6-14**). Using the chart, it can be seen that a small difference between torque values at the bolt head can result in a large difference in clamping forces. Because torque is actually force used to turn a fastener against friction, the actual clamping forces can vary even at the same torque value. Up to about 25 ft-lb (35 Nm) of torque, the clamping force is fairly constant; however, above this point, variation of actual

Grade of Bolt	SD	BC	10K	12K		Socket or Wrench Size	
Minimum Tensile Strength (Psi)	71,160	113,800	142,000	170,674			
Grade Marking on Head	SD	BC	10K	12K			
Metric						Metric	
Bolt Diameter (mm)		TORQUE (In Foot Pounds)				Bolt Head (mm)	Nut (mm)
6	5	6	8	10		10	10
8	10	16	22	27		14	14
10	19	31	40	49		17	17
12	34	54	70	86		19	19
14	55	89	117	137		22	22
16	83	132	175	208		24	24
18	111	182	236	283		27	27
22	182	284	394	464		32	32
24	261	419	570	689		36	36

Figure 6-14 Metric bolt and nut torque specifications.

clamping forces at the same torque value can be as high as 200 percent. This is due to variations in thread conditions or dirt and oil in some threads. Up to 90 percent of the torque is used up by friction, leaving 10 percent for the actual clamping. The result could be that some bolts have to provide more clamping force than others, distorting the cylinder bores. For this reason *only* lubricate a fastener's threads if the vehicle manufacturer recommends doing so.

To compensate and correct for these factors, many manufacturers use **torque-to-yield** bolts. The yield point of identical bolts does not vary much. A bolt that has been torqued to its yield point can be rotated an additional amount without any increase in clamping force. When a set of torque-to-yield fasteners is used, the torque is actually set to a point above the yield point of the bolt. This ensures that the set of fasteners will have an even clamping force.

Manufacturers vary on specifications and procedures for securing torque-to-yield bolts. Always refer to the service manual for exact procedures. In most instances, first a torque wrench is used to tighten the bolts to their yield point. Next, the bolt is turned an additional amount as specified in the service manual.

The graph in **Figure 6-15** indicates that a bolt can be elongated considerably at its yield point before it reaches its failure point. Also notice that the clamp load is consistent between the proof load and the failure point of the bolt. Bolts that are torqued to their yield points have been stretched beyond their elastic limit and require replacement whenever they are removed or loosened.

Torque-to-yield is a stretch-type bolt that must be tightened to a specific torque and then rotated a certain number of degrees.

Liquid Thread Lockers

Thread lockers are used to keep fasteners from loosening due to vibration or torsional twisting. They are commonly anaerobic, meaning they won't cure in the outside air but will cure between fastener threads in the absence of air. Liquid thread lockers come in

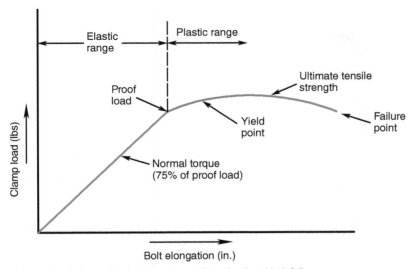

Figure 6-15 Relationship between proper clamp load and bolt failure.

different grades or strengths. "High strength" means heat is needed to break the "lock" on the fastener to take it back apart. Common engine bolts using thread locker are crankshaft pulley, flywheel, and camshaft bolts. See service information to determine which fasteners require locking compound.

GASKETS, SEALS, SEALANTS, AND ADHESIVES

Gaskets

There are three basic engine gasket classifications (**Figure 6-16**):

1. *Hard gaskets.* These include gaskets made of metals or metals covering a layer of clay/fiber compound. Head gaskets, exhaust manifold, intake manifold, and exhaust gas recirculation (EGR) valve gaskets are examples of this type of gasket.
2. *Soft gaskets.* These gaskets are made of rubber, cork, paper, and rubber-covered metal. Common usages include valve covers, oil pans, thermostat housings, water pumps, timing covers, and some intake manifolds.
3. *Liquid gasket makers.* These are usually a type of liquid material used to form gaskets. Silicones and anaerobics are examples of liquid gasket makers.

Cylinder Head Gaskets. Perhaps the cylinder head gasket is subjected to the greatest demands. It must be capable of sealing combustion pressures up to 2,700 psi (18,616 kPa) and withstanding temperatures over 2,000°F (1,100°C). In addition, it must be able to seal coolant and oil under pressure. To complicate the matter further, the head gasket must be able to perform these functions while accommodating a shearing action resulting from expansion rates of the metals. Shearing results from the difference in thermal expansion between the cylinder head and the block. As the two metals expand and contract at different rates and amounts, a scrubbing stress is created (**Figure 6-17**). The head gasket cannot move during these times. A large factor in determining what type of head gasket to use in a specific engine is the metals the head and the block are constructed from. The head and the block deck surface finish is mostly determined by what type of head gasket is being used. Changing the head gasket design without surface finish consideration can lead to early gasket failure.

The head gasket consists of a core, facing, and coating. The core is usually made from solid or clinched steel. The facing is usually constructed of graphite and rubber fiber. These materials allow sufficient compression to conform to minor surface irregularities.

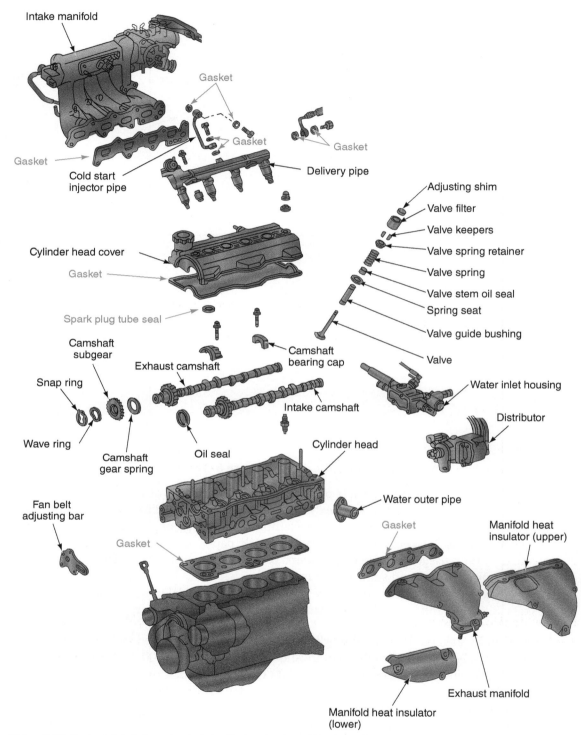

Figure 6-16 Seals and gaskets are located at critical seal points within the engine.

The coating may be Teflon or silicone based and works with the facing to seal minor surface irregularities and resist shearing. Some head gaskets use an elastomeric sealing bead to increase clamping forces around fluid passages (**Figure 6-18**).

One of the most recent developments in head gaskets is rubber-coated embossed (RCE) steel shim gaskets. A rubber coating is bonded to a steel shim gasket. The coating protects the shim from corrosive elements in the cooling and lubrication systems and

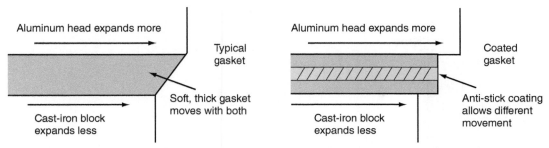

Figure 6-17 The coating on the gasket material allows for different expansion rates between metals and still provides a good seal without distorting the gasket.

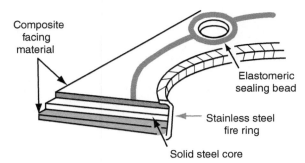

Figure 6-18 Sealing beads work to prevent fluid leakage.

provides good friction reduction. Because of their construction, RCE gaskets are also referred to as multilayer steel (MLS) gaskets. There are many advantages to the multilayer design. First, it has a uniform thickness that prevents bore distortion during cylinder head installation. Also, the MLS gasket is very resilient in that once compressed, it "bounces" back for a good seal.

Most cylinder head gaskets have a metal **fire ring** around the combustion chamber opening (**Figure 6-19**). The fire ring protects the gasket material from the high temperature it is exposed to. Also, the fire ring increases the gasket thickness around the cylinder bore so that it uses up to 75 percent of the clamping force to form a tight seal against combustion pressure losses.

Intake Manifold Gaskets. A leak in the connection of the intake manifold and cylinder head creates a vacuum leak that allows unmeasured air into the engine. This can cause the mixture entering the combustion chamber to be too lean (not enough fuel), resulting in rough idle, detonation, or both. The **intake manifold gasket** is designed to provide a good seal under the changing temperatures it will be subject to (**Figure 6-20**).

Many intake manifold gaskets are constructed of a solid or perforated steel core with a fiber facing. Some manufacturers use gaskets made from a rubber silicone bonded to steel or high-temperature plastic. On this style, the silicone provides the actual sealing. Some V-type engines use a **valley pan** style of intake manifold gasket. In addition, V-type engines use end strip seals to seal the connection between the ends of the manifold and the block. **Strip seals** are generally made from molded rubber or cork-rubber. Many fuel-injected engines use a plenum that requires a gasket between it and the intake manifold (**Figure 6-21**).

With the advancement of plastics and composites, some manufacturers are designing intake manifolds from these products. Some of these do not use the typical intake manifold gasket; instead, they use a series of O-rings to seal each port (**Figure 6-22**).

A **fire ring** is a metal ring surrounding the cylinder opening in a head gasket.

The **intake manifold gasket** fits between the manifold and cylinder head to seal the air-fuel mixture or intake air.

Valley pans prevent the formation of deposits on the underside of the intake manifold.

A **strip seal** provides a seal between the flat surfaces on the front and back of the engine block and the intake manifold.

Figure 6-19 This head gasket has a sealing bead and fire rings around the cylinders.

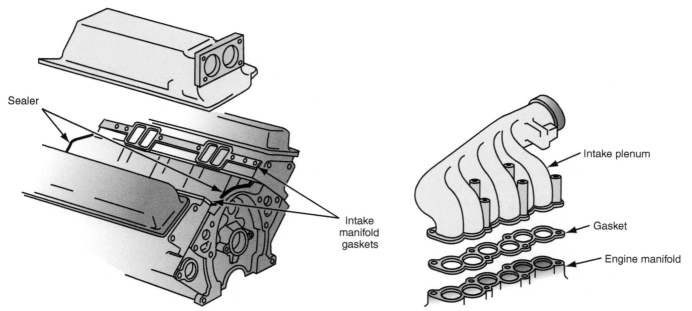

Figure 6-20 Intake manifold gasket location.

Figure 6-21 Intake plenum gasket location.

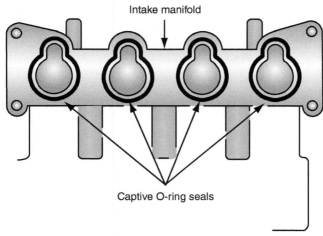

Figure 6-22 Some intake manifolds use captive O-rings to seal against the cylinder head.

Exhaust Manifold Gaskets. The **exhaust manifold gasket** must prevent leakage under extreme temperatures (**Figure 6-23**). Leaks in this connection disrupt the flow of exhaust gases and can result in poor engine performance. In addition, exhaust leaks can lead to burned valves and objectionable noises. The use of exhaust manifold gaskets by engine manufacturers has declined in recent years with improved machining processes and better materials; however, when the exhaust manifold is removed, it may become warped and require the use of a gasket. Most technicians opt to install an exhaust manifold gasket even if the engine was not originally equipped with one.

Valve Cover Gaskets. **Valve cover gaskets** are common locations for external oil leakage. This is largely due to the wide spacing between the attaching bolts. The wide spacing allows the stamped steel cover to distort easily and provides less clamping forces (**Figure 6-24**).

To seal properly, the valve cover gasket must be highly compressible yet have good torque retention. To perform these tasks, a variety of materials are used to construct valve cover gaskets. Some of the most common are synthetic rubber, cork-rubber, and molded rubber. Synthetic rubber gaskets seal by deforming instead of compressing. The synthetic rubber has a tendency to "remember" its original shape and will attempt to return to it. When the valve cover is tightened against the gasket, the gasket deforms. The gasket attempts to return to its original shape and pushes back against the valve cover, creating a

The **exhaust manifold gasket** seals the connection between the cylinder head and the exhaust manifold.

Valve cover gaskets seal the connection between the valve cover and the cylinder head. The gasket is not subject to pressures, but must be able to seal hot, thinning oil.

Valve covers may be called rocker arm covers.

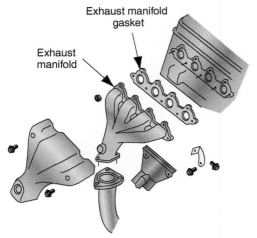

Figure 6-23 Exhaust manifold gasket location.

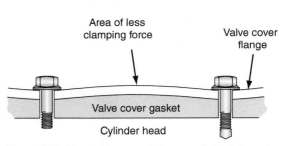

Figure 6-24 The valve cover gasket can easily be deformed.

tight seal. One drawback to this gasket is it can be difficult to install. Cork-rubber gaskets compress very well and provide good sealing. Molded rubber gaskets are the easiest to install and provide the best sealing. Sealers or adhesives should not be used on molded rubber gaskets. In addition to the valve cover gasket, many overhead camshaft engines use molded rubber semicircular plugs (**Figure 6-25**).

Oil Pan Gaskets. **Oil pan gaskets** can be multipiece or a single-unit molded gasket, depending on the crankcase design. The most common type of multipiece gasket uses two side pieces made of cork and rubber and two end pieces made of synthetic rubber (**Figure 6-26**). Many modern engines use a single-unit molded gasket (**Figure 6-27**).

Miscellaneous Gaskets. Many additional gaskets are used on the engine. Each of these must be replaced during the assembling process. Additional gasket applications include the following:

- Exhaust gas recirculation (EGR) valve
- Water pump

> **Oil pan gaskets** are used to prevent leakage from the crankcase at the connection between the oil pan and the engine block.

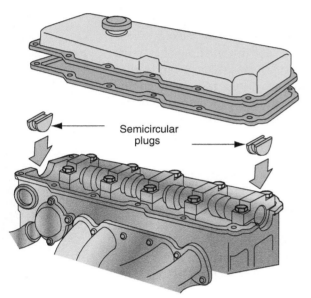

Figure 6-25 Camshaft plugs work with the valve cover gasket to seal oil.

Semicircular plugs

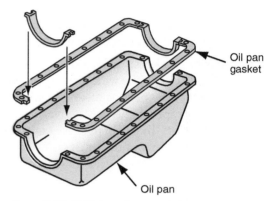

Oil pan gasket

Oil pan

Figure 6-26 Multipiece oil pan gasket.

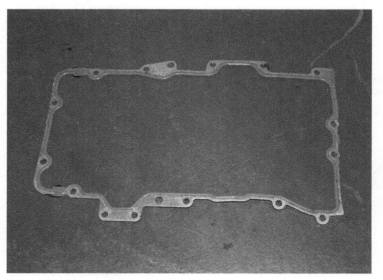

Figure 6-27 A one-piece oil pan gasket.

- Air cleaner
- Fuel injector mounting
- Timing cover
- Exhaust pipe

Each of these gaskets is designed for a particular purpose. The materials used must perform many of the tasks previously discussed, but under unique circumstances. Always refer to the service manual for the proper torque and sequence when installing these gaskets.

Seals

The most common seals in the engine are the **front main seal** (also called the timing cover seal) and the **rear main seal**. The front main seal seals around the harmonic balancer to prevent oil leakage from the front of the crankshaft, while the rear main seal prevents leakage from the rear of the crankshaft. Most modern engines use a type of **lip seal** to perform these functions. The lip seal is generally constructed of butyl rubber or Neoprene. The rubber seal is attached to a metal case that is driven into the bore (**Figure 6-28**). The seal lip will use the pressure of the fluid between the seal and the case to force the lip tight against the shaft. To assist in providing a tighter seal, some lip seals use a garter spring behind the lip. If the seal is installed in a location where the front of the seal may be exposed to dirt, a dust lip may be formed into the seal to deflect dirt away from the outside of the seal. When installing a new lip seal, the lip always faces the hydraulic pressure side or the fluid side.

The front main and oil pump housing oil seals are generally one-pipe lip seals with a steel outer ring (**Figure 6-29**). To aid in sealing during installation, apply silicone sealer to the outer diameter of the metal shell. Neoprene seals should always be lubricated prior to installation. This will prevent seal damage from overheating during initial startup. If the surface of the harmonic balancer is damaged, a **service sleeve** can be installed to provide a new sealing surface.

There are three common seal designs for rear main seals: rope-type, two-piece molded synthetic rubber, and one-piece rubber with steel ring. The rope-type rear main seal is found on many older engines. These are difficult to replace without removing the crankshaft. They were replaced with molded two-piece seals. Rope-type seals are also two pieces. The lower piece is installed into the bearing cap, while the upper piece is installed into the block.

> A **front main seal** is a lip-type seal that prevents leaks between the timing gear cover and the front of the crankshaft.

> The **rear main seal** is a seal installed behind the rear main bearing on the rear main bearing journal to prevent oil leaks.

> **Lip seals** are formed from synthetic rubber and have a slight raise (lip) that is the actual sealing point. The lip provides a positive seal while allowing for some lateral movement of the shaft.

> **Service sleeves** (sometimes referred to as speedy-sleeves) press over the damaged sealing area to provide a new, smooth surface for the seal lip.

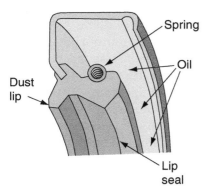

Figure 6-28 The common lip seal has an outer case, rubber sealing lips, and a gator spring. Hydraulic pressure works to cause the lip seal to ride tight against the shaft.

Front main seal

Figure 6-29 The front main seal fits into the timing cover.

A BIT OF HISTORY

In the year 2000, sales of light trucks surpassed car sales. The light truck market consists of these vehicles.

1. Minivans—17.9 percent
2. Full-sized vans—5.5 percent
3. Full-sized pickups—26.6 percent
4. Compact pickups—14.7 percent
5. Full-sized sport-utility vehicles (SUVs)—8 percent
6. Compact SUVs—23.3 percent
7. Small SUVs—3.9 percent

Adhesives, Sealants, and Liquid Gasket Maker

A number of chemical sealing materials are available, which are designed to reduce parts ordered or inventoried and increase the probability of a good seal. Proper use of these chemical materials is required to ensure good results.

To assist in gasket material removal, spray or brush-on gasket remover solvent is available. Adhesives are designed to bond different types of gasket materials in place prior to installation. These would be used to hold the gasket in place during installation of parts that may require a lot of wiggling into place. Some gaskets must be coated with a sealer prior to installation. This is true of a few MLS cylinder head gaskets. Some sealants can damage gasket coatings: Be sure to check the instructions for the gasket you are installing.

In the 1980s, many manufacturers began switching to **sealants** and liquid gasket makers instead of molded gaskets at many joint connections throughout the engine. When applying these materials, care must be taken to ensure proper application. Bead size, continuity, surface prep, and location are factors affecting proper sealing. If the bead is too thin, a leak can result. A bead that is too wide can result in spillover and clog the oil pump or oil galleries. Also, the manufacturer may require different types of sealers at specified locations in the engine. For this reason, always refer to the service manual for the correct type and usage of sealers.

There are two basic types of sealants:

- Aerobic
- Anaerobic

Most aerobic sealants and liquid gasket makers are silicone compounds and work well with metals and plastics. The most common type of liquid gasket maker is **room-temperature vulcanizing (RTV)** compound. This type of gasket maker forms a seal by absorbing moisture from the air. RTV gasket makers are capable of filling gaps up to 0.25 in. Before attempting to use RTV, the sealing surfaces must be thoroughly cleaned. RTV begins to set in about 10 minutes, requiring the components to be assembled quickly. Some RTV suppliers use different colors to denote the temperature range the RTV is capable of withstanding. Use the correct RTC for the expected temperature. RTV cannot be used on high-temperature exhaust system components. Before using any adhesives, sealants, or liquid gasket makers on engine components, make sure the package label states "Sensor Safe." Otherwise the O_2 sensors could end up failing and needing replacement due to silicon contamination.

Anaerobic sealants and liquid gasket makers are best used between machine metal parts like transmission case halves. These liquids cure after the components have been assembled. Anaerobic gasket makers are capable of filling gaps up to 0.015 in. (0.4 mm) and withstand temperatures up to 350°F (177°C). Some engines using a one-piece main

Sealants are commonly used to fill irregularities between the gasket and its mating surface. Some sealants are designed to be used in place of a gasket.

Aerobic sealants require the presence of oxygen to cure.

Anaerobic sealants will only cure in the absence of oxygen.

Room-temperature vulcanizing (RTV) compound is a type of engine sealer that may be used in place of a gasket on some applications.

bearing require the use of special anaerobic sealants that do not cure until a specified temperature is reached. This ensures proper maintenance of oil clearances as the main bearing bolts are torqued, without fear of the sealant setting too soon.

Liquid gasket makers can also be made from elastomeric rubber. They do an excellent job of sealing for many miles. Parts that use this type of gasket maker can be a challenge to disassemble.

SUMMARY

- Metals are divided into two basic groups: ferrous and nonferrous. Ferrous metals are those containing iron. Nonferrous metals contain no iron.
- Alloys are mixtures of two or more metals.
- Iron containing very low carbon (between 0.05 and 1.7 percent) is called steel.
- Composites are human-made materials using two or more different components tightly bound together. The result is a material consisting of characteristics that neither component possesses on its own.
- Powder metallurgy is the manufacturing of metal parts by compacting and sintering metal powders.
- Metric and standard bolts are used on today's automobiles.

- Gaskets are used to prevent gas, coolant, oil, or pressures from escaping between two mating parts.
- Seals are used to seal between a stationary part and a moving one.
- Sealants are used to seal parts and gaskets.
- Liquid gasket makers have replaced many of the gaskets on engines.
- Examples of sealants and liquid gasket makers are anaerobics and aerobic (RTV).
- Cylinder head gaskets must seal cylinder pressures up to 2,700 psi (18,616 kPa) and withstand temperatures over 2,000°F (1,100°C).
- Aerobic sealants require the presence of oxygen to cure. Anaerobic sealants will only cure in the absence of oxygen.

REVIEW QUESTIONS

Short-Answer Essays

1. Describe the advantage of aluminum alloy engine components compared to cast iron or steel components.

2. List the methods available to locate cracks in castings.

3. Describe the purpose of torque-to-yield bolts.

4. Explain the bolt grade marking system.

5. Define the following terms used with bolts:
   Head
   Diameter
   Pitch
   Length
   Grade

6. Describe what an anaerobic sealant is.

7. What is the function of a seal?

8. Describe the demands that a head gasket is subjected to.

9. Explain the advantages of rubber-coated embossed or multilayer steel head gaskets.

10. List three different types of valve cover gaskets.

Fill-in-the-Blanks

1. _____ metals are those containing iron. _____ metals contain no iron.

2. _____ _____ is the metal's resistance to being pulled apart.

3. _____ is the process of heating the metal to remove its hardness.

4. Most aluminum blocks are fitted with cylinder _____.

5. Bolt diameter is the measure across the _____ area.

6. _____ sealants require the presence of oxygen to cure.

7. When installing a lip seal, the lip always faces the _____ _____ or _____ side.

8. Neoprene seals should always be _____ before installation.

9. Head gaskets must be able to seal _____ and _____ from the combustion chamber under pressure.

10. The most common lip seals used on a typical engine are the _____ _____ and the _____ _____.

Multiple Choice

1. Ceramic engine components have all these characteristics *except*:
   A. Light weight
   B. Heat resistance
   C. Non-brittleness
   D. Ability to isolate sound and vibration

2. Forged engine components have these advantages and applications:
   A. Intake manifolds may be forged.
   B. Crankshafts may be forged.
   C. Rocker arm covers may be forged.
   D. They provide a softer metal that is more easily machined.

3. Engine components may be cracked by:
   A. extreme temperature changes in a very short time.
   B. extremely low, uniform temperatures.
   C. an excessive amount of antifreeze in the coolant.
   D. engine operating temperature below normal.

4. The manufacturing process that yields the strongest finished part is:
   A. casting.   C. machining.
   B. forging.   D. stamping.

5. Each of the following statements is true *except*:
   A. Tempering increases the toughness of the metal.
   B. Annealing makes a metal easier to be machined.
   C. Sintering bonds the metal powder together.
   D. Case hardening makes the metal's grain coarser.

6. When installing intake manifold gaskets:
   A. End strips are used to seal the ends of an OHV intake manifold to the block.
   B. A valley pan–type intake gasket is used on in-line engines.
   C. A leaking intake manifold gasket will cause a rich air-fuel mixture.
   D. Silicone sealer should be placed on both sides of the intake manifold gasket before installation.

7. All of these statements about head gaskets are true *except*:
   A. A conventional head gasket contains a core, facing, and coating.
   B. The core is usually made from silicone.
   C. The facing is usually made from graphite or rubber fiber.
   D. The coating may be made from Teflon.

8. The fire ring in a head gasket
   A. compensates for movement between metals with different expansion rates.
   B. increases heat transfer to the cooling system.
   C. protects the gasket material from high combustion temperatures.
   D. seals coolant passages through the head gasket.

9. Which of the following statements is true for replacing valve cover gaskets?
   A. A sealant should be used with molded rubber valve cover gaskets.
   B. An anaerobic sealant should be used with cork valve cover gaskets.
   C. The exhaust manifold and the valve cover gasket are usually made of the same material.
   D. Some valve cover gaskets are made from a synthetic rubber.

10. When discussing exhaust manifold connections, each of the following statements is true *except*:
    A. Some engines do not use an exhaust manifold gasket.
    B. Some exhaust manifolds are sealed with RTV.
    C. A damaged gasket can result in poor engine performance.
    D. An exhaust manifold gasket is a hard gasket.

CHAPTER 7
INTAKE AND EXHAUST SYSTEMS

Upon completion and review of this chapter, you should understand and be able to describe:

- The purpose of the air filter.
- The design of the air filter.
- The three different materials used to manufacture intake manifolds.
- The purpose of the intake manifold.
- The advantages of aluminum and plastic intake manifolds compared to cast iron.
- The operation and advantages of intake manifolds with dual runners.
- Two different methods for operating the valves that open and close the intake manifold runners.

- How the engine creates vacuum.
- How vacuum is used to operate and control many automotive devices.
- The operation of exhaust system components, including exhaust manifold, gaskets, exhaust pipe and seal, muffler, resonator, and tailpipe, and clamps, brackets, and hangers.
- The benefits and operation of turbochargers and superchargers.
- The purpose and operation of the catalytic converter.

Terms To Know

Air filter
Airflow restriction indicator
Back pressure
Catalytic converter
Compressor wheel
Exhaust manifold
Intake air temperature (IAT) sensor

Intake manifold
Intercooler
Mass airflow (MAF) sensor
Muffler
Oil coking
Resonator
Roots-type supercharger
Supercharger

Tailpipe
Turbine wheel
Turbo boost
Turbo lag
Turbocharger
Variable-length intake manifold
Wastegate

INTRODUCTION

It's very simple—the better an engine can breathe, the more power it will make. The intake and exhaust system is the breathing apparatus for the engine. To maximize volumetric efficiency, the intake and exhaust manifolds must be carefully designed. Both must allow adequate airflow to provide strong engine operation. A plugged air filter can seriously restrict airflow, causing the engine to run with reduced power or even not to start. A restricted exhaust system can cause the same symptoms. The engine must be able to freely take air in and easily push air out to breathe well. The intake system must be well sealed so that a strong vacuum is formed. This ensures that air will flow into the cylinder when the throttle and intake valves are open.

Some engines used a forced induction system to increase airflow into the engine. A turbocharger or supercharger pressurizes the air above atmospheric pressure to force more air in when the intake valves open. This improvement in volumetric efficiency can increase engine power dramatically.

AIR INDUCTION SYSTEM

The intake system consists of the air inlet tube, an air filter, ducting, a throttle bore, and the intake manifold (**Figure 7-1**). The opening and closing of the throttle plate in the throttle bore controls airflow into the engine. Think of the throttle plate in a gasoline engine as a restrictive device that limits and controls the amount of air that will enter the engine. Some systems use a throttle cable to open and close the throttle plate, while others employ "drive by wire" technology. In these systems, the Powertrain Control Module (PCM) responds to an input from the apps (accelerator pedal position sensors) and controls a motor at the throttle plate to open it to the correct angle. The intake system should provide airflow at a high velocity and low volume when the engine is operating at low rpms. This ensures good fuel atomization and mixing when the injector sprays into the airstream in the manifold near the intake valve. The intake system should deliver a high volume of air when the engine is running at high rpms, to fill the combustion chamber as much as possible in a short time. This keeps power output high even at higher engine speeds. Volumetric efficiency will naturally fall at higher rpms, but the design of the intake and exhaust systems can reduce the losses.

The intake system pulls in fresh air from outside the engine compartment. Ducts direct this air through the air filter and into the throttle body assembly to power the engine. The air filter is placed below the top of the engine to allow for aerodynamic body designs. Electronic sensors measure airflow, temperature, and density. The air induction system performs the following functions:

- Filters the air to protect the engine from wear
- Silences air intake noise
- Heats or cools the air as required
- Provides the air the engine needs to operate
- Monitors airflow temperature and density for more efficient combustion and a reduction of hydrocarbon (HC) and carbon monoxide (CO) emissions
- Operates with the positive crankcase ventilation (PCV) system to burn the crankcase fumes in the engine

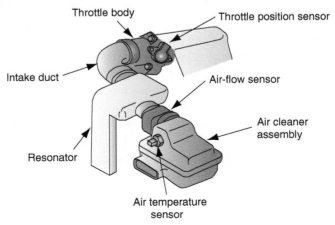

Figure 7-1 A late-model intake air distribution system.

AIR INTAKE DUCTWORK

Ductwork is used to direct the air into the throttle body. Cool outside air is drawn into the air cleaner assembly, and on some engines, warm air from around the exhaust is also brought in for cold engine operation. Many air ducts include a resonator chamber that looks like an irregularly shaped protrusion on the side of an intake duct. This is used to reduce the rumble of intake air as it rushes through the ductwork. Air then flows through the air filter, where particulate matter that could damage the engine is removed. Another leg of ducting connects the air filter housing to the throttle bore. Many vehicles have a **mass airflow (MAF) sensor** in this ducting to measure the mass of air flowing into the engine. Most engines also use an intake air temperature (IAT) sensor in the air cleaner housing of intake ductwork (**Figure 7-2**). As the air temperature increases, the air density decreases. The PCM will reduce the amount of fuel delivered, as needed to ensure excellent fuel economy and low emissions.

The **mass airflow (MAF) sensor** measures the mass of airflow entering the engine and sends an input to the PCM reflecting this value. This is a primary input for the PCM's control of fuel delivery.

Be sure that the intake ductwork is properly installed and all connections are airtight, especially those between the MAF sensor or remote air cleaner and the throttle plate assembly. If outside air can sneak in through a leak in the ductwork, it is not filtered. Vehicles that use a MAF sensor measure the air coming into the engine from the air filter. The PCM uses this input as the primary indicator of how much fuel should be injected. If air is drawn into the engine beyond the MAF sensor through a leak in the duct, it will not be measured. The PCM will deliver less fuel, and the engine will run too lean. Generally, metal or plastic air ducts are used when engine heat is not a problem. Special paper/metal ducts are used when they will be exposed to high engine temperatures.

AIR CLEANER/FILTER

The primary function of the **air filter** is to prevent airborne contaminants and abrasives from entering the engine. Without proper filtration, these contaminants can cause serious damage and appreciably shorten engine life. All incoming air should pass through the filter element before entering the engine (**Figure 7-3**).

The **air filter** prevents dirt from entering the engine.

Intake Air Temperature
Sensor

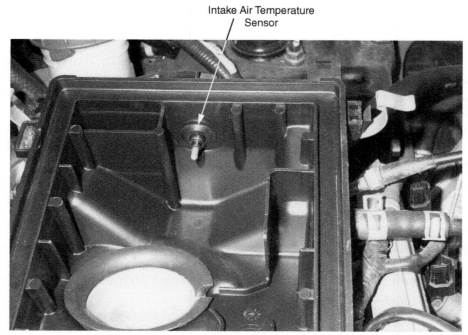

Figure 7-2 The intake air temperature sensor helps the PCM determine the correct fuel delivery and spark timing.

Figure 7-3 The air filter is located in the air filter housing within the intake ductwork.

Air Filter Design

Air filters are basically assemblies of pleated paper supported by a layer of fine mesh wire screen. The screen gives the paper some strength and also filters out large particles of dirt. A thick plastic-like gasket material normally surrounds the ends of the filter. This gasket adds strength to the filter and serves to seal the filter in its housing. If the filter does not seal well in the housing, dirt and dust can be pulled into the airstream to the cylinders. In most air filters, the air flows from the outside of the element to the inside as it enters the intake system.

The shape and size of the air filter element depends on its housing; the filter must be the correct size for the housing or dirt will be drawn into the engine (**Figure 7-4**). On today's engines, air filters are either flat (**Figure 7-5**) or round (**Figure 7-6**). Air filters must be properly aligned and closed around the filter to ensure good airflow of clean air.

Some air cleaners have an **airflow restriction indicator** mounted in the air cleaner housing (refer to **Figure 7-7**). If the air filter element is not restricted, a window in the side of the restriction indicator shows a green color (or it may be clear). When the air filter element is restricted, the window in the airflow restriction indicator is orange and the words "Change Air Filter" appears. The air filter must then be replaced. After the air filter is replaced, a reset button on top of the airflow restriction indicator must be pressed to reset the indicator so it displays green in the window.

Some air cleaners have a combined MAF sensor and **intake air temperature (IAT) sensor** attached to the air outlet on the air cleaner housing. A duct is connected from the MAF sensor to the throttle body. The MAF sensor must be attached to the air cleaner so air flows through this sensor in the direction of the arrow on the sensor housing (**Figure 7-8**).

An **airflow restriction indicator** is a meter that shows the amount of air flow past a port. This gives the technician an indicator that the air filter may be not flowing as well as it could because it is dirty.

The **intake air temperature (IAT) sensor** sends an analog voltage signal to the PCM in relation to air intake temperature.

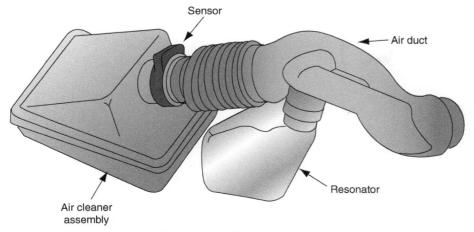

Sensor

Air duct

Resonator

Air cleaner
assembly

Figure 7-4 The air cleaner assembly is shaped to fit in the crowded engine compartment.

Figure 7-5 Typical flat air cleaner element.

Figure 7-6 Typical round air filter for a late-model vehicle.

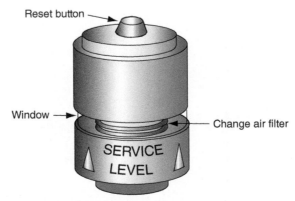

Reset button

Window

Change air filter

SERVICE
LEVEL

Figure 7-7 Airflow restriction indicator.

If the airflow is reversed through the MAF sensor, the powertrain control module (PCM) supplies a rich air-fuel ratio and increased fuel consumption. Other air cleaners contain a separated IAT sensor (**Figure 7-9**).

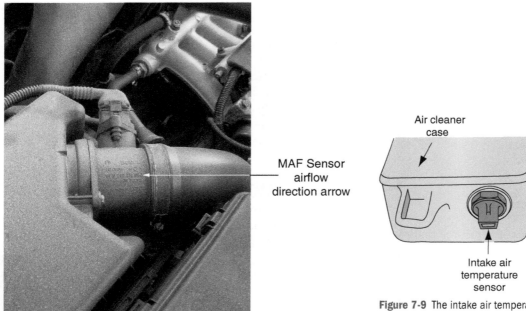

Figure 7-8 The mass airflow sensor must be installed in the proper direction.

Figure 7-9 The intake air temperature sensor may be mounted in the air cleaner housing.

INTAKE MANIFOLD

An **intake manifold** is a cast iron, aluminum, or plastic component with internal passages that distributes air or an air-fuel mixture from the throttle body or carburetor to the cylinder head intake ports.

The **intake manifold** distributes the clean air as evenly as possible to each cylinder of the engine. Fuel is injected at the very end of the intake manifold runners near the combustion chamber.

Modern intake manifolds for engines with port fuel injection are typically made of die-cast aluminum or plastic (**Figure 7-10**). These materials are used to reduce engine weight. A plastic manifold transfers less heat to the intake air, and this results in a denser air-fuel mixture. Plastic manifolds are also lower in cost to produce and transfer less vibration, making them quieter. Because intake manifolds for port-injected engines only deliver air to the cylinders, fuel vaporization and condensation are not design considerations. These intake manifolds deliver air to the intake ports, where it is mixed with the fuel delivered by the injectors (**Figure 7-11**). A primary consideration of these manifolds is the delivery of equal amounts of air to each cylinder.

Figure 7-10 This die-cast aluminum manifold feeds the five-cylinder engine through short runners off of the plenum.

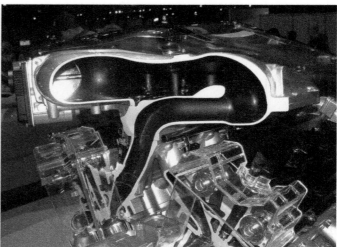

Figure 7-11 In a port fuel injection engine, the intake manifold delivers air to the intake ports.

Intake manifolds also serve as the mounting point for many intake-related accessories and sensors. Some include a provision for mounting the thermostat and thermostat housing. In addition, connections to the intake manifold can provide a vacuum source for the exhaust gas recirculation (EGR) system, automatic transmission vacuum modulators, power brakes, and/or heater and air-conditioning airflow control doors. Many of these vacuum-operated devices are found only on older vehicles. On modern vehicles they are electronically controlled. When you compare an older engine to a modern one, one of the first things you will notice is the absence of many vacuum-controlled components that would operate from the intake manifold's vacuum. Other devices located on or connected to the intake manifold include the manifold absolute pressure (MAP) sensor, knock sensor, various temperature sensors, and EGR passages.

Most engines cannot produce the amount of power they should at high speeds, because they do not receive enough air. This is the reason why many race cars have hood scoops. With today's body styles, hood scoops are not desirable, because they increase aerodynamic drag. However, to get high performance out of high-performance engines, more air must be delivered to the cylinders at high engine speeds. There are a number of ways to do this; increasing the air delivered by the intake manifold is one of them.

INTAKE MANIFOLD TUNING

Intake manifolds can be designed to improve the volumetric efficiency of the engine. This process is called intake manifold tuning. As the air rushes through the intake manifold into the cylinders, the opening and closing of the intake valves cause the airflow to pulse. If the intake manifold is divided into individual runners for each cylinder, the pulsing effect of the airflow pushes or rams more air into the cylinder. By adjusting or tuning the length of the individual runners, the manufacturer can design an intake manifold to supply the amount of air required by the particular engine. Smaller diameter, longer intake manifold runners ram more air into the cylinders at lower engine rpm. To push more air into the cylinders at higher rpm when air pulsing is faster, the intake manifold runners are designed larger in diameter and shorter. Runners must be curved to avoid sharp bends that create more airflow restriction.

Manufacturers have also developed manifolds that deliver more air only at high engine speeds. The design of manifolds is intended to keep air moving quickly. At low rpm, narrower diameter, longer runners keep the smaller volume of air moving quicker. At high rpm, shorter, wider runners offer less restriction and allow the air to keep moving quickly.

One such system (**Figure 7-12**) uses a **variable-length intake manifold**. At low speeds, the air travels through only part of the manifold on its way to the cylinders. When the engine speed reaches about 3,700 rpm, two butterfly valves open, forcing the air to take a longer route to the intake port. This increases the speed of the airflow, as well as increasing the amount of air available for the cylinders. As a result, more power is available at high speeds without decreasing low-speed torque and fuel economy and without increased exhaust emissions.

Another approach is the use of two large intake manifold runners for each cylinder (**Figure 7-13**). Separating the intake runners from the intake ports is an assembly that has two bores for each cylinder. There is one bore for each runner. Both bores are open when the engine is running at high speeds. However, a butterfly valve in one set of the runners is closed by the PCM when the engine is operating at lower speeds. This action decreases airflow speed and volume at lower engine speeds and allows for greater airflow at high engine speeds. The butterfly valves in the intake manifold runners may be operated by a vacuum actuator, and the PCM operates an electric/vacuum solenoid that turns the vacuum on and off to the actuator. In other intake manifolds, the valves that open and close some of the runners are operated electrically by the PCM, much like a solenoid (**Figure 7-14**).

A PCM-operated electric solenoid-type valve that opens and closes some of the intake manifold runners may be called an intake manifold tuning valve (IMTV).

The **variable-length intake manifold** has two runners of different lengths connected to each cylinder head intake port.

Runner valve open

Runner valve closed

Figure 7-12 A variable-length intake manifold.

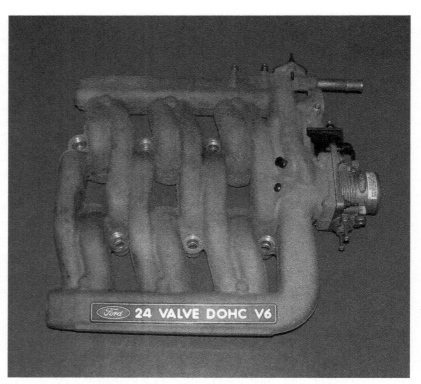

Figure 7-13 This manifold provides a short and a long runner for each cylinder. Both are opened at higher engine rpms to improve engine breathing.

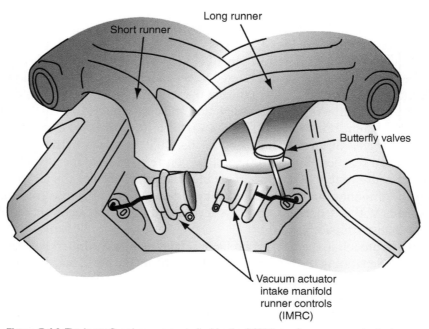

Figure 7-14 The butterfly valves are controlled by the PCM through a vacuum actuator to open the second set of runners at high rpms.

Combining a variable intake manifold with a variable valve timing system will increase the power output of a vehicle significantly. In many cases it will also reduce the emissions output and make the engine more efficient. Many vehicles produced today have such a combination of systems.

VACUUM BASICS

Vacuum refers to any pressure that is lower than the earth's atmospheric pressure at any given altitude. Atmospheric pressure appears as zero on most pressure gauges. This does not mean there is no pressure; rather, it means the gauge is designed to read pressures greater than atmospheric pressure. All measurements taken on this type of gauge are given in pounds per square inch and should be referred to as psig (pounds per square inch gauge). Gauges and other measuring devices that include atmospheric pressure in their readings also display their measurements in psi; however, these should be referred to as psia (pounds per square inch absolute). There is a big difference between 12 psia and 12 psig. A reading of 12 psia is less than atmospheric pressure and therefore would represent a vacuum, whereas 12 psig would be approximately 26.7 psia. When referring to the pressure in the intake manifold, we use manifold absolute pressure (MAP) as the measure. MAP is a reading of absolute pressure in the manifold. The MAP reading at idle is typically 7–9 psi. Because vacuum is defined as any pressure less than atmospheric, vacuum is any pressure less than 0 psig or 14.7 psia. Remember that vacuum is simply a difference in the pressures of two areas. Vacuum is not normally measured using negative psi numbers. Vacuum had to originally be measured in a tube filled with mercury, a metal that is in its liquid state at room temperatures. The mercury would then be drawn by the vacuum up a tube, and the amount drawn up was measured in inches. Today, we do not use mercury tubes but rather a calibrated gauge that reads in inches (or millimeters) of mercury (in. Hg or mm Hg). Other units of measurement for vacuum are kilopascals and bars. Normal atmospheric pressure at sea level is about 1 bar or 100 kilopascals.

An engine in good condition should develop 16–18 in. Hg vacuum at idle. The amount of low pressure produced by the piston during its intake stroke depends on a number of things. Basically it depends on the cylinder's ability to form a vacuum and the intake system's ability to fill the cylinder. When there is high vacuum (16–18 Hg [381 to 559 mm Hg]), we know the cylinder is well sealed and not enough air is entering the cylinder to fill it. At idle, the throttle plate is almost closed and nearly all airflow to the cylinders is stopped. This is why vacuum is high during idle. Because there is a correlation between throttle position and engine load, it can be said that load directly affects engine manifold vacuum. Therefore, vacuum will be high whenever there is no, or low, load on the engine.

VACUUM CONTROLS

Engine manifold vacuum is used to operate and control several devices on an engine. Prior to the mid-1960s, vacuum was used only to operate the windshield wipers and a distributor vacuum advance unit. Since then, the use of vacuum has become extensive. Some emission control output devices operate on a vacuum. This vacuum is usually controlled by solenoids that are opened or closed, depending on electrical signals received from the PCM. Engine vacuum is used to control operation of certain accessories, such as air conditioner/heater systems, power brake boosters, speed-control components, automatic transmission vacuum modulators, and so on. A vacuum solenoid either blocks or flows vacuum to actuate components or control the flow of vacuum. The evaporative emissions control system and the EGR system often use electrically controlled vacuum solenoids (**Figure 7-15**).

Figure 7-15 This vacuum solenoid is controlled by the PCM and uses the pull of vacuum to draw fuel vapors into the intake manifold.

TURBOCHARGERS

Another method of improving volumetric efficiency and engine performance is to compress the intake air before it enters the combustion chamber. This can be accomplished through the use of a turbocharger or a supercharger. While at idle the engine still develops the same amount of vacuum; under boost the supercharger or turbocharger delivers roughly 8–15 psi of pressure above atmospheric pressure. When the throttle opens, the greater pressure difference increases airflow into the cylinders. This increases the density of the air charge in the cylinder producing more power.

A turbocharger or supercharger changes the effective compression ratio of an engine, simply by packing in air at a pressure that is greater than atmospheric; for example, an engine that has a compression ratio of 8:1 and receives 10 pounds (69 kPa) of boost will have an effective compression ratio of 10.5:1.

A **turbocharger** is a blower or special fan assembly that uses the expansion of hot exhaust gases to turn a turbine and compress incoming air. In a typical car engine, a turbocharger may increase the horsepower approximately 20 percent. A typical turbocharger consists of the following components:

- Turbine wheel
- Shaft
- Compressor wheel
- Wastegate valve
- Actuator
- Center housing and rotating assembly (CHRA)

Basic Operation

The **turbine wheel** and the compressor wheel are mounted on a common shaft. Both wheels have fins or blades, and each wheel is encased in its own spiral-shaped housing.

A **turbocharger** uses the expansion of exhaust gases to rotate a fan-type wheel, which increases the pressure inside the intake manifold.

A **turbine wheel** is a vaned wheel in a turbocharger to which exhaust pressure is supplied to provide shaft rotation.

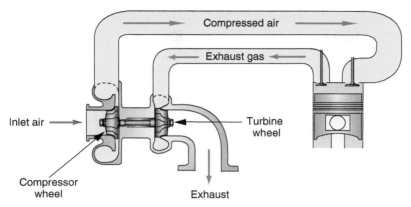

Figure 7-16 Exhaust gases spin the turbine wheel. The compressor wheel is connected to the turbine through a shaft and spins at the same rate to pressurize the intake charge.

The shape of the housing works to control and direct the flow of the gases. The shaft is supported on bearings in the turbocharger housing. The expelled exhaust gases from the cylinders are directed through a nozzle against the blades of the turbine wheel. When engine load is high enough, there is enough exhaust gas flow to cause the turbine wheel and shaft to rotate at a high speed. This action creates a vortex flow. Since the compressor wheel is positioned on the opposite end of the shaft, the compressor wheel must rotate with the turbine wheel (**Figure 7-16**).

The **compressor wheel** is mounted in the air intake system. As the compressor wheel rotates, air is forced into the center of the wheel, where it is caught by the spinning blades and thrown outward by centrifugal force. The air leaves the turbocharger housing and enters into the intake manifold. Since most turbocharged engines are port injected, the fuel is injected into the intake ports. The rotation of the compressor wheel compresses the air and fuel in the intake manifold, creating a denser air-fuel mixture. This increased intake manifold pressure forces more air-fuel mixture into the cylinders to provide increased engine power. The turbocharger must reach a certain rpm before it begins to pressurize the intake manifold. Some turbochargers begin to pressurize the intake manifold at 1,250 engine rpm and reach full boost pressure in the intake manifold at 2,250 rpm.

In a normally aspirated engine, air is drawn into the cylinders by the difference in pressure between the atmosphere and engine vacuum. A turbocharger pressurizes the intake charge to a point above normal atmospheric pressure. **Turbo boost** is the positive pressure increase created by the turbocharger.

> A **compressor wheel** is a vaned wheel mounted on the turbocharger shaft that forces air into the intake manifold.

> **Turbo boost** is a term used to describe the amount of positive pressure increase of the turbocharger; for example, 10 psi (69 kPa) of boost means the air is being induced into the engine at 24.7 psi (170 kPa). Normal atmospheric pressure is 14.7 psi (101 kPa); add the 10 psi (69 kPa) to get 24.7 psi (170 kPa).

 A BIT OF HISTORY

Turbochargers have been common in heavy-duty applications for many years, but they were not widely used in the automotive industry until the 1980s. Turbochargers had two traditional problems that prevented their wide acceptance in automotive applications. Older turbochargers had a lag, or hesitation, on low-speed acceleration, and there was the problem of bearing cooling. Engineers greatly reduced the low-speed lag by designing lighter turbine and compressor wheels with improved blade design. Water cooling combined with oil cooling provided improved bearing life. These changes made the turbocharger more suitable for automotive applications.

Turbocharger wheels rotate at very high speeds, in excess of 100,000 rpm. Engineers design and balance the turbocharger to run in excess of 150,000 rpm, about 25 times the maximum rpm of most engines. Due to this high-speed operation, turbocharger wheel balance and bearing lubrication are very important.

Boost Pressure Control. If the turbocharger boost pressure is not limited, excessive intake manifold and combustion pressure can destroy engine components. Also, if the amount of boost is too high, detonation knock can occur and decrease engine output. To control the amount of boost developed in the turbocharger, most turbochargers have a **wastegate** diaphragm mounted on the turbocharger. A linkage is connected from this diaphragm to a wastegate valve in the turbine wheel housing (**Figure 7-17**).

Under low to partial load conditions, the diaphragm spring holds the wastegate valve closed. This routes all of the exhaust gases through the turbine housing. Boost pressure from the intake manifold is also supplied to the wastegate diaphragm.

Under full load, when the boost pressure in the intake manifold reaches the maximum safe limit, the boost pressure pushes the wastegate diaphragm and opens the wastegate valve. This action allows some exhaust to bypass the turbine wheel, which limits turbocharger shaft rpm and boost pressure (**Figure 7-18**).

On some engines, the boost pressure supplied to the wastegate diaphragm is controlled by a computer-operated solenoid. In many systems, the PCM pulses the wastegate solenoid on and off to control boost pressure. Some computers are programed to momentarily allow a higher boost pressure on sudden acceleration to improve engine performance.

> A **wastegate** limits the maximum amount of turbocharger boost by directing the exhaust gases away from the turbine wheel. The wastegate valve is also referred to as a bypass valve.

Wastegate diaphragm

Figure 7-17 The wastegate diaphragm is operated by a PCM-controlled solenoid to protect the engine from dangerous overboosting.

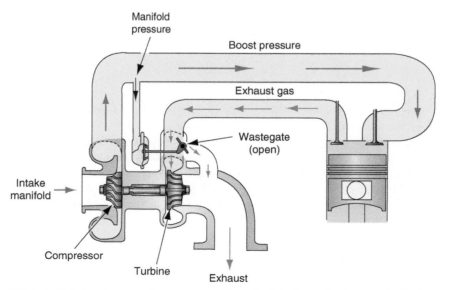

Figure 7-18 As boost pressure increases toward the safe limit, the wastegate opens to divert exhaust gases away from the turbine wheel.

Turbo Lag

Turbo lag occurs when the turbocharger compressor and turbine wheels are not spinning fast enough to create boost. It takes time to get the exhaust gases to bring the turbocharger wheels up to operating speed. The size and weight of the turbine and compression wheels, along with housing design, are factors that affect the amount of turbo lag. Smaller, lighter turbochargers result in less lag time but lower boost pressures. Some engines use two sequential turbochargers: a very small, lightweight one to help acceleration at low rpms and a bigger volume one to create plenty of boost to maximize power at higher rpms. Low-pressure turbochargers may provide boost of only 4–8 psi, but they reduce the lag time until the 8–15 psi from the full-size turbo kicks in. Also used on some diesel engines, variable nozzle turbine turbocharger systems have been developed to reduce the lag period (**Figure 7-19**).

> **Turbo lag** is a short delay period before the turbocharger develops sufficient boost pressures.

Turbocharger Cooling

Exhaust flow past the turbine wheel creates very high turbocharger temperature, especially under high engine load conditions. Many turbochargers have coolant lines connected from the turbocharger housing to the cooling system (**Figure 7-20**). Coolant circulation through the turbocharger housing helps cool the bearings and shaft. Full oil pressure is supplied from the main oil gallery to the turbocharger bearings and shaft to lubricate and cool the bearings. This oil is drained from the turbocharger housing back into the crankcase. Seals on the turbocharger shaft prevent oil leaks into the compressor or turbine wheel housings. Worn turbocharger seals allow oil into the compressor or turbine wheel housings, resulting in blue smoke in the exhaust and oil consumption. Some heat is also dissipated from the turbocharger to the surrounding air.

Some turbochargers do not have coolant lines connected to the turbocharger housing. These units depend on oil and air cooling. On these units, if the engine is shut off immediately after heavy-load or high-speed operation, the oil may burn to some extent in the turbocharger bearings. When this action occurs, hard carbon particles, which destroy the turbocharger bearings, are created. The coolant circulation through the turbocharger housing lowers the bearing temperature to help prevent this problem. When turbochargers that depend on oil and air cooling have been operating at heavy load or high speed, idle the engine for at least one minute before shutting it off. This action will help prevent turbocharger bearing failure.

Intercoolers

A disadvantage of the turbocharger (and supercharger) is that it heats the incoming air. The hotter the air, the less dense it is. As the air gets hotter, fewer air molecules can enter the cylinder on each intake stroke. Also, hotter intake air leads to detonation problems.

Figure 7-19 Variable nozzle turbine type turbocharger.

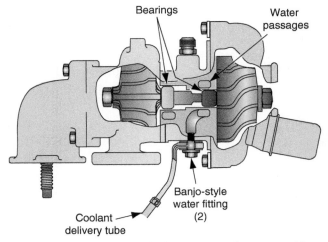

Figure 7-20 A water-cooled turbocharger uses coolant to control the temperature of the bearings and the shaft.

Figure 7-21 The intercooler cools the pressurized air to create a denser intake charge.

> The **intercooler** cools the intake air temperature to increase the density of the air entering the cylinders. It is also referred to as a charge air cooler.

To combat these effects, many turbocharger systems use an **intercooler** (**Figure 7-21**). The intercooler is like a radiator in that it removes heat from the turbocharger system by dissipating it to the atmosphere. The intercooler can be either air cooled or water cooled (**Figure 7-22**). Cooling the air makes it denser, increasing the amount of oxygen content with each intake stroke. The intercooler cools the air that leaves the turbocharger at about 100°F (38°C) before it enters the cylinders. For every 10°F (5.5°C) that the air is cooled, a power gain of about 1 percent is obtained. If the intercooler is capable of cooling the air by 100°F (38°C), then a 10 percent power increase is obtained.

Scheduled Maintenance

> **Oil coking** occurs after the engine is shut down and the turbocharger is still spinning. The oil is no longer circulating, and what remains inside the turbo can literally bake into a white, hard, caked restriction in the lines. Allowing the engine to idle for a minute before shutting the engine off allows the turbo compressor to wind down and prevents this.

Four main things will reduce the life of a turbocharger: lack of oil, **oil coking**, contaminants in the oil, and ingestion of foreign material through the air intake. To prevent premature turbocharger failure, engine oil and filters should be changed at the vehicle manufacturer's recommended intervals. The engine oil level must be maintained at the specified level on the dipstick. The air cleaner element and the air intake system must be maintained in satisfactory condition. Dirt entering the engine through an air cleaner will damage the compressor wheel blades. When coolant lines are connected to the turbocharger housing, the cooling system must be maintained according to the vehicle manufacturer's maintenance schedule to provide normal turbocharger life.

Turbocharged engines have a lower compression ratio than a normally aspirated engine, and many parts are strengthened in a turbocharged engine because of the higher cylinder pressure. Therefore, many components in a turbocharged engine are not interchangeable with the parts in a normally aspirated engine.

> **AUTHOR'S NOTE** During my experience in the automotive service industry, I encountered a significant number of turbocharged engines with low boost pressure. This low boost pressure was often caused by low engine compression. Because low engine compression reduces the amount of air taken into the engine, this problem also reduces turbocharger speed and boost pressure. Therefore, when diagnosing low boost pressure, always verify the engine condition before servicing or replacing the turbocharger.

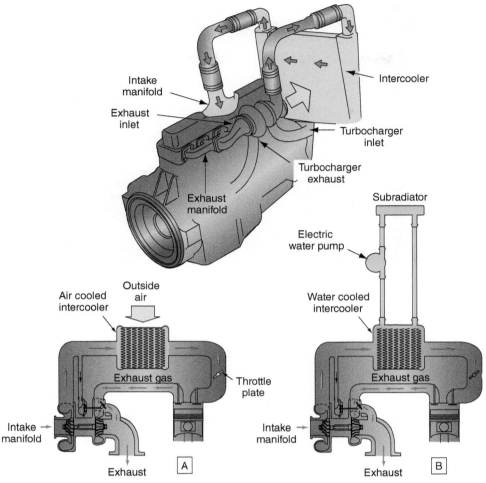

Figure 7-22 Intercoolers can be air cooled (A) or water cooled (B).

SUPERCHARGERS

After falling into disuse for a number of years (except for racing applications), the **supercharger** started to reappear on automotive engines as original equipment manufacturer (OEM) in 1989 (**Figure 7-23**). The supercharger is belt driven from the crankshaft by a ribbed belt (**Figure 7-24**). A shaft is connected from the crankshaft pulley to one of the drive gears in the front supercharger housing, and the driven gear is meshed with the drive gear. The rotors inside the supercharger are attached to the two drive gears.

The drive gear design prevents the rotors from touching; however, there is a very small clearance between the drive and driven gears. In some superchargers, the rotor shafts are supported by roller bearings on the front and needle bearings on the back. During the manufacturing process, the needle bearings are permanently lubricated. The ball bearings are lubricated by a synthetic-based, high-speed gear oil. A plug is provided for periodic checks of the front bearing lubricant. Front bearing seals prevent lubricant loss into the supercharger housing.

In a typical car engine, a supercharger increases horsepower approximately 20 percent compared to a normally aspirated engine with the same CID; for example, the General Motors normally aspirated 1997 3800 V-6 engine is rated at 200 HP. The supercharged version of this engine is rated at 240 HP. The supercharger will operate around 10,000 to 15,000 rpm. Unlike the turbocharger, the amount of boost the supercharger produces is a

The **supercharger** is a belt-driven air pump used to increase the compression pressures in the cylinders.

The supercharger may be called a blower.

Figure 7-23 A factory-installed supercharger.

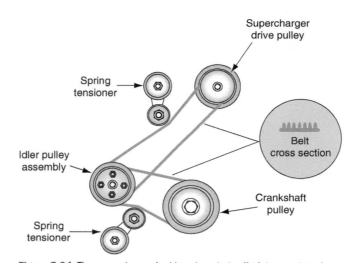

Figure 7-24 The supercharger is driven by a belt off of the crankshaft.

function of engine rpm (not load). The advantage of the supercharger is that it will produce more torque at lower engine speeds than a turbocharger. Also, there is no lag time associated with a supercharger. The disadvantage of the supercharger is that it consumes horsepower as it is driven.

Supercharger Operation

The **Roots-type supercharger** is a positive displacement supercharger that uses a pair of lobed vanes to pump and compress the air.

While there have been a number of supercharger designs on the market over the years, the most popular is the **Roots-type supercharger**. The pair of three-lobed rotor vanes (**Figure 7-25**) in the Roots supercharger is driven by the crankshaft. The lobes force air into the intake manifold. The helical design evens out the pressure pulses in the blower and reduces noise. It was found that a 60-degree helical twist works best for equalizing the inlet and outlet volumes. Another benefit of the helical rotor design is it reduces carryback volumes—air that is carried back to the inlet side of the supercharger because of the unavoidable spaces between the meshing rotors—which represents a loss of efficiency. Many aftermarket superchargers are a centrifugal type.

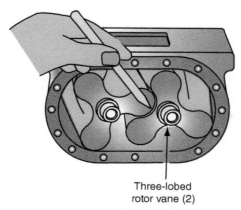

Three-lobed
rotor vane (2)

Figure 7-25 Common design of the rotor.

To handle the higher operating temperatures imposed by supercharging, an engine oil cooler is usually built into the engine lubrication system. This water-to-oil cooler is generally mounted between the engine front cover and oil filter.

Intake air enters the inlet plenum at the back of the supercharger, and the rotating blades pick up the air and force it out the top of the supercharger. The blades rotate in opposite directions, acting as a pump as they rotate. This pumping action pulls air through the supercharger inlet and forces the air from the outlet. There is a very small clearance between the meshed rotor lobes and between the rotor lobes and the housing (**Figure 7-26**).

Air flows through the supercharger system components in the following order:

1. Air flows through the air cleaner and MAF sensor into the throttle body.
2. Airflow enters the supercharger intake plenum.
3. From the intake plenum the air flows into the rear of the supercharger housing (**Figure 7-27**).
4. The compressed air flows from the supercharger to the intercooler inlet.
5. Air leaves the intercooler and flows into the intercooler outlet tube.
6. Air flows from the intercooler outlet tube into the intake manifold adapter.
7. Compressed, cooled air flows through the intake manifold into the engine cylinders.
8. If the engine is operating at idle or very low speeds, the supercharger is not required. Under this condition, airflow is bypassed from the intake manifold adapter through a butterfly valve to the supercharger inlet plenum (**Figure 7-28**).

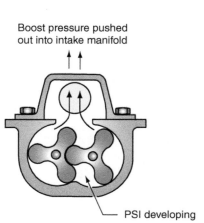

Boost pressure pushed
out into intake manifold

PSI developing

Figure 7-26 As the rotors mesh, they pump air out under pressure.

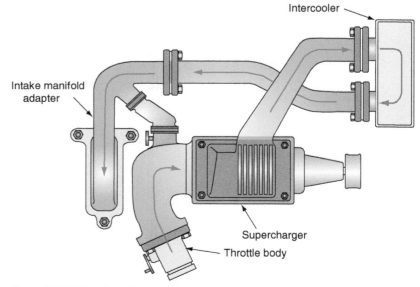

Intercooler

Intake manifold
adapter

Supercharger

Throttle body

Figure 7-27 Airflow through a supercharger.

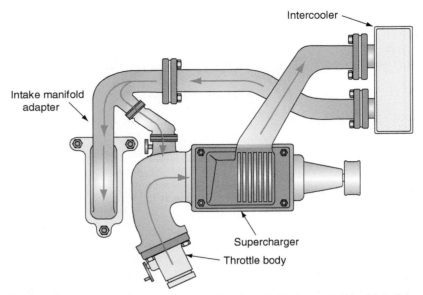

Figure 7-28 The bypass actuator can divert airflow from the intake manifold back into the supercharger plenum to reduce boost.

In one example, the bypass butterfly valve (**Figure 7-29**) is operated by an air bypass actuator diaphragm as follows:

1. When manifold vacuum is 7 in. Hg (23.6 kPa) or higher, the bypass butterfly valve is completely open and a high percentage of the supercharger air is bypassed to the supercharger inlet.

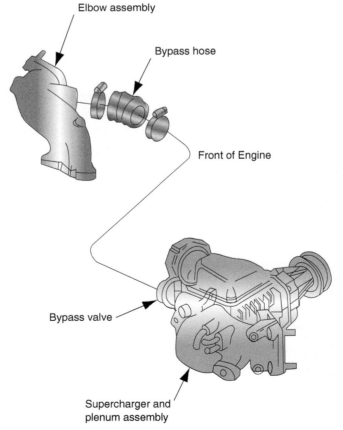

Figure 7-29 Supercharger bypass hose and intake elbow.

2. If the manifold vacuum is 3 to 7 in. Hg (10 to 23.6 kPa), the bypass butterfly valve is partially open and some supercharger air is bypassed to the supercharger inlet, while the remaining airflow is forced into the engine cylinders.
3. When the vacuum is less than 3 in. Hg (10 kPa), the bypass butterfly valve is closed and all the supercharger airflow is forced into the engine cylinders.

On some superchargers, the pulley size causes the rotors to turn at 2.6 times the engine speed. Since supercharger speed is limited by engine speed, a supercharger wastegate is not required. Belt-driven superchargers provide instant low-speed action compared to exhaust-driven turbochargers, which may have a low-speed lag because of the brief time interval required to accelerate the turbocharger shaft. Compared to a turbocharger, a supercharger turns at much lower speeds.

Friction between the air and the rotors heats the air as it flows through the supercharger. The intercooler dissipates heat from the air in the supercharger system to the atmosphere, creating a denser air charge. When the supercharger and the intercooler supply cooled, compressed air to the cylinders, engine power and performance are improved.

The compression ratio in the supercharged 3.8-liter engine is 8.2:1, compared to a normally aspirated 3.8-liter engine, which has a 9.0:1 compression ratio. The following components are reinforced in the supercharged engine because of the higher cylinder pressure:

- Engine block
- Main bearings
- Crankshaft bearing caps
- Crankshaft
- Steel crankshaft sprocket
- Timing chain
- Cylinder head
- Head bolts
- Rocker arms

Superchargers can be enhanced with electrically operated clutches and bypass valves. These allow the same computer that controls fuel and ignition to kick the boost on and off precisely as needed. This results in far greater efficiency than a full-time supercharger.

EXHAUST SYSTEM COMPONENTS

The various components of the typical exhaust system include the following:

- Exhaust manifold
- Exhaust pipe and seal
- Catalytic converter
- Muffler
- Resonator
- Tailpipe
- Heat shields
- Clamps, brackets, and hangers
- Exhaust gas oxygen sensors

All the parts of the system are designed to conform to the available space of the vehicle's undercarriage and yet be a safe distance above the road.

Exhaust Manifold

The **exhaust manifold** (**Figure 7-30**) collects the burnt gases as they are expelled from the cylinders and directs them to the exhaust pipe. Exhaust manifolds for most vehicles are made of cast or nodular iron. Many newer vehicles have stamped, heavy-gauge sheet metal or stainless steel units.

> The **exhaust manifold** is an exhaust collector that is mounted onto the cylinder head.

Figure 7-30 An exhaust manifold for a four-cylinder engine.

In-line engines have one exhaust manifold. V-type engines have an exhaust manifold on each side of the engine. An exhaust manifold will have three, four, or six passages, depending on the type of engine. These passages blend into a single passage at the other end, which connects to an exhaust pipe. From that point, the flow of exhaust gases continues to the catalytic converter, muffler, and tailpipe, and then exits at the rear of the car.

V-type engines may be equipped with a dual exhaust system that consists of two almost identical, but individual, systems in the same vehicle.

Exhaust systems are designed for particular engine–chassis combinations. Exhaust system length, pipe size, and silencer size are used to tune the flow of gases within the exhaust system. Proper tuning of the exhaust manifold tubes can actually create a partial vacuum that helps draw exhaust gases out of the cylinder, improving volumetric efficiency. Separate, tuned exhaust headers (**Figure 7-31**) can also improve efficiency by preventing the exhaust flow of one cylinder from interfering with the exhaust flow of another cylinder. Cylinders next to one another may release exhaust gas at about the same time.

Figure 7-31 Engine efficiency can be improved with tuned exhaust headers.

When this happens, the pressure of the exhaust gas from one cylinder can interfere with the flow from the other cylinder. With separate headers, the cylinders are isolated from one another, interference is eliminated, and the engine breathes better. The problem of interference is especially common with V-8 engines. However, exhaust headers tend to improve the performance of all engines.

Exhaust manifolds may also be the attaching point for the air injection reaction (AIR) pipe (**Figure 7-32**). This pipe introduces cool air from the AIR system into the exhaust stream. The added air in the exhaust stream during engine warm-up helps burn excess fuel in the exhaust and heat up the catalytic converter faster. Some exhaust manifolds have provisions for the exhaust gas recirculation (EGR) pipe. This pipe takes a sample of the exhaust gases and delivers it to the EGR valve. Also, some exhaust manifolds have a tapped bore that retains the oxygen sensor (**Figure 7-33**).

Exhaust Pipe and Seal

The exhaust pipe is metal pipe, made of aluminized steel, stainless steel, or zinc-plated heavy gauge steel, that runs under the vehicle between the exhaust manifold and the catalytic converter (**Figure 7-34**).

Catalytic Converters

A **catalytic converter** (**Figure 7-35**) is part of the exhaust system and a very important part of the emission control system. Because it is part of both systems, it has a role in both. As an emission control device, it is responsible for converting undesirable exhaust gases

> A **catalytic converter** reduces tailpipe emissions of carbon monoxide, unburned hydrocarbons, and nitrogen oxides.

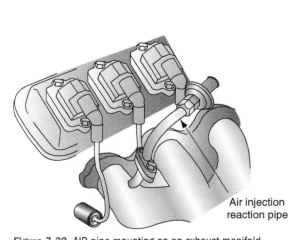

Figure 7-32 AIR pipe mounting on an exhaust manifold.

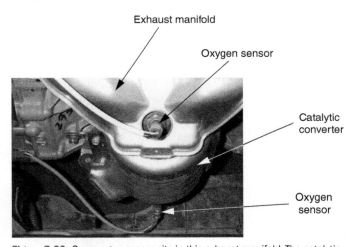

Figure 7-33 One oxygen sensor sits in this exhaust manifold. The catalytic converter mounts onto the exhaust manifold, and another oxygen sensor fits into the exhaust pipe.

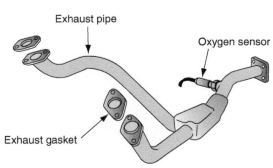

Figure 7-34 The front exhaust pipe for a V-6 engine.

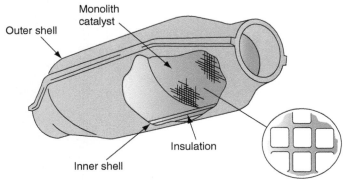

Figure 7-35 Monolithic-type catalytic converter.

into harmless gases. As part of the exhaust system, it helps reduce the noise level of the exhaust. A catalytic converter contains a ceramic element coated with a catalyst. A catalyst is a substance that causes a chemical reaction in other elements without actually becoming part of the chemical change and without being used up or consumed in the process.

Catalytic converters may be pellet type or monolithic type. A pellet-type converter contains a bed made from hundreds of small beads and is mostly used on older vehicles. Exhaust gases pass over this bed. In a monolithic-type converter, the exhaust gases pass through a honeycomb ceramic block. The converter beads or ceramic block are coated with a thin coating of cerium, platinum, palladium, rhodium, or any combination of these and are held in a stainless-steel container. Modern vehicles are equipped with three-way catalytic converters, which means the converter reduces the three major exhaust emissions, hydrocarbons (HC), carbon monoxide (CO), and oxides of nitrogen (NO_x). The converter oxidizes HC and CO into water vapor and carbon dioxide (CO_2) and reduces NO_x to oxygen and nitrogen.

Many vehicles are equipped with a mini-catalytic converter that either is built into the exhaust manifold or is located next to it (**Figure 7-36**). These converters are used to clean the exhaust during engine warm-up and are commonly called warm-up converters. Some older catalytic converters have an air hose connected from the AIR system to the oxidizing catalyst. This air helps the converter work by making extra oxygen available. The air from the AIR system is not always forced into the converter; rather, it is controlled by the vehicle's PCM. Fresh air added to the exhaust at the wrong time could overheat the converter and produce NO_x, something the converter is trying to destroy.

OBD II regulations call for a way to inform the driver that the vehicle's converter has a problem and may be ineffective. The PCM monitors the activity of the converter by comparing the signals of an HO_2S located at the front of the converter with the signals from an HO_2S located at the rear (**Figure 7-37**). If the sensor outputs are the same, the converter is not working properly and the malfunction indicator lamp (MIL) on the dash will light.

Converter Problems

The converter is normally a trouble-free emission control device, but three things can damage it. One is leaded gasoline. Lead coats the catalyst and renders it useless. The difficulty of obtaining leaded gasoline has reduced this problem. Another is overheating.

Figure 7-36 A warm-up converter may be used before the primary converter.

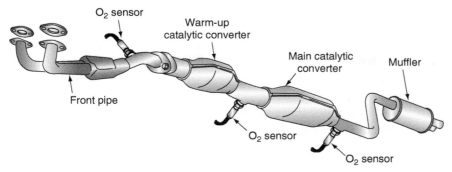

Figure 7-37 Exhaust system for an OBD II vehicle.

If raw fuel enters the exhaust because of a fouled spark plug or other problem, the temperature of the converter quickly increases. The heat can melt the ceramic honeycomb or pellets inside, causing a major restriction to the flow of exhaust. A plugged converter or any exhaust restriction can cause damage to the exhaust valves due to excess heat, loss of power at high speeds, or stalling after starting (if totally blocked). The last is damage or poisoning due to silicone contamination from coolant or sealant.

MUFFLERS

The **muffler** is a cylindrical or oval-shaped component, generally about 2 feet (0.6 meters) long, mounted in the exhaust system about midway or toward the rear of the car. Inside the muffler is a series of baffles, chambers, tubes, and holes to break up, cancel out, or silence the pressure pulsations that occur each time an exhaust valve opens.

Two types of mufflers are commonly used on passenger vehicles (**Figure 7-38**). Reverse-flow mufflers change the direction of the exhaust gas flow through the inside of the unit. This is the most common type of automotive muffler. Straight-through mufflers permit exhaust gases to pass through a single tube. The tube has perforations that tend to break up pressure pulsations. They are not as quiet as the reverse-flow type.

There have been several important changes in recent years in the design of mufflers. Most of these changes have been centered at reducing weight and emissions, improving fuel economy, and simplifying assembly. These changes include the following:

1. *New materials.* More and more mufflers are being made of aluminized and stainless steel. Using these materials reduces the weight of the units as well as extends their lives.
2. *Double-wall design.* Retarded engine ignition timing that is used on many small cars tends to make the exhaust pulses sharper. Many cars use a double-wall exhaust pipe to better contain the sound and reduce pipe ring.

> The **muffler** is a device mounted in the exhaust system behind the catalytic converter that reduces engine noise.

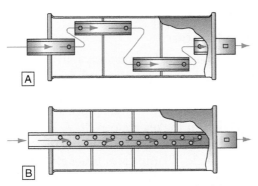

Figure 7-38 (A) Reverse-flow muffler, (B) Straight-through muffler.

3. *Rear-mounted muffler.* More and more often, the only space left under the car for the muffler is at the very rear. This means the muffler runs cooler than before and is more easily damaged by condensation in the exhaust system. This moisture, combined with nitrogen and sulfur oxides in the exhaust gas, forms acids that rot the muffler from the inside out. Many mufflers are being produced with drain holes drilled into them.

4. *Back pressure.* Even a well-designed muffler will produce some **back pressure** in the system. Back pressure reduces an engine's volumetric efficiency, or ability to breathe. Excessive back pressure caused by defects in a muffler or other exhaust system part can slow or stop the engine. However, a small amount of back pressure can be used intentionally to allow a slower passage of exhaust gases through the catalytic converter. This slower passage results in more complete conversion to less harmful gases. Also, no back pressure may allow intake gases to enter the exhaust.

> **Back pressure** reduces an engine's volumetric efficiency.

Resonator

On some older vehicles, there is an additional muffler, known as a **resonator** or silencer. This unit is designed to further reduce or change the sound level of the exhaust. It is located toward the end of the system and generally looks like a smaller, rounder version of a muffler.

> A **resonator** is a device mounted on the tailpipe to reduce exhaust noise further.

Tailpipe

The **tailpipe** is the last pipe in the exhaust system. It releases the exhaust fumes into the atmosphere beyond the back end of the car.

> The **tailpipe** conducts exhaust gases from the muffler to the rear of the vehicle.

Heat Shields

Heat shields are used to protect other parts from the heat of the exhaust system and the catalytic converter (**Figure 7-39**). They are usually made of pressed or perforated sheet metal. Heat shields trap the heat in the exhaust system, which has a direct effect on maintaining exhaust gas velocity.

  A BIT OF HISTORY

In the 1930s Charles Nelson Pogue developed the Pogue carburetor. This carburetor used exhaust heat to vaporize the fuel before it was mixed with the air entering the engine. Charles Pogue was issued several patents on this carburetor and claimed greatly increased fuel mileage. Mr. Pogue never did sell his invention to any of the car manufacturers, but rumors were repeated for years about the fantastic fuel economy supplied by this carburetor.

Clamps, Brackets, and Hangers

Clamps, brackets, and hangers are used to properly join and support the various parts of the exhaust system. These parts also help isolate exhaust noise by preventing its transfer through the frame (**Figure 7-40**) or body to the passenger compartment. Clamps help secure exhaust system parts to one another. The pipes are formed in such a way that one slips inside the other. This design makes a close fit. A U-type clamp usually holds this connection tight (**Figure 7-41**). Another important job of clamps and brackets is to hold pipes to the bottom of the vehicle. Clamps and brackets must be designed to allow the exhaust system to vibrate without transferring the vibrations through the car.

Many different types of flexible hangers are available. Each is designed for a particular application. Some exhaust systems are supported by doughnut-shaped rubber rings

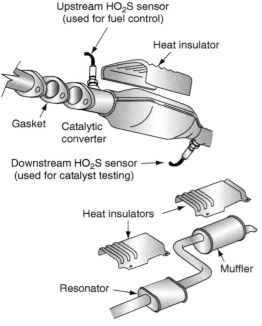

Figure 7-39 Typical locations of heat shields in an exhaust system.

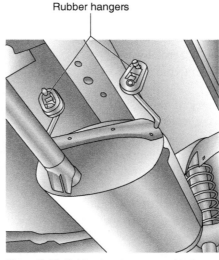

Figure 7-40 Rubber hangers are used to keep the exhaust system in place without allowing it to contact the frame.

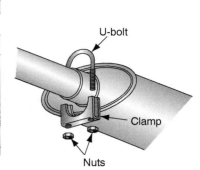

Figure 7-41 A U-clamp is often used to secure two pipes that slip together.

between hooks on the exhaust component and on the frame or car body. Others are supported at the exhaust pipe and tailpipe connections by a combination of metal and reinforced fabric hanger. Both the doughnuts and the reinforced fabric allow the exhaust system to vibrate or move without breakage that could be caused by direct physical connection to the vehicle's frame.

Some exhaust systems are a single unit in which the pieces are welded together by the factory. By welding instead of clamping the assembly together, car makers save the weight of overlapping joints as well as that of clamps.

AUTHOR'S NOTE During my experience in the automotive service industry, I encountered several cases where restricted exhaust or intake systems were misdiagnosed and confused with ignition system or fuel system defects. Defective fuel system or ignition system components may cause a loss of engine power and reduced maximum speed, but these symptoms are accompanied by cylinder misfiring, engine surging, or both. When the exhaust or intake system is restricted, the maximum speed is reduced, but the engine does not misfire or surge.

SUMMARY

- The air induction system allows a controlled amount of clean, filtered air to enter the engine. Cool air is drawn in through a fresh air tube. It passes through an air cleaner before entering the throttle body.
- The air intake ductwork conducts airflow from the remote air cleaner to the throttle body mounted on the intake manifold.

- The air cleaner/filter removes dirt particles from the air flowing into the intake manifold, to prevent these particles from causing engine damage.
- The intake manifold distributes the air or air-fuel mixture as evenly as possible to each cylinder. Intake manifolds are made of cast iron, plastic, or die-cast aluminum.

- The vacuum in the intake manifold operates many systems such as emission controls, brake boosters, heater/air conditioner, cruise controls, and more. A diagram of emission system vacuum hose routing is located on the underhood decal. Loss of vacuum can create many drivability problems.
- Turbochargers and superchargers create forced induction systems that improve the volumetric efficiency and power of the engine.
- Turbochargers use the heat energy in exhaust gases to spin the turbine and pressurize intake air on the compressor side.
- A turbocharger wastegate regulates boost pressure based on engine load to prevent engine-damaging overboost conditions.
- A belt-driven supercharger supplies air to the intake manifold under pressure to increase engine power.

- A vehicle's exhaust system carries away gases from the passenger compartment, cleans the exhaust emissions, and muffles the sound of the engine. Its components include the exhaust manifold, exhaust pipe, catalytic converter, muffler, resonator, tailpipe, heat shields, clamps, brackets, and hangers.
- The exhaust manifold is a bank of pipes that collects the burned gases as they are expelled from the cylinders and directs them to the exhaust pipe.
- The catalytic converter reduces HC, CO, and NO_x emissions.
- The muffler consists of a series of baffles, chambers, tubes, and holes to break up, cancel out, and silence pressure pulsations.

REVIEW QUESTIONS

Short-Answer Essays

1. Explain the operation of an airflow restriction indicator.

2. Explain the purposes of the intake manifold.

3. Explain the advantages of plastic intake manifolds.

4. Explain the operation of an intake manifold with dual runners at low and high engine speeds.

5. Explain how a turbocharger or supercharger creates more engine power.

6. Describe basic turbocharger operation.

7. What is the purpose of the intercooler?

8. Describe the advantages of tuned exhaust manifolds compared to conventional exhaust manifolds.

9. Explain two catalytic converter operating problems.

10. When referring to a turbocharger system, describe what oil coking is, how it is caused, and how to prevent it.

Fill-in-the-Blanks

1. Without proper intake air filtration, contaminants and abrasives in the air will cause severe _____ damage.

2. If the air filter is restricted, the airflow restriction indicator window appears _____ in color.

3. When an intake manifold has dual runners, both runners are open at _____ engine speeds.

4. The exhaust flows over the _____ wheel in a turbocharger.

5. Turbocharged or supercharged engines have _____ compression ratios compared to naturally aspirated engines.

6. The air flows through the intercooler _____ it flows through the supercharger.

7. In place of cast- or nodular-iron exhaust manifolds, newer vehicles have manifolds manufactured from stamped, heavy-gauge sheet metal or _____ _____.

8. A tuned exhaust manifold prevents exhaust flow from one cylinder from interfering with the _____ _____ from another cylinder.

9. A catalytic converter may be overheated by a(n) _____ air-fuel ratio.

10. The exhaust pipe between the engine and the catalytic converter may be a _____ _____ design in order to be quieter.

Multiple Choice

1. The intake manifold is typically made of.

   A. plastic.   C. cast iron.

   B. stainless steel.   D. fiber composites.

2. An engine should develop _____ in. Hg vacuum at idle.

   A. 5   C. 18

   B. 8   D. 25

3. A variable-length intake manifold is designed to increase.

   A. compression.   C. intake turbulence.

   B. airflow at idle.   D. airflow at high rpm.

4. The catalytic converter reduces _____ emissions.

   A. HC, CO_2, O_2 C. CO, CO_2, NO_x

   B. HC, CO, NO_x D. N, HC, O_2

5. *Technician A* says that a turbocharger may spin at speeds of 100,000 rpm.

   Technician B says that oil coking is a common cause of turbocharger failure.

   Who is correct?

   A. A only   C. Both A and B

   B. B only   D. Neither A nor B

6. On a turbocharged engine, when the opens, boost pressure is reduced.

   A. Compressor bypass

   B. Turbine dump valve

   C. Wastegate

   D. Intercooler router

7. A supercharger typically uses _____ to pressurize the intake charge.

   A. two rotors

   B. four rotors

   C. a compressor pump

   D. three helix gears

8. A mini-converter is used.

   A. on small engines where a normal converter will not fit properly.

   B. on engines that used leaded fuels.

   C. in conjunction with EGR systems to supply clean exhaust for the cylinders.

   D. to reduce emissions during engine warm-up.

9. A restricted exhaust system can cause.

   A. stalling.

   B. backfiring.

   C. loss of power.

   D. acceleration stumbles.

10. All of these statements about an intake manifold dual runner system are true *except*:

    A. Both runners to each cylinder are open at 1,400 engine rpm.

    B. The butterfly valves in one set of runners may be operated by a vacuum actuator.

    C. The butterfly valves may be operated electrically by the PCM.

    D. One runner to each cylinder is open at 900 engine rpm.

CHAPTER 8
ENGINE CONFIGURATIONS, MOUNTS, AND REMANUFACTURED ENGINES

Upon completion of this chapter, you should understand and be able to describe:

- The different engine configurations.
- The methods of mounting the engine.
- The function of different engine mounts.
- The common attaching points for the engine.
- The common accessories found on the engine assembly.
- The benefits and limitations of a remanufactured engine.

Terms To Know

Crate engine	Engine mount	Remanufactured engine
Differential	Longitudinally mounted	Transverse-mounted engine
Engine cradle	Mid-mounted engine	

INTRODUCTION

When performing major engine mechanical repairs, you will be confronted with different engine configurations. The engine may be mounted transversely or longitudinally, or it may be a mid-engine, which is mounted in the rear. You will need to understand the operation of motor mounts as part of your assurance that the engine will perform smoothly. When overhauling or replacing an engine, you will have to remove many engine accessories and brackets; this requires some familiarity with these components. In many cases, you, your shop management, or your customer may choose to install a **remanufactured engine**. You will need to understand the benefits and limitations of choosing a "reman" or **crate engine** versus you overhauling the engine in-house.

ENGINE CONFIGURATIONS

The most common configuration for a front-wheel-drive vehicle is to have a **transverse-mounted engine** (**Figure 8-1**). The engine is mounted sideways in the engine compartment so that the front of the engine is mounted toward one side of the vehicle. This is a popular design, in part because it can reduce the size of the engine compartment. The transaxle is attached to the side of the engine and faces the other side of the vehicle. The transaxle contains the **differential** assembly, located in the rear of the vehicle on rear-wheel-drive vehicles. The compact size of the transaxle lowers overall vehicle weight. This is always a goal of vehicle manufacturers as they try to improve fuel economy. A transverse engine is often removed from the bottom of the engine compartment with the transaxle attached. This often requires parts of the **engine cradle** (suspension cradle) to be removed (**Figure 8-2**). The engine and the suspension components are attached to the engine cradle, so disassembly of parts of the suspension may be required to remove the engine.

Figure 8-1 This transverse-mounted engine has many accessories and attachments that would have to be removed during an engine overhaul or an engine swap.

Cradle toward the
rear of the vehicle

Cradle
mount

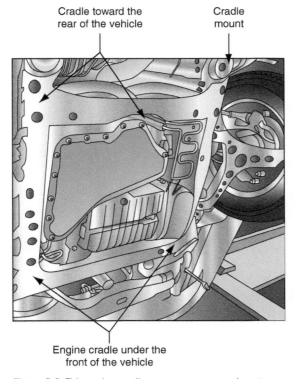

Engine cradle under the
front of the vehicle

Figure 8-2 This engine cradle comes out as one piece to remove the engine through the bottom. The suspension will have to be disconnected.

Many rear-wheel-drive and four-wheel-drive vehicles install their engines longitudinally (**Figure 8-3**). A few front-wheel-drive vehicles also mount their engines in this fashion, though it is not common. With a **longitudinally mounted** engine, the front of the engine faces the front of the vehicle, and the transmission mounts on the back of the engine. The engine and transmission are bolted together and mounted to the frame through a rear engine mount (**Figure 8-4**). A longitudinal engine may be removed from the top through the hood opening or through the bottom of the engine compartment.

A **longitudinally mounted** engine is placed in the vehicle lengthwise, usually front to rear.

Figure 8-3 A longitudinal engine mounted in an all-wheel-drive SUV.

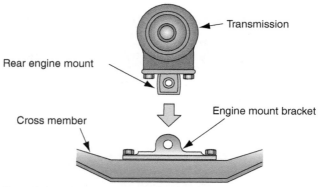

Figure 8-4 The engine and transmission use a rear engine mount to help stabilize the components.

A few vehicles mount their engines in the rear of the vehicles. Some are located between the passenger compartment and the rear suspension. These are called **mid-mounted engines** (or simply mid-engines) and are typically mounted transversely. They are used in compact, high-performance, rear-wheel-drive sports cars. The front of the vehicle can be low and sloped to maximize aerodynamics. The weight of the powertrain near the center of the vehicle places the center of gravity near the vehicle midpoint. This improves steering and handling for fine sports car performance. The most rare configuration is an engine mounted at the true end of the vehicle.

A **mid-mounted engine** is located between the rear of the passenger compartment and the rear suspension. It is also called a mid-engine.

ENGINE MOUNTS

An **engine mount** attaches the engine to the chassis and absorbs the engine vibrations.

Engines usually have two or three **engine mounts** to attach the engine to the frame and to isolate the vibration of the engine from the chassis and passenger compartment (**Figure 8-5**). There may be additional mounts that hold the transmission or transaxle and also help support the engine. The engine will have a top-engine mount that may look quite different from the one shown in Figure 8-4. Some mount at the very front of the engine and attach to the front radiator cross member (**Figure 8-6**). A lower engine mount on a transverse-mount engine may attach to either the back of the engine or the front of the transaxle (**Figure 8-7**).

Engine
mount

With
bushings

Figure 8-5 This top-engine mount has two bushings to support the engine vertically and horizontally.

Figure 8-6 This top-engine mount attaches to the front cross member and the front of the engine.

Figure 8-7 This lower engine mount actually attaches at the front of the transaxle, very close to the engine.

The rubber bushing has voids

Figure 8-8 The voids in the bushings allow a looser attachment to the case. This allows the engine to vibrate and the mount to isolate those vibrations from the chassis.

Types of Engine Mounts

Many engine mounts are simply rubber bushings in a metal casing. The rubber bushing absorbs the engine vibrations. The rubber is attached to the housing with some slack to allow movement within the case (**Figure 8-8**). The rubber may have voids to allow more movement within the mount.

Hydraulic engine mounts use two chambers filled with fluid and an orifice in between to dampen the vibration of the engine. It takes some time and energy to displace the fluid from one chamber to the other as the engine moves. This helps absorb the engine vibration. They look and function very similarly to suspension shock absorbers (**Figure 8-9**).

Some hydraulic engine mounts are electronically controlled. When the engine is idling, the engine mounting should be very soft to allow the engine to run as smoothly as possible during its low-load condition. Once the engine is driving under a load, stiffer dampening must occur to isolate the harsher movements from the chassis. To achieve that exceptionally smooth idle, more and more vehicles are using electronically controlled hydraulic engine mounts.

In one example, the two hydraulic chambers are connected by a large and a small orifice. When the PCM determines that the engine is idling, it actuates a vacuum motor that opens the normally closed larger orifice. This allows more fluid flow and softer dampening to absorb the shaking of the engine at idle. Only the smaller valve is open during engine-loaded conditions to provide the stiffer dampening needed to isolate engine vibration.

REMANUFACTURED ENGINES

It is a trend in today's automotive industry to replace seriously damaged engines with remanufactured or new units rather than overhaul the existing engine in the shop. Remanufactured engines are available from vehicle manufacturers and other suppliers (**Figure 8-10**). They will come with the internal components of the engine fully assembled and adjusted. Beyond that, the engines you receive may vary significantly. You can purchase a crate engine with or

Figure 8-9 This hydraulic engine mount looks very much like a shock absorber. Its function is quite similar as well.

Figure 8-10 The crate engine comes fully assembled, less the accessories. Some engines come more complete than others.

without cylinder heads. Some will come with new oil and water pumps and covers. In other cases, it is your responsibility to replace these components as warranted by the vehicle. Many crate engines will come as relatively bare engines with no covers or new accessories. All sensors, brackets, hoses, and tubing will be transferred from the old engine to the new.

There are several advantages to installing a remanufactured engine. A crate engine comes fully assembled and carries a written warranty. That means that if the "new" engine suffers damage within the warranty period, the supplier will provide a new unit. If a technician rebuilt an engine in the shop and it came back within a few thousand miles and another engine mechanical problem, the shop would be responsible for making it right for the customer. This would cost the shop the technician's labor and the cost of all the new parts to repair the damage.

It is also often more cost-effective to install a reman engine rather than overhaul one in-house. A supplier that provides crate engines has a production shop of sorts. The technicians who work there overhaul engines every day and all day. They are very efficient at this. The supplier is also able to source parts in high volume for significant discounts.

 A BIT OF HISTORY

The idea of supplying remanufactured engines rather than overhauling existing ones is relatively new. In the 1980s, for example, most shops were rebuilding all their engines. Technicians were well practiced at the fine measuring required for professional overhauls. Many shops were still performing cylinder head service or "valve jobs" in-house as well. Now several shops have let much of the engine overhaul machining tools deteriorate or have sold them to machine shops where most of the engine machining is now done. Many "old-timers" and young enthusiasts alike miss the more frequent opportunities to build an engine.

AUTHOR'S NOTE Many students are disappointed to learn that engine overhaul is not a common task in most general automotive repair shops. If engine overhaul is something you are interested in, you can make that a priority when looking for work. There are still many shops out there that specialize in engine repair. You can also consider working for an engine manufacturer.

A technician in a shop that overhauls an engine a few, or even a dozen, times a year will take longer to do the same work. The shop may not be able to source the discount parts that are available, especially to a high-volume consumer. The machining work that may need to be done at a local machine shop also adds cost and time to the overhaul.

Sometimes when you are beginning the disassembly of an engine, it becomes apparent that the damage is so extensive that a remanufactured engine would definitely be cheaper (**Figure 8-11**). A hole in the engine block from a piston that let loose would make that engine cheaper to replace than to repair. You may not make a decision until you have taken the engine apart and evaluated it. Other shops simply follow the policy of installing crate engines whenever possible.

There are reasons not to install remanufactured engines. For some vehicles, reman engines are not readily available or may be cost prohibitive (**Figure 8-12**). It is difficult to locate remanufactured engines for many low-volume import engines, for example. The condition of the vehicle and the desire of the customer may well warrant the cost of a thorough in-house overhaul. In some cases, the customer may want a new engine from the manufacturer despite the cost. This may be either a truly new or a remanufactured engine, as made available by the particular manufacturer. In other cases, the vehicle may

Figure 8-11 After finding piston and ring pieces in the oil pan, this shop decided that a crate engine would be more cost-effective.

Figure 8-12 This high-mileage engine needs some major mechanical work. A reman engine is not readily available. The customer likes the vehicle so much that he or she has chosen to give the engine a thorough in-house overhaul.

be in such poor condition that it is not worth installing a new crate engine. The customer could also be in a financial situation that does not allow him or her to purchase a new engine. As we will discuss during our short block service discussions in a later chapter, you may be able to perform satisfactory, though minimal, repairs on a high-mileage engine. This sort of work may buy the customer the time he or she needs to be able to purchase a new engine or a new vehicle. By the time you complete your study of this text, you should feel confident in your ability to make these sorts of judgment calls.

SUMMARY

- Most front-wheel-drive engines are mounted transversely.
- Many rear-wheel-drive engines are mounted longitudinally.
- Some sports cars have mid-mounted engines.
- Engine mounts absorb the engine vibrations.
- Many engine mounts are either rubber or hydraulic.

- Electronically controlled hydraulic engine mounts can produce a very smooth idle.
- The engine has mechanical attachments to many other systems.
- Remanufactured engines are available for many popular engines.
- Crate engine may be a cost-effective alternative to an in-house overhaul.

REVIEW QUESTIONS

Short-Answer Essays

1. List the types of engine configurations.

2. Explain the advantages of a transverse-mounted engine.

3. Explain the advantages of a mid-mounted engine.

4. Explain the purposes of engine mounts.

5. Describe the operation of the hydraulic engine mount.

6. Explain the operation of the electronically controlled engine mount.

7. Explain the benefit of the electronically controlled engine mount.

8. Explain two examples when you would be likely to install a remanufactured engine.

9. Describe two benefits of a remanufactured engine.

10. Describe why you might choose to overhaul an engine in-house.

Fill-in-the-Blanks

1. A _____ engine is mounted between the rear of the passenger compartment and ahead of the rear suspension.

2. A transverse engine assembly allows better _____ because it is lighter.

3. A longitudinally mounted engine is usually used on four-wheel-drive and _____ vehicles.

4. The two most common types of engine mountings are the longitudinal and the _____.

5. The _____ engine mounts work like a shock absorber.

6. Rubber engine mounts have _____ to allow more movement within the mount.

7. _____ _____ hydraulic engine mounts can become firmer when the engine is driving under a load.

8. Remanufactured engines come with a _____.

9. A remanufactured engine is also called a(n) _____ engine.

10. The decision to install a remanufactured engine or overhaul an engine in-house may be made by _____, the shop management, or the _____.

Multiple Choice

1. The most common type of engine installation in a front-wheel-drive vehicle is.
   A. longitudinal.   C. mid-mount.
   B. transverse.   D. rear.

2. The most common type of engine installation in a rear-wheel-drive vehicle is.
   A. rear.   C. transverse.
   B. mid-mount.   D. longitudinal.

3. *Technician A* says that a transverse engine is usually coupled to a transaxle.
   Technician B says that most front-wheel-drive vehicles use longitudinally mounted engines.
   Who is correct?
   A. A only   C. Both A and B
   B. B only   D. Neither A nor B

4. *Technician A* says that a mid-engine is naturally faster than a longitudinally mounted engine.
   Technician B says that a mid-engine installation can improve vehicle handling.
   Who is correct?
   A. A only   C. Both A and B
   B. B only   D. Neither A nor B

5. A rubber engine mount has voids to make the mount.
   A. softer.   C. last longer.
   B. stiffer.   D. resist twisting.

6. A hydraulic engine mount moves fluid through a.
   A. pintle.   C. canister.
   B. solenoid.   D. orifice.

7. *Technician A* says that an electronically controlled hydraulic engine mount can improve idle quality.
   Technician B says that these mounts can help isolate engine vibration during hard acceleration.
   Who is correct?
   A. A only   C. Both A and B
   B. B only   D. Neither A nor B

8. *Technician A* says that it may be a shop policy to install a remanufactured engine rather than overhaul an engine in-house.
   Technician B says that some engines are damaged badly enough that it is not cost-effective to repair them.
   Who is correct?
   A. A only   C. Both A and B
   B. B only   D. Neither A nor B

9. A crate engine may come with a new _____ installed.
   A. A/C compressor   C. Both A and B
   B. Power-steering   D. Neither A nor B
      pump

10. A reason to install a remanufactured engine may be.
    A. cost.
    B. warranty.
    C. time savings.
    D. all of the above.

CHAPTER 9
CYLINDER HEADS

Upon completion and review of this chapter, you should understand and be able to describe:

- The design of a valve, including an explanation of the stem, head, face, seat, margin, and fillet.
- The reasons for burned valves.
- The causes of valve channeling.
- The different types of valve seats.
- The reasons for adding Stellite to valve seats.
- Valve seat recession and its causes.

- The purposes of valve stem seals, and the different types of these seals.
- The results of worn valve stems and guides.
- The different combustion chamber designs.
- The combustion process.
- How combustion chamber design can affect exhaust emissions and engine performance.

Terms To Know

Burned valve	Margin	Swirl chambers
Chamber-in-piston	Off-square	Tumble port
Channeling	Pentroof combustion	Turbulence
Eddy currents	chambers	Valve float
End gases	Poppet valves	Valve guides
Face	Quenching	Valve job
Fillet	Skin effect	Valve seats
Free-wheeling engine	Squish area	Valve seat insert
Head	Stellite	Valve seat recession
Hemispherical chamber	Stem necking	Valve spring
Integral valve seats	Stratified charge	Valve stem
Interference engine	Surface-to-volume (S/V) ratio	Wedge chamber

INTRODUCTION

Proper combustion requires the engine to "breathe" properly. The air intake system directs air to the throttle body. From the throttle body, the intake air is directed through the intake manifold, through cylinder head, and past the intake valves into the combustion chambers. Once in the combustion chambers, the air-fuel mixture is ignited, and the resulting power output is affected by chamber design. Finally, the spent gases are expelled from the combustion chambers, past the exhaust valves, and out through the exhaust manifold and exhaust system.

The intake and exhaust manifolds, cylinder head, valves, and rocker arms must all work together to move the air-fuel mixture in and out of the combustion chambers and expel the exhaust gases. In this chapter, we discuss cylinder heads and related components such as valves, valve seats, and valve guides. Discussion of the typical system failures and their causes is also included. Today's technician must be able to identify a failed component and determine the cause of the failure. Although lack of lubrication and overheating are the most common causes of component failure, all engines will exhibit some form of normal wear between moving parts. A technician must be able to classify component wear as normal or abnormal. If abnormal wear is determined, the cause must be identified and corrected before the engine is reassembled.

CYLINDER HEADS

On most engines, the cylinder head contains the valves, valve seats, valve guides, valve springs, and the upper portion of the combustion chamber (**Figure 9-1**). In addition, the cylinder head has passages to allow coolant and oil flow through the head. Many modern engines have overhead camshafts (OHCs). In these cases, the cylinder head also houses the camshaft(s) and all of the valvetrain components.

The cylinder head is attached to the block above the cylinder bores. This mating surface must be flat and have the correct surface finish. The cylinder head gasket is installed between the cylinder head and the block. It has the difficult task of sealing in the extreme forces of combustion and preventing internal or external leaks of oil or coolant.

The cylinder head is made of cast iron or aluminum alloy. It is common for engine manufacturers to use aluminum alloys to cast the head and mate it to a cast-iron block. The difference in thermal expansion between the cylinder head and block creates radial stress that must be withstood by the head gasket. The gasket must be able to withstand scrubbing or even shearing during uneven expansion and contraction while maintaining its seal. The head gasket must be able to seal the combustion chamber even when one component expands differently than the other. Head-gasket failures can cause serious drivability concerns and engine damage. When the head gasket fails, it frequently allows coolant to enter the combustion chamber and also admits combustion gases into the cooling system. The burning coolant causes white, sweet-smelling smoke to be emitted from the tailpipe and will often cause misfire in the affected cylinder. In extreme cases, it can hydraulically lock the engine and cause extensive piston and cylinder damage. In most cases, the coolant passing across the oxygen sensors will destroy them. Many technicians replace the oxygen sensors after a head-gasket failure or at least warn the customer that the sensors may fail in the near future. The combustion gases in the cooling system cause

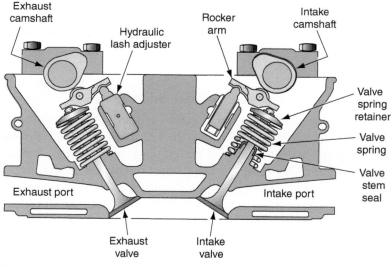

Figure 9-1 Typical OHC cylinder head components.

excessive pressure and air pockets in the system that frequently result in engine overheating and boiling over. Other times the symptoms may include erratic heater operation and engine temperature, consistent or intermittent overheating, rough running, misfiring (especially at start-up), excessive pressure in the cooling system, bubbles in the coolant overflow and coolant overflow from the tank, coolant loss without an external leak, or low compression on two adjacent cylinders. A blown head gasket can also allow coolant to mix in the oil, creating a fluid in the lubrication system that closely resembles a coffee milkshake. The oil's lubricating and corrosion protection properties are diminished. Driving the vehicle with this condition can cause extensive engine damage.

A cracked cylinder head can often cause the same symptoms as a blown head gasket. Occasionally, the crack in the head gasket or cylinder head is very small and will cause less dramatic symptoms. When the engine is cool, the metals in either of the suspect parts contract and enlarge the crack. Coolant may leak into the affected cylinder and foul the spark plug or cause small amounts of white smoke to exit the tailpipe. A classic symptom of a cracked cylinder head (or a small crack in the head gasket) is misfire on a cylinder accompanied by a little white smoke for about one minute after start-up. The white smoke also smells like coolant, not regular exhaust. Make sure that you do not confuse this white smoke with the normal condensation cloud that is emitted through the exhaust on a very cold day. Then the engine heat can cause enough expansion in the metal to seal the gap, and the engine may run perfectly for the rest of the day.

Prompt repair of a vehicle exhibiting any of these symptoms is essential to prevent more severe engine damage. Overheating typically causes failure of the head gasket or a crack in the cylinder head. Often, the white smoke will billow from the tailpipe, making diagnosis straightforward. In other cases, very little smoke will be visible, and further testing will be needed to determine the cause of the overheating or misfire.

The combination of an aluminum cylinder head and a cast-iron block with coolant circulating between them presents the problem of electrolysis. This is the chemical action producing a small electrical charge in the coolant. It exists when two dissimilar metals are touching the same fluid. As the coolant ages, it becomes acidic and the chemical additives that prevent electrolysis are used up. This is when electrolysis accelerates. Electrolysis will eat away at the aluminum components in the cooling system, including the cylinder head. Proper cooling system maintenance reduces the impact of electrolysis on aluminum components.

Shop Manual
Chapter 9, page 382

Intake and Exhaust Ports

The intake and exhaust ports cast into the cylinder head are carefully machined to match the flow of air from the intake manifold and out to the exhaust manifold. Some cylinders have Siamese ports, meaning that two cylinders share the same opening. This is not an ideal design but may be used to save space or cost. Intake ports may use a hump in them to increase air velocity, thus increasing volumetric efficiency. When the intake ports are on one side of the head and the exhausts on the other, the head is called a crossflow cylinder head design. This adds to the volumetric efficiency of the head, especially during valve overlap. As the intake valve opens, the incoming air is drawn into the moving air across the chamber from the exhaust flow (**Figure 9-2**). Non-crossflow designs have a much harder time filling the combustion chamber fully with air. Exhaust airflow out of the chamber may actually provide resistance to incoming air as it is pushing out in the opposite direction of the incoming air. The proximity of the hot exhaust gases to the intake charge also has a tendency to heat (and therefore reduce the density of) the intake charge, reducing the volumetric efficiency.

Camshaft Bore

The camshaft(s) on an overhead-cam engine has a bore machined into the head to house the camshaft. The top halves of the bores are caps or a cap housing that bolts to the head (**Figure 9-3**). On some engines, the bores may be fixed, and you can slide the camshaft

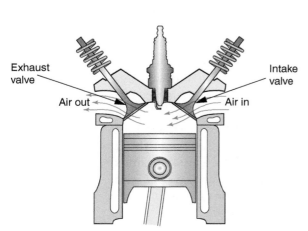

Figure 9-2 The momentum of airflow helps draw in a fresh charge; in a crossflow chamber this is particularly effective.

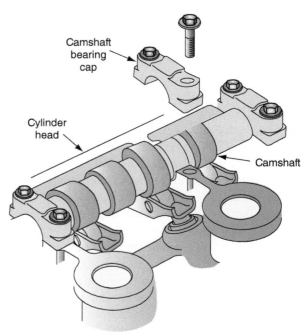

Figure 9-3 These soft aluminum camshaft caps serve as the cam bearing surface. In this case, there are no actual cam bearings or bearing inserts.

AUTHOR'S NOTE Many enthusiasts port and polish their heads during a performance overhaul. A die grinder can be used to polish the surface of the intake and exhaust passages and enlarge the actual ports to better match the opening of the manifold. This is called port matching. Some enthusiasts continue to enlarge the ports beyond the opening. This is called porting. Care must be taken when porting because removing metal from this area makes the walls thinner. Manufacturers have spent countless hours designing the flow through these ports, however; so take care not to change the basic flow design.

into place from either the front or rear of the head. Overhead camshafts sometimes use bearings. More often the integral bearing surface is the soft aluminum of the head. Oil passages lead into these bores to provide essential lubrication. If the cam bores are damaged, the head usually requires replacement, though in some cases the bores are machined oversize and inserts can be installed.

Valves

Modern automotive engines use **poppet valves** (**Figure 9-4**). The poppet valve is opened by applying a force that pushes against its tip. The camshaft lobes work to open the valves through the components of the valvetrain. Closing of the valve is accomplished through spring pressure.

The valve as a whole has many parts (**Figure 9-5**). The large-diameter end of the valve is called the **head**. The angled outer edge of the head is called the **face** and provides the contact point to seal the port. The seal is made by the valve face contacting the valve seat in the cylinder head. The valve seat is press fit into the cylinder head or is an integral part of the head. To provide a positive seal, the valve face is cut at an angle. The angle is usually 45°, though some manufacturers may use a 30° angle on some valves and seats. The angle causes the seal to wedge tighter when the valve is closed and allows free flow of the air-fuel

Poppet valves control the opening or passage by linear movement.

The **head** is the enlarged part of the valve.

The **face** is the tapered part of the section of the valve head that makes contact with the valve seat.

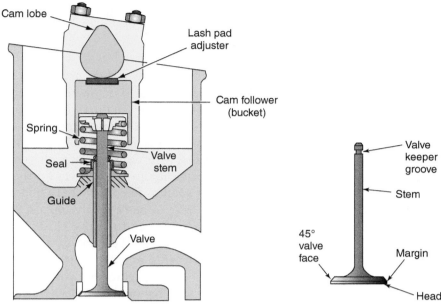

Figure 9-4 Typical poppet valve configuration using an overhead camshaft.

Figure 9-5 The parts of the valves.

mixture when the valve is open. The valve face and seat must both be precisely machined to the correct angle and width to provide good sealing. Pits, corrosion, or carbon on either surface can result in leakage of compression and combustion forces.

The area between the valve face and the valve head is referred to as the **margin**. The margin allows for machining of the valve face and for dissipation of heat away from the face and stem. The margin adds structure to the valve. If it is cut too thin during refinishing, the valve will burn or become misshapen and leak cylinder pressure. The **valve stem** guides the valve through its linear movement. The stem also has valve keeper grooves machined into it close to the top. These grooves are used to hold keepers that retain the valve springs. The stem rides in a valve guide located in the cylinder head.

The head of the intake valve is typically larger than the exhaust valve's. This is because maximizing intake valve size is considered more important than the exhaust valve. There is only so much area in the combustion chamber to locate the valves, and the larger the intake valve, the smaller the exhaust valve has to be. Smaller exhaust valves are able to work efficiently because the exhaust is being forced out of the chamber by the upward movement of the piston and by the heat generated by combustion; thus they are less sensitive to restrictions, whereas the air-fuel charge is being pulled into the cylinder by vacuum alone, and restrictions will reduce the engine's efficiencies. Larger intake valves reduce the restrictions and improve volumetric efficiency.

Most valves are constructed of high-strength steel; however, some manufacturers are experimenting with ceramics. Most valve heads are constructed of 21-2N and 21-4N stainless steel alloys; 21-2N alloy is commonly used in original equipment exhaust valves, and 21-4N is a higher grade of stainless steel containing more nickel. One of the primary concerns in valve construction and selection of alloys is the operating temperatures to which they are subjected. Combustion chamber temperatures can range from 1,500°F to 4,000°F (82°C to 220°C). The types of material used to construct the valve must be capable of withstanding these temperatures and dissipating the heat rapidly. For every 25°F (14°C) reduction in valve temperature, the valve's burning durability is doubled. There are several alloys available for valve construction that will increase the valve's burning durability. An alloy that is being used by many manufacturers is Inconel. Inconel is constructed from a nickel base with 15 to 16 percent chromium and 2.4 to 3.0 percent titanium.

The **margin** is the material between the valve face and the valve head.

The **valve stem** guides the valve through its linear movement.

Shop Manual
Chapter 9, page 390

Valve construction design is also an important aspect for controlling heat. A valve can be constructed as one piece or two piece. One-piece valves run cooler since the weld of a two-piece valve inhibits heat flow up the stem. Two-piece valves allow the manufacturer to use different metals for the valve head and stem.

In addition to alloy selection, manufacturers design the valve and cylinder head to dissipate heat away from the valve (**Figure 9-6**). Most of the heat is dissipated through the contact surface of the face and seat. The heat is then transferred to the cylinder head. About 76 percent of the heat is transferred in this manner. Most of the remaining heat is dissipated through the stem to the valve spring and on to the cylinder head. To increase the transfer of heat through the stem, some stems are filled with sodium. Turbocharged, supercharged, or other high-performance engines may use sodium-filled valves for their excellent heat dissipation quality and decreased weight. A sodium-filled valve has a hollow stem that is partially filled with metallic sodium (**Figure 9-7**). When the valve opens, the sodium splashes down toward the head and picks up heat. When the valve closes, the sodium moves up the stem, taking heat away from the hottest area of the valve, the head. The heat is transferred from the stem to the guide and then dissipated in the coolant passages around the guide. Sodium-filled valves should not be machined; if you were to cut through the valve, the sodium could explode. These valves should be replaced when reconditioning the head.

Many manufacturers use chrome-plated stems to provide additional protection against wear resulting from initial engine starts when oil is not present on the valve stem. In addition, chrome works to protect against galling when cast-iron guides are used. If the valve stem is reground, the chrome plating will be removed. In this case, either the valve will have to be replated or a bronze guide must be used.

The **fillet** is the curved area between the stem and the head. It provides structural strength to the valve. The fillet shape also affects the flow of the air-fuel mixture around the valve. A valve with a steeply raked fillet (called a tulip valve) has better flow than a valve with a flatbacked fillet. Fillet shape is only a factor while the valve is opening; after the valve is completely open, the fillet shape has very little effect on airflow.

Although not normally considered a wear area, the fillet must be thoroughly cleaned. Carbon will build up in this area, causing the flow to be disrupted. Excessive carbon in this location can be an indication of worn valve stem seals or guides. A wire brush is usually required to remove carbon buildup on the fillet.

One-piece valves do not weld the head to the stem like two-piece valves. The advantage of two-piece valves is that two different metals can be used to construct the head and the stem.

The **fillet** is the curved area between the stem and the head.

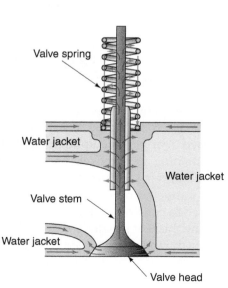

Figure 9-6 Heat must be removed from the valve.

Figure 9-7 The hollow valve is partially filled with sodium. When the valve closes, the sodium splashes up to help take heat away from the head and into the stem.

Aluminized Valves. Aluminized valves provide increased valve life by reducing the effects of corrosion. Small particles of aluminum are fused to the valve head, causing a reaction that makes the surface corrosion resistant. Aluminized valves should not be reconditioned, because this process removes the coating from the head.

Determining Valve Malfunctions

It is not enough to be able to recognize a failed valve. It is very important to determine the cause of the malfunction to prevent a repeat failure. Different types of failures are usually caused by identifiable conditions.

Deposits on Valves. Excessive carbon buildup on valve stems is a relatively common problem on modern vehicles (**Figure 9-8**). This can be caused by excessive valve stem temperatures due to increased clearance between the valve guide and the valve stem. Deposits may increase as the valve seals wear and allow more oil into the guide, which can then bake onto the back of the valve. The use of improper or contaminated oil also leads to the formation of carbon deposits on the valve stems.

The carbon buildup shown on the valve in Figure 9-8 affects engine performance. First, the mass of carbon restricts airflow into the combustion chamber and the loss of air intake reduces power. Second, the weight of the carbon commonly causes **valve float** at higher rpms. Valve float occurs when the springs can no longer hold the valve closing consistent with the profile of the cam lobe. Instead of closing swiftly with the drop of the camshaft lobe, the valve floats over the ramp and hangs open at higher rpm, when the momentum is greater. This causes engine misfire. Another common symptom of carbon buildup affects starting in the morning or when the engine is cold. The engine will typically start and stall two or three times and then start and run roughly for a few seconds. What happens is that the fuel injected into the intake is absorbed by the sponge-like carbon during the first couple of starting attempts. Then when the carbon becomes saturated with fuel, the vehicle starts and runs roughly, until the excess fuel is drawn out of the carbon.

> **Valve float** is a condition that allows the valve to remain open longer than intended. It is the effect of inertia on the valve.

Burned Valves. A **burned valve** is identifiable by a notched edge beginning at the margin and running toward the center of the valve head (**Figure 9-9**). The most common cause of a burned valve is a leak in the sealing between the valve face and seat. At the location of the leak, the valve will not cool efficiently. This is because the heat from the valve head cannot be transferred to the valve seat and cylinder head. As the area increases in temperature, it begins to warp. The warpage results in less contact at the valve seat and compounds the problem of heat transfer. The resultant heat buildup melts the edge of the valve.

> **Burned valves** are actually valves that have warped and melted, leaving a groove across the valve head.

Figure 9-8 Carbon buildup on the neck of the valve interferes with breathing.

Figure 9-9 Burned valves are characterized by a groove from the face toward the fillet.

Exhaust leaks allow cool air to flow past the valve. This cools the valve too fast, changes its structure, and reduces its burning durability.

Valve cupping is a deformation of the valve head caused by heat and combustion pressures.

Channeling is referred to as local leakage caused by extreme temperatures developing at isolated locations on the valve face and head.

In a **free-wheeling engine**, valve lift and angle prevent valve-to-piston contact if the timing belt or chain breaks.

When a timing belt or chain breaks, the pistons and crankshaft will quickly stop moving, but the valvetrain will stop moving almost immediately. If an engine is designed so that the valves could possibly touch the piston when this happens (because a valve is fully open and the piston is at TDC), it is called an **interference engine**.

Since heat transfer is dependent upon full contact of the valve face to the seat, anything preventing this contact can result in a burned valve. This includes weak valve springs, carbon buildup, worn valve guides, and valve clearance adjusted too tight. Additional causes of burned valves include preignition, exhaust system leaks, and failure of the cooling system.

Not all overheating conditions result in burned valves. The metal of the valve head may become soft and form into a cup shape (**Figure 9-10**). The force applied by the springs to seat the valve, combined with combustion pressures, deforms the valve head.

Channeling. Under normal conditions, the fillet area of the valve should be the hottest area (**Figure 9-11**). The temperature around the valve edge should be the coolest. If the valve does not seat properly, the temperature is not transferred to the cylinder head. The area not contacting the seat becomes the hottest area of the valve head (**Figure 9-12**). The area with the hottest temperature will melt away, developing a channel. **Channeling** usually occurs just prior to valve burning. Channeling is generally due to four main causes:

1. Flaking off of deposits from the valve
2. Extensive valve face peening due to foreign material lodged between the valve face and seat
3. Corroded valve face
4. Thermal stress of the cylinder head, resulting in radial cracking

Metal Erosion. Metal erosion is identified by small pits in the valve face that usually cannot be removed by valve grinding (**Figure 9-13**). This erosion is generally caused by coolant entering the combustion chamber. The chemical reaction etches away the face material. This condition can also occur if some chemicals enter the combustion chamber. These chemicals can come in the form of coolant system conditioners, oil additives, and fuel additives.

Breakage. Breakage of the valve can occur due to valve and piston contact. In a **free-wheeling engine**, valve lift and angle prevent valve-to-piston contact if the timing belt or chain breaks. In an **interference engine**, this can be the result of timing chain or belt breakage. Interference designs are commonly used because they offer improved fuel efficiency, increased power output, and lower emissions. However, if the valve timing components fail, the valve can contact the top of the piston or another valve. This contact may cause the valve to break or be drawn into the combustion chamber. If this occurs, major engine damage results. If the engine you are working on has experienced this type of failure, inspect the head and block carefully for cracks and other damage. The head must usually be replaced when this type of failure occurs.

Figure 9-10 Excessive heating of the valve can lead to cupping.

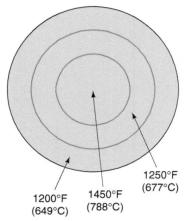

1250°F (677°C)

1200°F (649°C)   1450°F (788°C)

Figure 9-11 Normal valve temperatures as heat is dissipated through the face and seat.

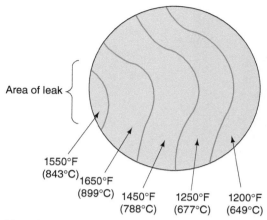

Figure 9-12 Improper face and seat contact results in improper heat dissipation.

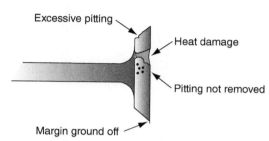

Figure 9-13 Metal erosion of the valve.

Other than valve-to-piston contact, valves may also experience fatigue and impact breaks. A fatigue break is a gradual breakdown of the valve due to excessive heat and pressure. Impact breakage can result from the valve face being forced into the valve seat with excessive force. Impact breakage can be the result of excessive valve clearance, valve springs with excessive pressure, loss of valve keepers, or any condition that can result in **stem necking**.

Necking is identified by a thinning of the valve stem just above the fillet (**Figure 9-14**). The head pulls away from the stem, causing the stem to stretch and thin. This condition is caused by overheating or excessive valve spring pressures. In addition, necking can be caused by exhaust gases circling around the valve stem. The corrosive nature of these gases and the action resulting from the swirl eat away the metal of the valve stem. Necking resulting from this action is caused by anything resulting in increased exhaust temperatures, including the following:

- Improper ignition timing
- Improper air-fuel mixtures
- Overloading
- Improper gear ratio in the final drive
- Exhaust restriction

An additional cause of valve breakage is referred to as **off-square** seating or tipping. If the valve does not seat properly, it may flex as it attempts to set into the valve seat. This flexing weakens the stem and eventually may lead to its breakage (**Figure 9-15**).

Stem necking is a condition of the valve stem in which the stem narrows near the bottom.

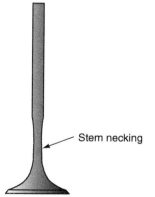

Figure 9-14 A valve with a necked stem must be replaced.

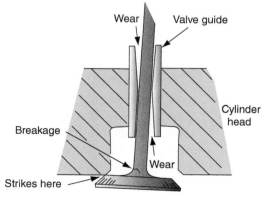

Figure 9-15 Tipping of the valve in the guide can keep the valve from seating properly, resulting in breakage near the fillet.

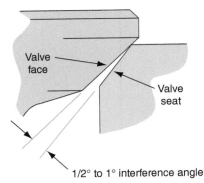

Figure 9-17 The valve face contacts the valve seat to seal the combustion chamber. The face and seat may be cut at different angles to create an interference angle. This allows the valve to seal with greater seating pressure.

Figure 9-16 A valve with a scored stem must be replaced to prevent premature failure of the valve seal.

Off-square seating can be caused by worn valve guides, distortion of the valve seat, improper rocker arm alignment, and weak valve springs.

Valve Stem Wear. When visually inspecting the valve stem, look for galling and scoring (**Figure 9-16**). Either of these can be caused by worn valve guides, carbon buildup, or off-square seating. Also look for indications of stress. A stress crack near the fillet of the valve indicates excessive valve spring pressures or uneven seat pressure. Stress cracks near the top of the stem can be caused by excessive valve lash, worn valve guides, or weak valve springs.

Inspect the tip of the valve stem for wear and flattening. These conditions can be caused by improper rocker arm alignment, worn rocker arms, or excessive valve lash. Separation of the tip from the stem can be caused by these same conditions.

Valve Seats

When the valve closes, it must form a seal to prevent loss of compression and combustion pressures. The valve face rests against the valve seat to accomplish this task (**Figure 9-17**). In addition to sealing compression pressures, the seat and face provide a path to dissipate heat from the valve head to the cylinder head.

There are two basic types of **valve seats**: integral and removable inserts. **Integral valve seats** are machined into the cylinder head. The seat is induction hardened to provide a durable finish. Most aluminum cylinder heads (and many cast-iron cylinder heads) use **valve seat inserts** that are pressed into a recessed area (**Figure 9-18**). Common materials used in seat insert construction include cast iron, bronze, and steel alloys. Cast-iron inserts are used to repair cast-iron heads not originally equipped with seat inserts. Bronze inserts are typically used in aluminum cylinder heads. Bronze inserts are claimed to have better heat transfer, thus prolonging valve life. Steel alloys are available in several grades of stainless steel and with several grades of hardness. It has been commonly thought the harder the valve seat the better. The material used for the valve's seat should be matched to the material used in its face; otherwise, less than recommended heat transfer and increased wear can

Off-square seating or tipping refers to the seat and valve stem not being properly aligned. This condition causes the valve stem to flex as the valve face is forced into the seat by spring and combustion pressures.

Shop Manual
Chapter 9, page 396

The **valve seat** provides the mating surface for the valve face.

Integral valve seats are part of the cylinder head.

Valve seat inserts are pressed into a machined recess in the cylinder head.

occur. One of the hardest seats is made of **Stellite**. The Stellite is applied to the valve seat to protect against oxidation and corrosion, and for added wear resistance. The seat is also induction hardened to provide a durable finish. Seat hardening is done by an electromagnet using induction to heat the valve seat to a temperature of about 1,700°F (930°C). This hardens the seat to a depth of about 0.060 in. (1.5 mm). The greatest advantage of integral seats is their ability to transfer heat. Integral seats run about 150°F (83°C) cooler than seat inserts.

The valve seat must be finished properly and cut to the specified width. Typically, the intake seats are approximately 1/16 in. and the exhaust seats about 3/32 in. Always check the manufacturers' specifications when refinishing valve seats. The exhaust seat is a little wider because it has to dissipate more heat after being open during the end of the power stroke. The seat width is critical to proper cooling and sealing. The seat should be narrow enough to provide a high seating force from the valve spring. If the seat is too wide, the valve will not seat with enough force to chip away any carbon developing on the seat. It can also allow the valve to run cooler than designed, which can allow carbon to develop on the seat. The seat must be wide enough to allow extreme heat to dissipate during the short time that the valve is closed. If the seat is too narrow, the valve will burn or cup over time because it will run too hot.

Seat Recession. Over time, valve seats have a tendency to recede further down into the head. This places the valves further into the head so that when the valves open there is not as much open area for air to flow; power is reduced. You can inspect for this before disassembling a cylinder head; simply look at each of the valves and make sure they are at the same level in the head. This measurement is called the installed height. There are three common causes of **valve seat recession** (**Figure 9-19**). The first is due to the pressure and heat in the combustion chamber causing local welding of the valve face and valve seat. When the valve is opened again, it forces a breaking of this weld, resulting in the removal of metal. In the past, this action of local welding was controlled by adding lead to the fuel. The lead would be the subject of the welding and breaking instead of the valve face and seat. To counteract the damage to valves and valve seats as a result of the loss of lead, most manufacturers use very hard valve seats.

The second leading cause of valve recession is sustained high engine speeds. The third cause is weak valve springs. Both high engine speeds and weak valve springs cause valve recession in the same manner. Under these two conditions, the valve rotates faster than normal. This results in excessive grinding between the valve face and seat. If the valve face is a harder material than the seat, the seat will recede.

> **Stellite** is a hard-facing material made from a cobalt-based material with a high chromium content.

> **Valve seat recession** is the loss of metal from the valve seat, causing the seat to recede into the cylinder head.

Figure 9-18 A valve seat insert.

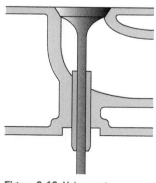

Figure 9-19 Valve seat recession.

Shop Manual
Chapter 9, page 396

Valve guides support and guide the valve stem through the cylinder head.

The solution for seat recession is to install hard valve seats and use high-quality valves. In addition, spring dampeners should be used to reduce chatter.

Valve Guides

Valve guides may be integral to the head or they may be pressed in. Cast-iron cylinder heads typically use integral valve guides. Holes are drilled down through the head where the valves will ride. On aluminum heads, the holes are bored oversized and guide inserts are pressed into the head. The guide inserts are usually cast iron or bronze (**Figure 9-20**). While bronze is a softer material, it has a natural lubricating quality that helps it resist wear.

The guide ensures that the valve rides straight up and down and contacts the valve seat accurately. When valve guides wear, the valve can wobble in the head and the valve face-to-seat sealing is compromised. The wobbling of the stem also reduces valve seal contact and distorts the seals. Guide wear allows leakage past the valves and seals and reduces compression and power. Wear can also cause stress cracks and breakage near the fillet. The guides are machined to provide just the right amount of clearance between the guide and the stem of the valve. The clearance is typically 0.001 to 0.003 in., though some manufacturers may specify more clearance. If the fit is too loose, the valve will wobble and provide poor seating, and if it is too tight, the valve can stick. The small clearance allows a thin film of oil to lubricate the valve and guide. Too much clearance allows oil to be pulled past the intake valve seal as the valve wobbles, resulting in oil consumption and bluish gray smoke out of the tailpipe. This is most noticeable at start-up, during extended idling, or on deceleration when vacuum is high. The guides have coolant passages around them, helping the valves dissipate heat.

Valve Seals

To control the amount of oil allowed to travel down the valve stem to provide lubrication, valve stem seals are used. These seals are constructed of synthetic rubber or plastic and are designed to control the flow of oil, not prevent it. There are three basic types of valve stem seals: O-ring, umbrella, and positive seal.

The O-ring seal is basically a square-cut seal made of rubber (**Figure 9-21**). The valve stem has a groove machined near the top to accept the O-ring. The O-ring forms a seal between the valve stem and the spring retainer. A shield is also used to deflect oil away from the valve stem.

Umbrella seals fit over the valve stem (**Figure 9-22**). The seal provides a tight fit around the valve stem and a loose fit around the guide. This design allows the seal to move up and down with the valve.

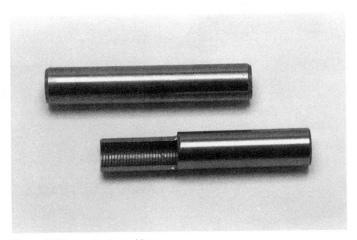

Figure 9-20 Insert valve guides.

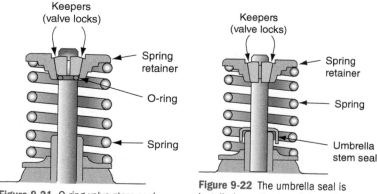

Figure 9-21 O-ring valve stem seal.

Figure 9-22 The umbrella seal is installed over the top of the valve stem.

Figure 9-23 Positive valve seal designs.

Positive seals are generally constructed of rubber with a Teflon insert (**Figure 9-23**). The design fits the valve guide snugly with either a light spring, spring clip, or press fit. Positive seals are considered by many to be the best type of valve stem seal. It is possible to replace umbrella or O-ring seals with positive seals. If the original guide was not cut for a positive seal, it can be cut to accept the seal.

CYLINDER HEAD COMPONENT RELATIONSHIPS

All components of the cylinder head associated with valve opening and seat (the valve, spring, guide, and seat) must work together to accomplish their designed tasks. Wear, damage, or problems in one component will have an effect on the other components. If the valve stem or guide is worn, there will be excessive movement of the valve inside the guide (**Figure 9-24**). This movement will translate to improper seating and possible burning of the valve. Figure 9-24A shows oil leaking past a worn intake valve guide. Atmospheric pressure pushes the oil into the air-fuel mixture. When the intake valve opens and the mixture is drawn into the cylinder, the oil is burned with the air-fuel charge. This results in oil consumption and blue smoke from the tailpipe. Figure 9-24B shows an exhaust valve with a worn guide allowing oil to be sucked by the guide due to venturi action. This too will result in oil consumption and blue smoke.

A valve that is not seating properly will cause poor engine performance. Both compression and combustion gases will be lost through the leakage. Lost compression reduces

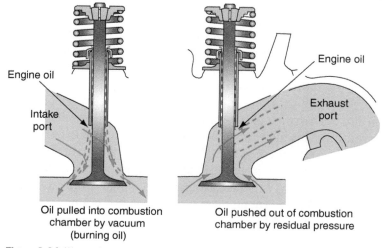

Engine oil

Engine oil

Intake port

Exhaust port

Oil pulled into combustion chamber by vacuum (burning oil)

Oil pushed out of combustion chamber by residual pressure

Figure 9-24 Worn valve guides can translate to improper valve sealing and oil consumption.

A **valve job** includes refinishing or replacing the valves and seats and any other worn components in the valvetrain.

A **valve spring** connects to the valve by a keeper and retainer, and pulls the valve back into the valve seat, thus putting the valve in a closed position where no fuel or air can pass through it.

Wedge chamber design locates the spark plug between the valves in the widest portion of the wedge in the cylinder head.

Quenching is the cooling of gases as a result of compressing them into a thin area. The quench area has a few thousandths of an inch clearance between the piston and combustion chamber.

The **squish area** of the combustion chamber is the area where the piston is very close to the cylinder head. The air-fuel mixture is rapidly pushed out of this area as the piston approaches TDC, causing a turbulence and forcing the mixture toward the spark plug. The squish area can also double as the quench area.

Turbulence is the movement of air inside the combustion chamber.

A high **surface-to-volume (S/V) ratio** results in higher HC emissions. The S/V ratio is a mathematical comparison between the surface area of the combustion chamber and the volume of the combustion chamber. A typical S/V ratio is 7.5:1.

the potential power of combustion, and when combustion gases leak past a valve, they are not exerting force on top of the piston to produce power. A small leak by the valve seat will rapidly turn into a burned valve. You will feel poor valve sealing as misfire and a loss of power in the engine. You may also hear popping in the intake or exhaust system as the pressurized combustion gases are allowed in through a leaking valve.

Thorough reconditioning of the cylinder head components is often called a **valve job**. It consists of repairing or replacing all of the necessary components in the valvetrain to provide trouble-free operation for extended mileage, typically 100,000 miles or more. A complete valve job requires that the cylinder head(s) be removed from the engine. If the valve seals and/or **valve springs** are faulty, a repair may be made with the cylinder head still installed on the engine. Replacing these valve seals with the head installed was common with older engines where the seals did not last very long.

AUTHOR'S NOTE During the cylinder head reconditioning process, all valvetrain components must be machined or replaced so they meet the engine manufacturer's specifications. An unsuccessful cylinder head reconditioning job may be defined as one that did not provide a reasonable mileage interval before some valvetrain component failed. It has been my experience that unsuccessful cylinder head reconditioning jobs are usually caused by failure to make sure that all valvetrain components are within manufacturer's specifications. For example, the technician may have done an excellent job of reconditioning the valves, seats, and guides, but the condition and measurements on the valve springs were overlooked.

COMBUSTION CHAMBER DESIGNS

The size, shape, and design of the combustion chamber affect the engine's performance, fuel efficiency, and emission levels. Manufacturers can use several different combustion chamber designs to achieve desired engine efficiency results.

Wedge Chamber

The most common method of creating a **wedge chamber** is to use a flat-topped piston and cast a wedge into the cylinder head (**Figure 9-25**). Wedge chamber design locates the spark plug between the valves in the widest portion of the wedge. The air-fuel mixture is compressed into the **quench** and **squish areas**, resulting in a turbulence that mixes the air-fuel mixture. The squish area squishes the air-fuel mixture into the center of the combustion chamber as the piston moves up toward Top Dead Center (TDC). This added **turbulence** helps mix the air-fuel mixture, cools it, and helps promote a faster burn once the ignition system begins combustion. The quench area cools the end gases in the chamber to prevent them from igniting before the intended flame front engulfs them. The squish and quench area of the combustion chamber gives it a lot of surface area compared to the volume. This high **surface-to-volume (S/V) ratio** of the combustion chamber has a negative impact on emissions. A higher S/V ratio contributes to greater hydrocarbon emissions as the fuel clings to the relatively cooler walls and exits the combustion chamber unburned in the form of hydrocarbons. The wedge combustion chamber design does not promote high volumetric efficiency. Air does not easily reach the hidden pocket of the wedge and the added turbulence reduces airflow, particularly at higher rpm. Another design method is to use flat chambers in the cylinder head and to offset the cylinder bores. This results in the piston approaching the cylinder head at an angle and produces the same basic functions of a wedge chamber.

Hemispherical Chamber

The **hemispherical chamber** is designed in a half-circle with the spark plug located in the center of the dome (**Figure 9-26**). The valves are located on an incline of 60° to 90° from each other.

The hemi-head design provides an even flame front because the spark plug is centered. The design also allows for easy "breathing" of the engine because the valves are located across from each other. During the period of valve overlap, the stream of airflow out the exhaust valve helps pull air in through the intake valve. The area inside the hemi chamber is wide open and fills easily. The hemi design has a very low S/V ratio. A disadvantage to this design is that the combustion chamber creates little turbulence. While this helps volumetric efficiency at higher rpm, it can allow the hot gases to ignite without a spark, causing uncontrolled combustion. To prevent this, some manufacturers may use a domed piston.

Placing the crown of the piston this close to the cooler cylinder head prevents the gases in this area from igniting prematurely.

The term *hemi* may be used for an engine with **hemispherical combustion chambers.**

Pentroof Chamber

Pentroof combustion chambers are a very common design for today's multivalve engines. They were first used in the 1980s and have grown in popularity since. The inverted-V shape of the chamber provides a low S/V ratio, resulting in lower emissions (**Figure 9-27**). These chambers are similar to the hemispherical head and share most of the advantages of that design. The pentroof chamber allows for larger valves and provides some turbulence.

Pentroof combustion chambers are a common design for multi-valve engines using two intake valves with one or two exhaust valves and provide good S/V ratio.

Swirl Chamber

To improve combustion efficiencies, **swirl chambers** are designed with a curved port that causes the air-fuel mixture to swirl in a corkscrew fashion (**Figure 9-28**). The compression of this swirling mixture causes a thorough mixing of the gases, resulting in increased fuel economy and reduced exhaust emissions. Location of the valve and design of the chamber causes the swirl action.

Swirl chambers create an airflow that is in a horizontal direction.

Chamber-in-Piston

The **chamber-in-piston** design locates the combustion chamber in the piston head instead of the cylinder head (**Figure 9-29**). The advantage of this chamber design is that the piston remains hot enough to provide proper vaporization of the air-fuel mixture. This design is used on many diesel engines.

A **chamber-in-piston** is a piston with a dish or depression in the top of the piston.

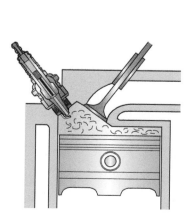

Figure 9-25 Wedge combustion chamber design.

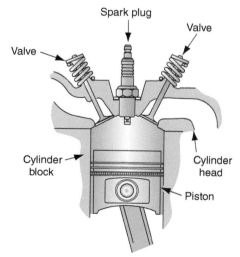

Figure 9-26 Hemispherical combustion chamber design. Notice the domed piston.

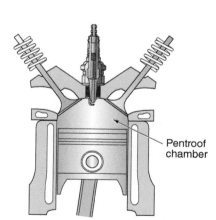

Figure 9-27 Pentroof combustion chamber design.

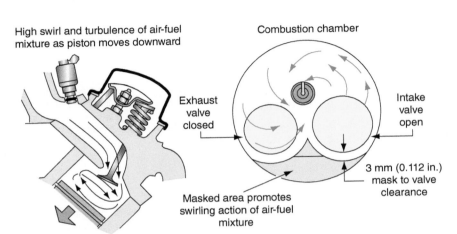

Figure 9-28 Swirl combustion chamber design.

High swirl and turbulence of air-fuel mixture as piston moves downward

Combustion chamber

Exhaust valve closed

Intake valve open

3 mm (0.112 in.) mask to valve clearance

Masked area promotes swirling action of air-fuel mixture

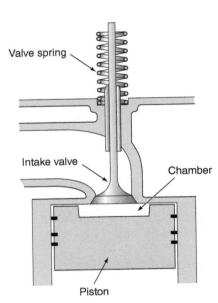

Figure 9-29 Chamber-in-piston combustion chamber design.

Valve spring

Intake valve

Chamber

Piston

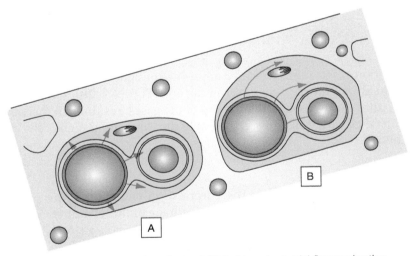

Figure 9-30 (A) Standard uniform-flow and (B) fast-burn tangential-flow combustion chambers.

Fast Burn

A faster combustion can be achieved by directing the incoming air-fuel mixture tangentially through the intake valve (**Figure 9-30**). This results in a turbulence that mixes the gases. An additional benefit of this design is the compression ratio can be increased without an increase in octane requirements. This is because the fast-burn design has less potential for knock.

Tumble Port

Tumble port combustion chambers use a modified intake port design, resulting in a tumbling vortex mixture burning technique (**Figure 9-31**). This design uses the principle of **eddy currents** by using a volume of air that runs against the main current to provide a tumbling condition. Benefits include increased fuel economy and reduced hydrocarbon emissions.

Tumble port combustion chambers use a modified intake port design.

An **eddy** is a current that runs against the main current.

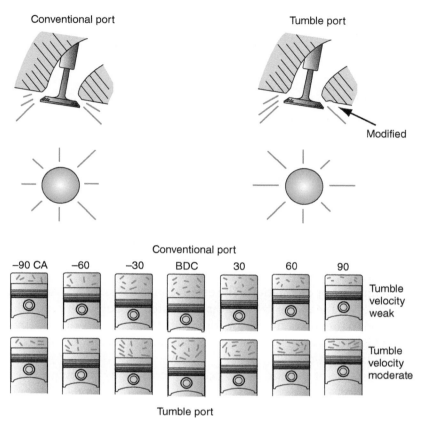

Figure 9-31 Comparison of conventional and tumble port designs.

Stratified-Charge Chambers

The **stratified-charge** combustion chamber is actually two chambers contained within the cylinder head (**Figure 9-32**). The smaller chamber (prechamber) is located above the main chamber and has its own intake valve. A very rich air-fuel mixture is delivered to the prechamber, where the spark plug is located. The rich mixture is very easy to ignite. At the same time, a very lean mixture is delivered to the main chamber. At the completion of the compression stroke, the spark plug fires to ignite the rich mixture in the precombustion chamber. The burning rich mixture will ignite the lean mixture in the main combustion chamber.

The first manufacturer to mass-produce the stratified-charge engine was Honda. It used the engine from 1975 to 1987 with the compound vortex controlled combustion

> A **stratified-charge** engine has two combustion chambers. A rich air-fuel mixture is supplied to a small auxiliary chamber, and the very lean air-fuel mixture is supplied to the main combustion chamber.

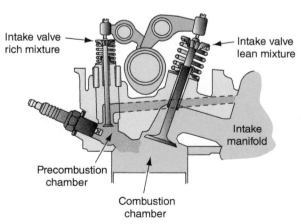

Figure 9-32 Stratified-charge combustion chamber.

(CVCC) design. A three-barrel carburetor was used. Two barrels supplied the lean mixture for the main combustion chamber, while the third barrel supplied the prechamber with the rich mixture.

THE COMBUSTION PROCESS

Combustion is the chemical reaction between fuel and oxygen that creates heat and pressure. It is a closely controlled burning of the air and fuel. Spark ignition occurs before top dead center on the compression stroke. The hot, compressed air-fuel mixture is ignited, and a flame front develops. For normal combustion to occur, the air-fuel mixture must be delivered in the proper proportions and mixed well, the spark must be timed precisely, and the temperatures inside the combustion chamber must be controlled. The flame can then move quickly and evenly (propagate) across the combustion chamber, harnessing the power of the fuel as heat. Pressure builds steadily as the gases expand from heat. The peak of this pressure develops around 10° ATDC (After Top Dead Center) to push the piston down on the power stroke.

The heat produced inside the combustion chamber must be controlled to reduce the possibility of preignition or detonation. The expansion of gases forces the unburned gases to be packed tighter and made hotter. Heat is dissipated through the spark plug case and threads in addition to the top of the piston, the walls of the combustion chamber, and the engine coolant (**Figure 9-33**). The design of the combustion chamber has a huge impact on the process of combustion.

> The unburned gases are called **end gases**.

The heat absorption of the **end gases** can be increased by creating some turbulence within the combustion chamber and by proper design of the squish and quench areas (**Figure 9-34**). As the gases swirl around, they come into contact with the walls of the combustion chamber or the top of the piston. Turbulence prevents the gases from becoming stagnant. Stagnant gases will only dissipate heat from the outer edges, allowing the center to become very hot.

Turbulence also causes the flame to burn faster. The turbulence can be promoted by piston top design and the squish area of the chamber. The squish area causes the gases to be squirted out of an area of the chamber as the piston compresses the gases.

As the gases burn, the end gases are packed into the quench area of the chamber. This thins the gases and allows them to dissipate heat faster. The quench area design affects the emission output of the engine. As previously discussed, to prevent the end gases from

Figure 9-33 During proper combustion, the flame spreads steadily across the chamber and heats all of the gases.

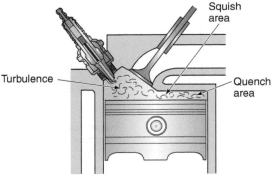

Squish area

Turbulence

Quench area

Figure 9-34 Quenching and turbulence are used to cool the temperatures of the end gases to prevent autoignition.

igniting on their own, heat is removed from them. This unburned mixture has a thickness of about 0.002 in. to 0.010 in. (0.1 mm to 0.3 mm). This thin layer of gases never burns, because the temperature is too low, and it is expelled into the exhaust system as unburned hydrocarbons. The larger the combustion chamber, the more the surface area for the formation of the **skin effect**.

> **Skin effect** is a small layer of unburned gases formed around the walls of the combustion chamber.

Manufacturers have reduced hydrocarbon emissions by redesigning the quench area. Reduction of hydrocarbons can be accomplished by increasing the height of the quench area. This increases the temperature of the end gases and reduces the thickness of the skin effect. Another means is by eliminating undesirable quench areas. This is done by removing sharp corners and pockets. Cylinder head gasket redesign and improved fit have eliminated many of these undesired quench areas.

Finally, skin effect has been reduced by lowering the compression ratio of the engine. This is not desirable from a performance point of view, but it was necessary to reduce emissions. Most engine manufacturers reduced the output of their engines during the late 1970s to the mid-1980s. Better understanding of the combustion process, chamber designs, fuels, and emission controls has made a return of higher compression engines possible.

A BIT OF HISTORY

With the great amount of moving parts within the engine, it is truly surprising how well it works. Unfortunately, engines requiring carbon-based fuels pollute the air and use a nonreplenishable fuel source. Today, most manufacturers are attempting to design a reliable, cost-efficient electric vehicle. This idea is not new. During the early years of the "horseless carriage," over one-third were driven electrically.

Preignition

Preignition occurs when a flame is ignited in the combustion chamber prior to the spark plug firing. Preignition results in a light knocking or rattling sound during acceleration at operating temperature. It is usually the result of hot spots in the chamber. This can be caused by excessive combustion chamber pressures. An excessively high compression ratio causes the temperature of the fuel charge to be raised above its flash point. A change in compression ratio can occur anytime there is a decrease in combustion chamber volume. This can be caused by excessive carbon buildup. Other things that change compression ratio include piston design and milling of the cylinder head, the block, or both. Another cause of preignition is the use of excessively low octane fuel. The octane required to properly operate the engine is based on the compression ratio of the engine. The higher the ratio, the higher the octane required.

The result of preignition is high pressures within the combustion chamber before the piston reaches TDC. This pressure forces the piston to reverse direction too soon, wasting useful combustion energy. It is possible to blow a hole in the top of the piston near the outer edge as a result of preignition (**Figure 9-35**).

Detonation

Normal combustion takes about 3 milliseconds (3/1,000 of a second). Abnormal combustion, detonation, is more like an explosion, occurring as fast as 2 millionths of a second (2/1,000,000 of a second). The explosive nature of detonation can cause serious engine damage. Detonation occurs when combustion pressures develop so fast that the heat and pressure will "explode" the unburned fuel in the rest of the combustion chamber. Before the primary flame front can sweep across the cylinder, the end gases ignite in an uncontrolled burst (**Figure 9-36**). The dangerous knocking results from the violent explosion.

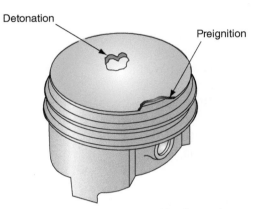

Figure 9-35 Piston damage caused by abnormal combustion.

Figure 9-36 End gases explode before the flame can expand across the chamber. The hot pressure spike knocks the piston so hard that it can blow a hole in it.

Detonation causes piston and ring damage, bent connecting rods and worn bearings, top ring groove wear, blown head gaskets, and possibly complete engine failure.

Common causes of preignition and detonation are the following:

- Deposits in the cooling system around the combustion chamber
- Engine overheating
- Too hot a spark plug
- An edge of metal or gasket hanging into the combustion chamber
- Fuel with too low an octane rating
- An inoperative or clogged exhaust gas recirculation (EGR) system
- Too much ignition timing advance
- Lean air-fuel mixtures (when there is less-than-desired fuel mixed with air)
- Carbon buildup in the combustion chamber
- A faulty knock sensor
- Excessive boost pressure from a turbocharger or supercharger

Misfire

Misfire is another type of abnormal combustion. When an engine misfires, it means that one or more of the cylinders are not producing their normal amount of power. The cylinder(s) is unable to burn the air-fuel mixture properly and extract adequate energy from the fuel. The misfire may be total, meaning that a flame never develops and the air and fuel are exhausted out of the cylinder unburned. Hydrocarbon emissions increase dramatically. Misfire may also be partial when a flame starts but sputters out before producing adequate power due to a lack of fuel, compression, or good spark. When an engine is misfiring, the engine bucks and hesitates; the misfire is often more pronounced under acceleration. Technicians normally call this a miss or a skip. Extended misfire can destroy the catalytic converter by overloading it with fuel and causing it to overheat. Misfire can be caused by ignition or fuel-system faults as well as engine mechanical problems. Typical engine mechanical faults that cause misfire are as follows:

- Burned or leaking valves
- Valve-stem or valve-back carbon buildup
- Weak or broken valve springs

- Worn valve guides or lifters
- Broken, worn, or bent rocker arms or pushrods
- Worn camshaft(s)
- Improper valve timing
- Worn pistons or rings

A BIT OF HISTORY

The Honda Insight and the Toyota Prius were the first hybrid cars to be introduced in the North American market. As the "hybrid" name suggests, these cars have two power sources to supply torque to the drive wheels. These power sources are a mechanical power source from the engine and an electric power source from batteries and electric motors. These cars may be referred to as parallel hybrids or even series-parallel because power may be supplied to the drive wheels from the engine or from the electric motor. An onboard computer system controls the power source to the drive wheels. When the vehicle is decelerating, the electric motor acts as a generator to help charge the batteries.

These batteries do not require charging from an external source. Several other manufacturers have introduced hybrid vehicles after the success of the Toyota and Honda versions. The internal combustion engine technology in these vehicles has since been transferred over to other nonhybrid vehicles.

MULTIVALVE ENGINES

Many of today's engines are designed with more than two valves per cylinder. An engine may use multiple valves to increase volumetric efficiency. The easier an engine can breathe, the more efficient and powerful it can be. A multivalve design allows more air-fuel mixture to enter the combustion chamber faster and helps the engine expel the spent exhaust gases. The most common multivalve engine is the four-valve-per-cylinder design (**Figure 9-37**).

Other designs use an extra intake valve, either a three-valve or a five-valve combustion chamber. One such variation is the use of a smaller second intake valve used as a jet valve. The jet valve directs air from above the throttle plates into the combustion chamber. At low engine speeds, this air causes a swirling action in the combustion chamber (**Figure 9-38**). The resultant turbulence increases the efficiency of the burn. Since the air inlet to the jet valve is above the throttle plates, the amount of air entering the valve decreases as engine speed increases.

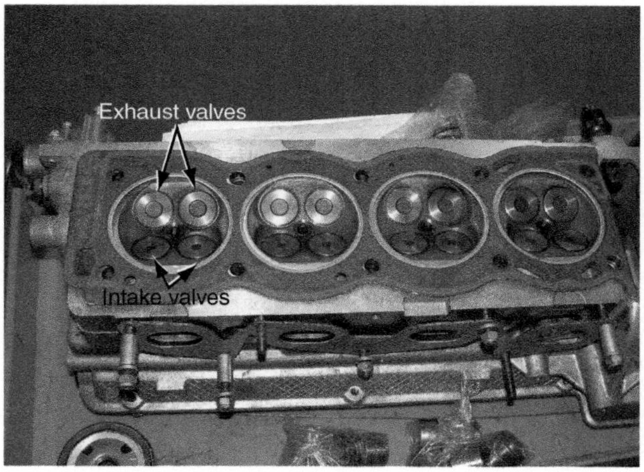

Figure 9-37 Multivalve cylinder head.

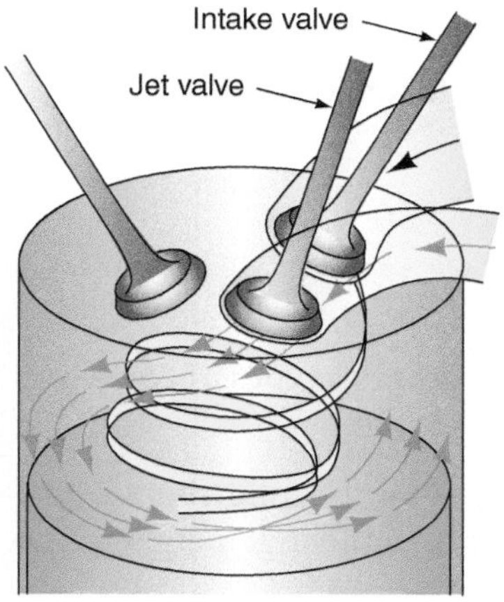

Figure 9-38 Some multivalve engines use a jet valve to increase the burn effectiveness at low engine speeds.

SUMMARY

- On most engines, the cylinder head contains the valves, valve seats, valve guides, valve springs, and the upper portion of the combustion chamber.
- The poppet valve is opened by applying a force that pushes against its stem. Closing of the valve is accomplished through spring pressure.
- The valve as a whole has many parts. The large-diameter end of the valve is called the head. The angled outer edge of the head is called the face and provides the contact point to seal the port. The area between the valve face and the valve head is referred to as the margin. The stem guides the valve through its linear movement.
- Manufacturers use several methods to dissipate the heat from the valve. First, the contact surface of the face and seat is used to transfer the heat to the cylinder head. The second method is through the stem, valve guide, valve spring, and on to the cylinder head.
- A burned valve is identifiable by a notched edge beginning at the margin and running toward the center of the valve head.
- Metal erosion is identified by small pits in the valve face and is generally caused by a chemical reaction in the combustion chamber.
- Breakage of the valve can occur due to valve and piston contact, valve springs that are too tight, loss of valve keepers, any condition that can result in stem necking, and off-square seating.
- The continual movement of the valve may result in wear on the portion of the stem traveling in the guide. As the stem wears, the oil clearance

increases, resulting in excessive oil being digested into the combustion chamber and the formation of deposits on the valve.
- The valve seat is located in the cylinder head and provides a seal when the valve face contacts it. It also provides a path to dissipate heat from the valve head to the cylinder head.
- There are two basic types of valve seats: integral and removable inserts. Integral valve seats are machined into the cylinder head. Most aluminum cylinder heads (and many cast-iron cylinder heads) use valve seat inserts that are pressed into a recessed area.
- Valve guides support and guide the valve stem through the cylinder head. There are two types of valve guides used: insert and integral. Insert guides are removable tubes that are pressed into the cylinder head; integral guides are machined into the head.
- Valve seals control the amount of oil allowed to travel down the valve stem to provide lubrication.
- The size, shape, and design of the combustion chamber affect the engine's performance, fuel efficiency, and emission levels.
- The wedge and the pentroof combustion chamber designs are popular on modern gasoline engines.
- Preignition is a form of abnormal combustion that occurs when a hot spot in the combustion chamber causes autoignition of the air-fuel mixture before the spark.
- Detonation occurs when the end gases autoignite after the spark. Detonation's damaging force can blow holes in the top of pistons.

REVIEW QUESTIONS

Short-Answer Essays

1. List the purpose of the following components:
   Cylinder head
   Intake valve
   Exhaust valve
   Valve seat

2. List and describe the common types of combustion chambers.

3. List the common causes for the following valve failures: burned valves, channeling, metal erosion, and breakage.

4. Explain the advantages of the pentroof combustion chamber.

5. List the causes of valve seat recession.

6. What are the differences between interference and free-wheeling engines?

7. List and describe the functions of the parts of a valve.

8. Explain why you should not refinish or machine sodium-filled valves.

9. List and describe the three common types of valve stem seals.

10. Describe the purpose of the valve guides and the difference between integral and insert types.

Fill-in-the-Blanks

1. The area between the valve head and the valve face is called the valve _____.

2. _____ control the flow of gases into and out of the engine cylinder.

3. Valve stem _____ control the amount of oil allowed to travel down the valve stem and provide lubrication.

4. When the valve closes, it must form a seal to prevent loss of compression. The _____ rests against the valve seat to accomplish this task.

5. _____ guides are removable tubes that are pressed into the cylinder head.

   _____ guides are machined into the head and are a part of the cylinder head.

6. Most of the heat is dissipated from the valve through the contact area between the valve face and _____ _____.

7. The most common cause of a burned valve is a leak in the sealing between the _____ _____ and seat.

8. Metal erosion is generally caused by _____ entering the combustion chamber.

9. Necking is identified by a _____ of the valve stem just above the _____.

10. An engine may use multiple valves to _____ volumetric efficiency.

Multiple Choice

1. The most common refinishing valve face angle is:
   A. 15°
   B. 30°
   C. 45°
   D. 60°

2. Valve stem seals.
   A. eliminate oil leakage into the valve guides.
   B. are not used on newer engines.
   C. keep combustion gases from corroding the valve stem.
   D. control oil flow to the valve stem.

3. Each of the following is a likely cause of valve burning *except*:
   A. A faulty thermostat
   B. A leak between the valve face and seat
   C. Worn valve guides
   D. A thick valve margin

4. *Technician A* says that the intake valve is often smaller than the exhaust valve.
   Technician B says that the exhaust valves can operate more efficiently because the air is under pressure when the exhaust valve opens.
   Who is correct?
   A. A only
   B. B only
   C. Both A and B
   D. Neither A nor B

5. Most of the valve's heat is dissipated through the.
   A. face.
   B. margin.
   C. stem.
   D. fillet.

6. The advantages of the pentroof combustion chamber include each of the following *except*:
   A. Good turbulence
   B. A high S/V ratio
   C. High volumetric efficiency
   D. Centrally located spark plug

7. The skin effect causes.
   A. premature valve burning.
   B. preignition.
   C. flaking of the Stellite seats.
   D. high HC emissions.

8. _____ occurs when another flame starts after the spark.
   A. Combustion
   B. Preignition
   C. Detonation
   D. Misfire

9. *Technician A* says that preignition is harmless.
   Technician B says that detonation can cause severe engine damage.
   Who is correct?
   A. A only
   B. B only
   C. Both A and B
   D. Neither A nor B

10. Each of the following is a likely cause of detonation *except*:
    A. Carbon buildup in the combustion chamber
    B. Engine overheating
    C. A lean air-fuel mixture
    D. Fuel with too high an octane rating

CHAPTER 10
CAMSHAFTS AND VALVETRAINS

Upon completion and review of this chapter, you should understand and be able to describe:

- The function of the valvetrain.
- The differences between overhead valve valvetrains and overhead camshaft valvetrains.
- The components of the valvetrain.
- The purpose and function of the camshaft.
- The relationship among the camshaft lobe design and lift, duration, and overlap.
- The benefits and limitations of valve overlap.
- The purpose of the lifters.
- The operation of hydraulic lifters.
- The purpose of pushrods.
- The common methods of mounting the rocker arms.
- Rocker arm geometry.

Terms To Know

Advanced camshaft
Camshaft bearings
Camshaft lift
Clearance ramp
Duration
Lifters
Lobe separation angle
Lobes

Pedestal-mounted rocker arm
Pump-up
Pushrods
Retarded camshaft
Rocker arm
Rocker arm ratio
Roller lifter

Shaft-mounted rocker arms
Stud-mounted rocker arm
Valve lift
Valve overlap
Valve spring
Zero lash

INTRODUCTION

Every component that works to open and close the valves is part of the valvetrain. Air must flow efficiently into and out of the cylinder head to supply maximum power. Proper breathing of the engine depends on the valves opening and closing at the correct time. All of the components that make up the valvetrain work together to perform this task. The camshaft(s) opens the valves for the correct amount of time and to the proper distance. The springs close the valves and must overcome tremendous momentum when the engine is spinning near the red line to hold the valves against the profile of the camshaft. The camshaft is mounted in the block in overhead valve engines. When the camshaft is mounted in the cylinder head, it is called an overhead camshaft engine. The other valvetrain components—pushrods, rocker arms, lifters, keepers, and valve rotators and retainers—will differ slightly depending on the engine design. Everything in the valvetrain is related and dependent upon each other; thus, a failure in the system can result in poor performance, engine noise, or mechanical failures. One stray movement in the valvetrain, a stretched component, or a snapped lock can throw the whole assembly into a wrecked heap of parts.

Shop Manual
Chapter 10, page 456

Valvetrain Components

The valvetrain components work together to open and close the valves. For proper engine operation, this function must be performed at the precise time without loss of motion. The components of the valvetrain vary depending on engine design. The common components include the following:

- Camshaft
- Lifters or lash adjusters
- Pushrods
- Rocker arms or followers
- Valves
- Valve springs
- Valve keepers
- Valve spring retainers
- Valve rotators
- Timing chain, belt, or gears
- Camshaft and crankshaft sprockets
- Variable valvetrain components

The valvetrain components work together to open and close the valves.

Shop Manual
Chapter 10, page 456

Camshafts

Simply stated, the camshaft is used to control valve opening (**Figure 10-1**). However, its function is actually more complicated. The camshaft also controls the rate of opening and closing, and how long the valve is open. The camshaft is precisely timed to the crankshaft to be sure the valves open at the correct time during their respective strokes. The camshaft spins at half the speed of the crankshaft to allow each valve to open only once during a complete engine cycle. One complete engine cycle means that the engine has rotated 720 degrees and that each of the cylinders has had one firing event. The camshaft may be driven by a gear, belt, or chain off the camshaft. Camshafts are designed differently depending on the type of lifter to be used with them. There are roller camshafts, hydraulic camshafts, and mechanical camshafts. The difference in the camshaft may be in the material used or in the shape of the lobe. Camshafts are usually constructed of nodular cast iron, although some high-performance engines use cast steel. Since roller lifter camshafts have the highest amount of load per square inch, most of these style camshafts are manufactured from forged steel billets.

Some late-model engines have camshafts made from a steel tube (**Figure 10-2**). The lobes are bonded to the steel tube, and a nose piece is friction welded to the tube.

Figure 10-1 Typical camshaft design.

Figure 10-2 A lightweight tubular camshaft used in a late-model V6 DOHC engine.

Camshaft end play is controlled by a thrust plate at the front of the camshaft. This type of camshaft is more rigid than a camshaft made from cast or forged steel. The increased camshaft rigidity reduces engine vibration.

The **lobes** of the camshaft push the valves open as the camshaft is rotated. There is usually one lobe for each intake and exhaust valve. The lobe design is a factor in determining how far the valve will open and for how long (**Figure 10-3**). The height of the lobe defines the amount of valve lift or opening.

The amount of **camshaft lift** affects the amount of **valve lift**. Additional factors of valve lift include rocker arm ratio, pushrod deflection, and lifter condition. Camshaft lift is a measurement of the height of the lobe above the base circle. When an overhead camshaft engine does not use rocker arms, the camshaft lobe opens the valve directly; camshaft lift is then the same as the valve lift. Valvetrains that use rocker arms use mechanical advantage in the rocker arm design to multiply camshaft lift. Valve lift is an important factor in determining how well the engine can breathe (ingest the air-fuel mixture and expel it). The valve is a restriction to airflow into the combustion chamber. This is true until the valve is lifted about 0.300 in. (8 mm) off its seat. At this point, the valve offers very little restriction to airflow. However, there is still an advantage to increasing valve lift over 0.300-in. (8-mm). A typical amount of valve lift is 0.4 in.–0.55 in. (10.16 mm–13.97 mm).

The longer the valve remains open after reaching the 0.300-in. (8-mm) lift, the better the airflow. Many manufacturers have increased the amount of lift in their camshafts to increase the amount of valve opening. Lift does not adversely affect fuel economy,

The **lobes** of the camshaft push the valves open as the camshaft is rotated.

Camshaft lift is the valve lifter movement created by the rotating action of the camshaft lobe. Typical lobe lift of original equipment in the block camshafts is between 0.240 in. and 0.280 in. (6 mm and 6.6 mm).

Valve lift is the total movement of the valve off of its seat, expressed in inches or millimeters.

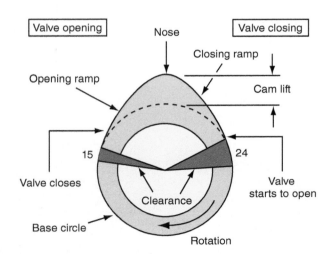

Figure 10-3 The lobe design determines how far the valve will open and for how long.

power, or emissions (as does excessive **duration**). For this reason, some manufacturers have maintained moderate duration of the camshafts and increased lift in an attempt to increase airflow. One of the limitations on maximum valve lift is the amount of travel of the valve spring. If the valve spring is required to travel a greater distance, its coils can bind.

The camshaft duration is determined by the length of the slope of the lobe beyond the base circle. The wider and longer the lobe is, the longer the valve stays open. Camshaft duration is measured in degrees of crankshaft rotation. Intake valve duration may differ from exhaust valve duration.

If a valve opens at top dead center (TDC) and closes at bottom dead center (BDC), the duration of the valve is 180 degrees. This is not practical for real-world engine operation. In actuality, the intake valve must open before TDC and close after BDC. The additional degrees of rotation must be calculated into the total duration degrees. If the intake valve opens 6 degrees before TDC and closes 12 degrees after BDC, total duration is 198 degrees of crankshaft rotation.

Since duration allows more time for the air-fuel mixture to fill the combustion chamber, duration has more effect on engine performance and efficiency than any other valve timing specification. There are instances when an engine may require longer durations to perform a specified function; for example, the high-performance engines used in most racing applications require a camshaft with higher durations. This camshaft allows the engine to breathe at higher rpm. The drawback is that the engine performs poorly at lower engine rpms. Variable valve timing technology allows a manufacturer to have the ability to design the engine for good performance, low emissions, and good economy. All of this can be done with only one design for the camshaft. By changing the timing of the camshaft, the lift and duration will change when compared to their relationship with the piston's position.

If a replacement camshaft is to be purchased, there are two methods used to measure duration. Society of Automotive Engineers (SAE) specifications are measured after 0.006 in. (0.15 mm) of lift on hydraulic lifter camshafts. Performance camshafts are usually measured for duration at 0.050-in. (1.25-mm) lift. An SAE-measured camshaft will have approximately 40 to 50 degrees more duration than a camshaft measured at 0.050-in. (1.25-mm) lift. This means that a camshaft with 220 degrees of duration at 0.006 in. has less-effective duration than a camshaft with 195 degrees of duration at 0.050 in.

The design of the lobe also determines how rapidly the valve opens. The steepness of the ramp dictates how quickly the valve opens and closes.

Clearance ramps are ground into the lobe to prevent the valvetrain components from hammering on each other. The clearance ramps are located between the base circle and the opening or closing ramps. The clearance ramps remove clearance in the valvetrain at a slower rate than the rate of lift (**Figure 10-4**). Most camshafts used with hydraulic lifters have a clearance ramp between 0.005 in. and 0.010 in. (0.13 mm and 0.25 mm). This ramp ensures that the lifter check valve is closed to provide a solid connection.

At the end of the exhaust stroke and the beginning of the intake stroke, both valves are open at the same time. This is called **valve overlap**. The amount of overlap is defined by the camshaft **lobe separation angle** (**Figure 10-5**). The larger or wider the separation angle, the less overlap. A tighter or lower separation angle means that the valves will be open together for a longer time, providing more overlap. A wider separation angle, used on most passenger cars, is 112 to 116 degrees. A high-performance engine or high-performance camshafts may have a tighter angle in the range of 112–116 degrees. The lobe separation angle or spread is determined by finding the lobe centers of the intake lobe and the exhaust lobe and averaging the two values. The lobe center can be found by using a degree wheel and a dial indicator (**Figure 10-6**). The degree wheel indicates the number of degrees after TDC where the maximum amount of valve lift is achieved for the intake

---

Camshaft **duration** is the amount of time the camshaft holds the valve open. It is measured in degrees of crankshaft rotation.

The area of the mechanical lifter camshaft from base circle to the edge comprises the **clearance ramp**.

When both the intake and exhaust valves are open at the same time an engine has **valve overlap**.

The **lobe separation angle** is the average of the intake and the exhaust lobe centers. It defines the amount of valve overlap. The tighter the angle (lower number of degrees), the more overlap the camshaft provides.

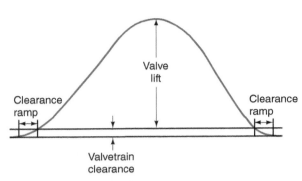

Valve
lift

Clearance
ramp

Clearance
ramp

Valvetrain
clearance

Figure 10-4 The clearance ramp removes valvetrain clearance at a slower rate than valve lift.

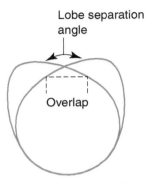

Lobe separation
angle

Overlap

Figure 10-5 The time, in crankshaft degrees, that both valves are open is overlap. This is defined by the lobe separation angle.

Figure 10-6 A degree wheel can be used to find the lobe centers.

An **advanced camshaft** has the intake valve open more than the exhaust valve at TDC. It is used to improve low-speed torque.

A **retarded camshaft** has the exhaust valve open more than the intake at TDC.

valve. The exhaust lobe centerline is determined in the same manner. The degree wheel is used to measure the number of degrees before or after TDC where the valve is fully opened. The lobe separation angle is the average of the two readings. For example, if the intake lobe center is 110 degrees after TDC and the exhaust lobe center is 120 degrees before TDC, the lobe separation angle is 115 degrees. If the intake lobe center is less than the lobe separation angle, the camshaft is advanced. An **advanced camshaft** has the intake valve(s) open more than the exhaust valve(s) at TDC. This type of camshaft is used to increase lower rpm performance for greater torque. **Retarded camshafts** are designed to have the exhaust valve open more than the intake at TDC.

Valve overlap helps the engine breathe better, particularly at higher rpm. The movement of airflow out of the exhaust valve helps draw the fresh air in through the intake valve. The momentum of airflow helps fully clean the spent exhaust gases out of the chamber even after TDC. This stream of airflow also helps pull the intake charge into the chamber. This is called scavenging. Too much overlap at a lower rpm, however, can allow some of the fresh air-fuel mixture to flow out of the exhaust valve and some exhaust to flow back into the intake. This is called reversion and leads to a rough idle, increased emissions, and reduced fuel economy. When you have exhaust gases in the intake, it reduces engine vacuum and counteracts the intended effects of overlap. Production vehicles use a wider separation angle to prevent the negative effects of reversion. Some engines using variable valve timing systems use two sets of camshaft lobes. At higher rpm, the camshaft lobes in use will provide greater overlap to improve volumetric efficiency and power. Some variable valve timing systems move the camshaft into a different position with regard to its timing. This will affect the performance and emissions of the engine significantly.

⚠ Caution

The Automotive Engine Rebuilders Association (AERA) recommends only a stock camshaft replacement on computer-controlled engines due to poor performance at some rpms. If you are replacing the camshaft on a computer-controlled engine with a non-OEM component, you may need to modify the computer, a few sensors, or both to make the engine operate correctly.

AUTHOR'S NOTE Aftermarket performance camshafts often produce an uneven or lopey idle accompanied by low vacuum. This is the result of reversion. If the engine is being designed for higher rpm performance, long overlap is essential, but be aware that you will sacrifice some low-rpm torque to achieve this goal. The fuel economy will suffer greatly, and emissions will typically be higher. Most camshaft upgrades are not legal for street use. They have a significant impact on emissions because of the long period of overlap and its resulting flow of intake charge out the exhaust. If you live in a state that performs tailpipe emissions testing, it is unlikely that a vehicle with a significantly modified camshaft will pass the emissions test. Any engine modification that increases the emissions of the vehicle is illegal for street use. Always refer to local laws and ordinances before performing modifications, especially if you are being paid to perform the services.

Some older designs of valvetrains used flat tappet camshafts. The camshaft lobes had a very slight taper of 0.0007 in. to 0.002 in. (0.0178 mm to 0.0508 mm) across their face (**Figure 10-7**). Along with offsetting the centerlines of the lifter and the lobe, this taper provided for lifter rotation. This was done to prevent premature wear, since the lifter's edge was not riding on the edge of the camshaft lobe. Rotating the lifter helped dissipate nearly 100,000 psi (690 MPa) load on the lifter. Roller lifters are commonly used today. Roller lifters are indexed and do not rotate.

Camshaft Bearings. The camshaft(s) rides in bearings or bearing surfaces on their journals. Oil pressure is fed through ports in the camshaft bearing bores. Like crankshaft main and rod bearings, camshaft bearings clearances are essential to good lubrication and adequate oil pressure. Camshafts that are installed within the engine block are usually fitted with full round bearings (**Figure 10-8**). These are one-piece bearings that are pressed into the block, and the journal of the camshaft is slid into place. A few overhead camshaft engines use split bearings with camshaft caps. Modern machining and oil improvements have made it possible for most manufacturers to eliminate the use of camshaft bearings in their overhead camshaft engines. In this case, the soft aluminum bore and caps serve as the bearing surface.

Camshaft bearings are required in most engines as a means of reducing friction between the camshaft and the cylinder block.

Shop Manual
Chapter 10, page 448

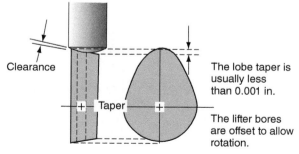

Figure 10-7 The taper of the camshaft lobes will provide lifter rotation to reduce wear.

Full round

Figure 10-8 Full-round camshaft bearings are used in engines with the camshaft in the block and in some overhead camshaft engines.

Lifters

Lifters are mechanical (solid) or hydraulic connections between the camshaft and the valves.

Lifters are mechanical (solid) or hydraulic connections between the camshaft and the valves. Lifters follow the contour of the cam lobes to lift the valve off its seat. Solid lifters provide a rigid connection, while hydraulic lifters use oil to absorb the resultant shock of valvetrain operation. Each lifter design has its advantages and disadvantages.

A solid lifter may be called a tappet or follower.

Solid lifters require a specified clearance between the components of the valvetrain. Periodic adjustment is required to maintain this clearance. If the clearance is too tight, there is no room for expansion due to heat. This can result in loss of power and value burning. If the clearance is too great, excessive noise, or valvetrain clatter, may be generated. In addition to noise, a change of 0.001 in. (0.03 mm) in valve lash may result in about a 3-degree change in duration. The lifter must be adjusted periodically to compensate for wear in the valve and valvetrain components. This is often recommended every 30,000 or 60,000 miles. Check the specific maintenance schedule. You will adjust the valve to have some lash between the rocker arm and the valve tip. This allows for a margin of error in valvetrain operation and also allows for the diminishing clearance as the valvetrain components heat up and expand. Many consumers see periodic valve adjustment as a disadvantage; it represents more costly maintenance.

A roller lifter is a normal lifter with a rolling pin on the bottom to decrease friction.

Hydraulic lifters use oil to automatically compensate for the effects of engine temperature and valvetrain wear. Some engine manufacturers use a roller-type hydraulic lifter, commonly called a **roller lifter**, in an effort to reduce friction (**Figure 10-9**). This type of lifter is fitted with a roller that rides on the camshaft lobe. Friction is reduced since the lifter body is not rubbing against the camshaft lobe (**Figure 10-10**). Roller lifters can also be designed as solid lifters for high-performance and heavy-duty diesel applications. Roller lifters are accompanied by a roller camshaft, which typically has steeper opening and closing ramps. Roller lifters do not need to rotate in the lifter bore, because they operate with lower friction.

Zero lash is the point where there is no clearance or interference between components of the valvetrains.

The hydraulic lifter automatically maintains **zero lash**. When the hydraulic lifter is resting against the base of the camshaft lobe, the valve is in its closed position (**Figure 10-11**). In this position, the plunger spring maintains a zero clearance in the valvetrain (**Figure 10-12**). Oil supplied by the oil pump is pushed through the oil feed holes of the lifter bore and into the lifter body. Oil is able to enter the lifter body only when the lifter is in this position, since the feed through the engine block is not aligned with the lifter body when it is on top of the lobe. The oil passes through the body and enters the inside of the plunger through an oil passage on its side. The check valve allows oil to flow from the plunger into the body, but not from the body to the plunger. Oil continues to flow into the lifter until it is full. Pressurized oil is trapped in the lifter by the closed check valve. The spring under the plunger pushes the lifter body toward the camshaft lobe and the plunger

Figure 10-9 A roller lifter from a late-model engine.

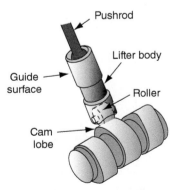

Figure 10-10 Roller lifters reduce friction and wear by using a roller on the camshaft.

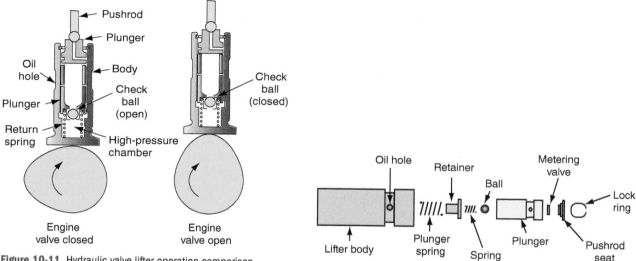

Figure 10-11 Hydraulic valve lifter operation comparison between valve open position and valve closed position.

Figure 10-12 The internal components of a hydraulic lifter.

toward the pushrod, removing any play in the valvetrain. The position of the check valve is determined by the pressure applied on the pushrod cap. This pressure is transferred from the closed valve through the pushrod to the cap. When the camshaft lobe attempts to raise the lifter body, the pressure on the cap and plunger is increased and the check valve closes, preventing the oil in the area below the plunger from escaping. Since the oil cannot be compressed, a rigid connection is formed, causing the plunger to raise the lifter body and open the valve.

As the temperature of the engine increases, parts begin to expand, reducing the amount of clearances. This causes the plunger of the lifter to be held lower in the lifter body, causing some of the oil that was trapped below the plunger to leak out from the lifter body. In addition, when the lifter is moved up the block bore by the camshaft lobe, the pressures of attempting to open the valves push the plunger down in the body. This results in leakdown. When the lifter follows the cam lobe back to the base, the oil feed passages align and the lifter is filled with oil again. This process of filling and leakdown automatically adjusts the hydraulic lifter to compensate for temperature changes and component wear. You can hear faulty lifters; they cause significant valvetrain clatter as the valvetrain components clash against each other after taking up the unintended clearance. You can hear the noise that they make when they first start up; this is normal. The clattering should disappear within a few seconds as oil pressure fills the body. When lifters wear, it takes longer for them to fill after start-up. Eventually they will become unable to hold oil as the camshaft tries to open the valve. This will cause noisy valvetrain clatter and loss of power because the valves will not open fully.

Solid lifters are often used in high-performance engines because they cannot **pump-up**. Since hydraulic lifters adjust automatically, they can overcompensate at higher engine speeds. At high engine rpm, valvetrain inertia causes the valves to open farther than what is accomplished by the cam lobe and rocker arm ratio. This valve float causes additional clearance to occur in the valvetrain. The lifter compensates for the additional clearance as it would under normal clearance corrections because the spring pushes the plunger up to take up the additional clearance. At this time, the body cavity fills with oil from the plunger cavity. The filling of the plunger cavity causes the lifter to be slightly longer than normal as it approaches the heel of the camshaft lobe. Again, the lifter compensates for this. On the next valve opening, the cavity is filled again and the cycle starts over. At high rpm, the lifter is unable to return to its normal size when it is

Lifter **pump-up** occurs when excessive clearance in the valvetrain allows a valve to float. The lifter attempts to compensate for the clearance by filling with oil. The valve is unable to close, since the lifter does not leak down fast enough.

approaching the heel of the lobe. The result is that the valve can be held open. When the engine speed is reduced, lifter operation returns to normal.

OHC Valvetrain Designs

Some overhead camshaft (OHC) engines use hydraulic lash adjusters (**Figure 10-13**). The lash adjuster works in the same manner as the hydraulic lifter. In some V6 DOHC valvetrains, the camshaft journals ride in cylinder head bores, and holders and holder plates retain the camshafts to the cylinder heads. The rocker arm shafts and rocker arms are mounted below the camshafts in the cylinder heads (**Figure 10-14**).

Cam followers are often used on overhead camshafts where the cam sits directly over the valves. The follower, often called tappet, is just a round cup that sits on top of the valve and provides a wider base for the camshaft to push on (**Figure 10-15**).

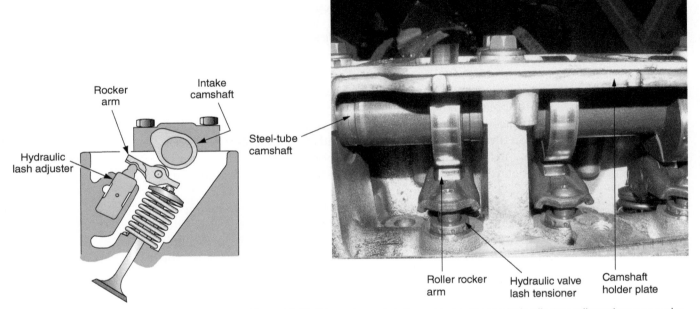

Figure 10-13 Some overhead camshaft engines use hydraulic valve lash adjusters.

Figure 10-14 This modern valvetrain uses hydraulic valve lash adjusters, roller rocker arms, and a tubular camshaft.

Figure 10-15 This follower sits directly under the camshaft and has a shim that can be changed to make valve clearance adjustments.

Often there are shims either under or on top of the followers that can be fitted to size to make the proper adjustment. These lifters also need periodic adjustment. There are a few manufacturers that use hydraulic followers. The hydraulic body sits under a cup (like a follower) and works just like a hydraulic lifter to constantly adjust valve clearance to zero.

Pushrods

Overhead valve (OHV) engines, with the camshaft located in the block, use **pushrods** to transfer motion from the lifter to the rocker arms (**Figure 10-16**). Besides being a link to the rocker arms, some pushrods are also used to feed oil from the lifter to the rocker arms (**Figure 10-17**). The oil is sent up the hollow portion of the pushrod.

Most pushrods are constructed from seamless carbon steel tubing. To decrease wear, some pushrods may have a hardened tip insert at each end. High-performance engines may use pushrods that are made of chromium-molybdenum tubular steel for increased strength. Chrome-moly pushrods may also have a thinner wall thickness to reduce weight and inertia.

The bottom of the pushrod fits into a socket in the lifter. The recesses of the socket are deep enough to prevent the pushrod from falling out during operation. The top of the pushrod fits into a small cup in the rocker arm.

Tolerances within the valvetrain generally allow some machining of the cylinder head without the need to correct pushrod length; however, if the deck surface of the cylinder head and/or the engine block is machined or if valve stem length is increased, it may be necessary to change the pushrods to maintain the correct valvetrain geometry. This is especially true on engines with nonadjustable valvetrains.

> Pushrods connect the valve lifters to the rocker arms.

Rocker Arms

The **rocker arm** is a pivoting lever used to transfer the motion of the pushrod to the valve stem. In overhead camshaft engines, the camshaft may operate the rocker arms directly (**Figure 10-18**). Rocker arms change the direction of the cam lift and spring closing forces, and they provide a leverage during valve opening. Most rocker arms are constructed of cast iron, cast aluminum, or stamped steel.

> The **rocker arm** is a pivoting lever used to transfer the motion of the pushrod to the valve stem.

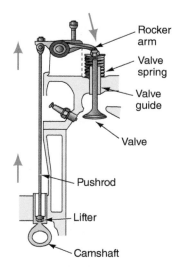

Figure 10-16 The valve lifter transfers force to the pushrod.

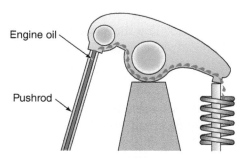

Figure 10-17 The pushrod can be used to deliver oil to the rocker arm.

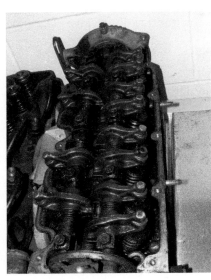

Figure 10-18 This OHC head uses rocker arms with integral hydraulic lifters to open the valves.

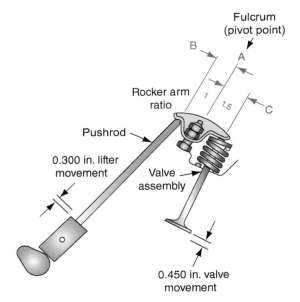

Figure 10-19 Rocker arms are designed to provide a mechanical advantage. The ratio is determined by comparing the distance from A to C to the distance from A to B. In this example, the ratio is 1.5:1.

The difference in the rocker arm dimensions from center to valve stem and center to pushrod is called the **rocker arm ratio**. It is a mathematical expression of the leverage being applied to the valve stem.

Shaft-mounted rocker arms are used in a mounting system that positions all rocker arms on a common shaft mounted above the cylinder head.

A **stud-mounted rocker arm** is mounted on a stud that is pressed or threaded into the cylinder head.

A **pedestal-mounted rocker arm** is mounted on a threaded pedestal that is an integral part of the cylinder head.

The design of the rocker arm is such that the side to the valve stem is usually longer than the side to the pushrod (**Figure 10-19**). This design gives the rocker arms a mechanical advantage. The **rocker arm ratio** works with the camshaft lobe to provide the desired lift. The ratio of the distance is calculated by comparing the distance from the rocker pivot to the point where the rocker hits the valve tip to the distance from the rocker pivot to the point where the rocker contacts the pushrod. Divide the longer distance to the valve tip by the shorter distance to the pushrod to find the ratio. For example, if the distance to the pushrod is 1.00 in. (25.4 mm) and the distance to the valve tip is 1.50 in. (38.1 mm), divide 1.50 in. by 1.00 in. (38.1 mm by 25.4 mm). The answer is 1.50 in. (1.5 mm); the rocker ratio is stated as 1.50:1 or 1.5 mm:1 mm. If the camshaft lift is 0.3 in. (7.62 mm), multiply that by the rocker ratio to find the valve lift: 0.3 in. × 1.50 in. = 0.45 in. (7.62 mm × 1.5 mm = 11.43 mm) of valve lift.

There are three common methods of mounting the rocker arm (**Figure 10-20**):

- Shaft mounted
- Stud mounted
- Pedestal mounted

Shaft-mounted rocker arms are located on a heavy shaft running the length of the cylinder head. The shaft is supported by stands. **Stud-mounted rocker arms** are used on overhead valve engines. Each rocker arm is mounted on a stud. A split ball is used as a pivot point for the rocker. **Pedestal-mounted rocker arms** are similar to the stud-mounted assemblies, except the rocker pivots on a split shaft.

Rocker Arm Geometry. When the valve is opened 50 percent of its travel, the rocker arm-to-stem contact should be centered on the valve stem (**Figure 10-21**). If the rocker arm is not centered, it will cause excessive side thrust on the stem. Rocker arm geometry does not usually need to be checked or corrected; however, if a different camshaft with a higher lift than the original camshaft is installed, rocker arm geometry will be affected. The higher lift causes the valve to move farther, causing the rocker arm tip to swing down

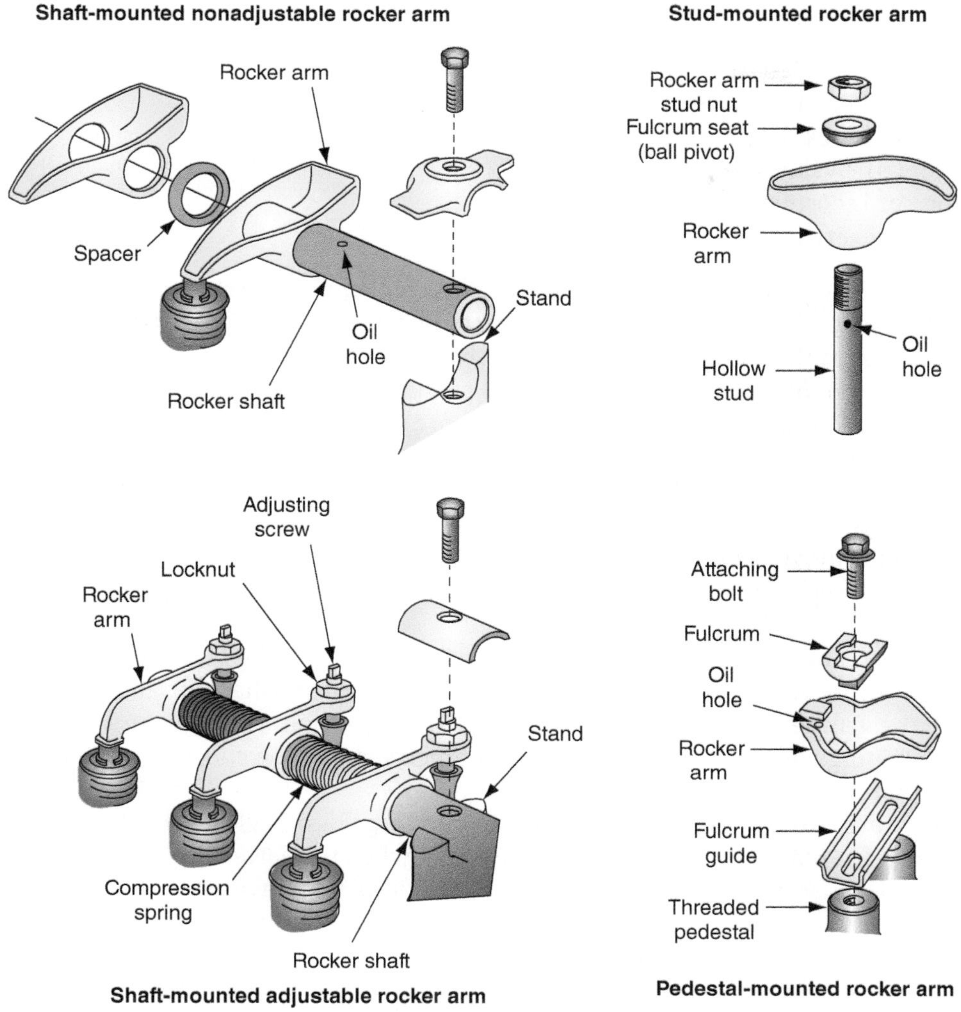

Shaft-mounted nonadjustable rocker arm

Rocker arm

Spacer

Oil hole

Rocker shaft

Stand

Stud-mounted rocker arm

Rocker arm stud nut

Fulcrum seat (ball pivot)

Rocker arm

Hollow stud

Oil hole

Adjusting screw

Locknut

Rocker arm

Compression spring

Rocker shaft

Stand

Shaft-mounted adjustable rocker arm

Attaching bolt

Fulcrum

Oil hole

Rocker arm

Fulcrum guide

Threaded pedestal

Pedestal-mounted rocker arm

Figure 10-20 Common methods of mounting the rocker arm.

Figure 10-21 This OHV engine uses a rocker arm shaft.

farther in its arc. As the tip travels a greater distance, the tip moves away from the stem center. Other causes of incorrect rocker arm geometry include the following:

- Cylinder head resurfacing
- Cylinder block deck resurfacing
- Sinking the valve seat
- Worn or incorrect pushrods
- Worn rocker arms

In engines that use pushrods, the angle of the rocker arm in relation to the pushrod is also important. Incorrect rocker arm geometry may result in the pushrod hitting the side of its passage in the cylinder head or the pushrod guide plate.

In addition, on engines that use hollow pushrods to supply oil to the upper engine, incorrect rocker arm geometry may prevent proper lubrication. These engines have a small hole in the recessed cup that accepts the pushrod tip. When the opening in the pushrod aligns with the hole in the rocker arm, oil is sprayed over the rocker arms and valve springs. The holes should align when the rocker arm is in the lash position. As the rocker arm is moved to open the valve, the holes are no longer aligned and oil flow stops. Incorrect rocker arm geometry may cause the holes not to align properly and result in too much or too little oil flow.

Any machining operations that alter the distance between the camshaft and the rocker arm may affect rocker arm geometry. Even changes to camshaft lift and duration can change the geometry. Usually the tolerances in the valvetrain will allow for some changes and machining without problems, but the geometry should be checked while rebuilding the engine.

Shop Manual
Chapter 10, page 453

The **valve spring** is a coil of specifically constructed metal used to force the valve to follow the profile of the camshaft and close fully. This provides a positive seal between the valve face and seat.

There are two basic functions of the **valve spring**: First, it closes the valve against its seat; second, it maintains tension in the valvetrain when the valve is open to prevent float. The spring must be able to perform these functions while withstanding temperature changes and vibrations. A valve spring is another vital component in the operation of the valvetrain. In today's high-speed, high-performance engines, the valve spring becomes a very significant factor in efficiency and power output. The valve spring must be able to handle the challenge of opening and closing the valves 50 times per second at 6,000 rpm. After miles of use, the spring may lose its tension or break. The reciprocating motion of the valve produces a considerable inertia effect. Worn or weak valve springs may not have sufficient tension for the lifter to follow the camshaft lobe, resulting in valve float.

Valve float causes the lifters to leave the camshaft lobe surface. In turn, this results in an abrupt seating of the valve and the lifter impacting the camshaft lobe. Valve float can result in excessive wear of the valve seat, lifter, camshaft, and valve. In addition, weak valve springs can prevent good heat transfer, resulting in burned valves and valve seat damage by bouncing the valve face against the seat. In some cases the piston can come in contact with the valve head. Weak valve springs usually have visible wear on their ends.

To accomplish its functions, the valve spring can be designed with different characteristics and features. The most common designs include dual springs, dampers, and variable-rate springs.

Dual springs are basically a set of two springs, with the smaller spring set inside the larger coil. The two springs have different diameters and lengths. The difference in sizes creates two different harmonic vibrations that cancel each other.

Some manufacturers use a damper inside the valve spring to reduce harmonic vibrations (**Figure 10-22**). The damper is a flat, wound coil designed to rub against the valve spring. The rubbing effect reduces the vibrations.

Variable-rate springs are designed with unequally spaced coils (**Figure 10-23**). This results in a spring that increases its rate of pressure as it is compressed. The closely wound coils are placed down toward the head. The more the spring is compressed, the more

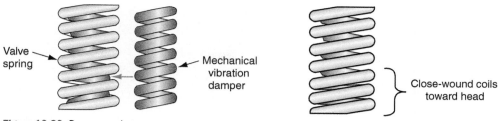

Figure 10-22 Damper spring.

Figure 10-23 A variable-rate valve spring.

pressure it exerts for the distance compressed. For example, a variable-rate spring may exert an additional 40 pounds of pressure the first tenth of an inch it is compressed; the next tenth of an inch of compression will result in an additional 50 pounds of pressure; the next tenth of an inch of compression will result in an additional 60 pounds of pressure, and so forth.

Along with different shapes and designs, different metals are used to construct valve springs. The steels used to construct a valve spring can vary from the very expensive H-11 tool steel used in Pro Stock drag racing to various grades of stainless steel. The type of steel determines the strength of the valve spring and its ability to transfer heat.

Heat is one of the major contributing factors to valve spring fatigue. Oil flowing over the valve spring is the only cooling method used. The majority of the heat generated in the valve spring is the result of friction between the coils of the spring. Some valve spring manufacturers attempt to reduce the amount of friction by coating the valve springs. Common coatings include Teflon and SDF-1. Teflon sheds the oil, while SDF-1 retains the oil by use of a wettable matrix. In tests, a coated spring runs about 20°F (11°C) cooler than an uncoated spring.

A BIT OF HISTORY

Quotes from a 1954 Buick owner's manual: 1. Every 5,000 miles the distributor points should be cleaned and spaced. The proper gap setting is 0.0125 in. to 0.0175 in. 2. The spark plugs should also be cleaned and gapped every 5,000 miles. Standard gap adjustment is 0.030 in. to 0.035 in. 3. At 5,000-mile intervals, the timing should be checked and adjusted if necessary. This should be done with a timing light or synchroscope. 4. The voltage and current regulator should be adjusted properly in order to avoid burned ignition points or burned-out light units and radio tubes. This recommended automotive service is much different from the service required on modern vehicles that do not have ignition points or timing adjustments; spark plug replacement intervals are often 100,000 miles, solid-state voltage regulators are not adjustable, and radios have transistors in place of tubes.

AUTHOR'S NOTE Many years ago, when I was a novice technician, I unbolted my stock overhead camshaft and replaced it with a "hot" performance camshaft. I adjusted the valves but left the replacement springs that came with the kit in my toolbox to install later. The camshaft combined with my new headers made a very noticeable improvement in engine power. Months later, when I was on a long and fast drive, my engine started misfiring. When I pulled over, I could hear clatter from under the valve cover. When I removed the valve cover, I saw the broken spring. It would have been easier to have installed the replacement springs in the shop than to replace the broken one in the back lot of someone else's garage in the rain.

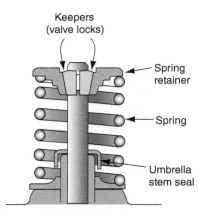

Figure 10-24 Valve keepers lock the valve spring retainer in place.

Figure 10-25 The exhaust valves on this engine are fitted with valve rotators to help clean the valve face and seat.

Valve Spring Retainers and Rotators

Valve springs are held in place by retainers locked onto the valves with keepers (**Figure 10-24**). The keepers lock onto the valve stem and are tapered so the retainer opening cannot fit over the keepers. The retainer may be a simple piece of steel to hold the spring in place. Other designs use a rotator built into the retainer. A rotator is used to turn the valve on its seat. It looks thicker than an ordinary retainer and is either ball or spring type. The ball type has small ball bearings that rotate on a ramped track in the assembly to turn the valve. The spring type has an unequal height spring to achieve the same effect. Often they are used only on the exhaust valves. It addition to helping clean the rotation, they help to more evenly dissipate heat to prevent hot spots, which cause burning and wear on the valve (**Figure 10-25**).

SUMMARY

- The camshaft is driven by the crankshaft and timed precisely to provide valve opening at the correct times.
- The valvetrain components work together to open and close the valves. The common components include the camshaft, lifters or lash adjusters, pushrods, rocker arms or followers, and valves.
- Simply stated, the camshaft is used to control valve opening, the rate of opening and closing, and how long the valve will be open.
- The lobes of the camshaft push the valves open as the camshaft is rotated. The height of the lobe determines the amount of valve lift or opening. The design of the lobe will also determine the duration of valve opening.
- Lift can be expressed in two methods. The first is cam lift, which is the measured lift of the cam lobe. The second is valve lift. This is the cam lift multiplied by the rocker arm ratio to provide a specification for the amount of valve movement.

- The camshaft lobes generally have a very slight taper, which works with offsetting the centerlines of the lifter to provide for lifter rotation.
- Lifters are mechanical (solid) or hydraulic connections between the camshaft and the valves. Lifters follow the contour of the cam lobes to lift the valve off its seat. Flat tappet lifters are designed to rotate in the lifter bore to reduce wear.
- Hydraulic lifters use oil to automatically compensate for the effects of engine temperature and valvetrain wear.
- Roller lifters create less friction than standard flat tappet lifters and use a different roller camshaft.
- OHV engines with the camshaft located in the block use pushrods to transfer motion from the lifter to the rocker arms.
- The rocker arm is a pivoting lever used to transfer the motion of the pushrod to the valve stem.

It changes the direction of the camshaft lift and spring closing forces, and it provides a leverage during valve opening.

■ OHC or DOHC engines have the camshafts mounted on the cylinder heads. The camshaft lobes may contact the rocker arms directly, and a lash adjuster may be positioned under one end of the rocker arm.

■ There are two basic functions of the valve spring: First, it closes the valve against its seat; second, it maintains tension in the valvetrain when the valve is open to prevent float.

■ To accomplish its functions, the valve spring can be designed with different characteristics and features. The most common designs include dual springs, dampers, and variable-rate springs.

REVIEW QUESTIONS

Short-Answer Essays

1. What is the function of the valvetrain?

2. Explain the purpose and function of the camshaft.

3. Describe the relationship among the camshaft lobe design and lift, duration, and overlap.

4. Explain the purpose of the lifters.

5. List and explain the common methods of mounting the rocker arms.

6. List the three types of valve springs.

7. What is meant by the term *duration* as it relates to the camshaft?

8. If the rocker arm ratio is 1.5:1 and the cam lobe lift is 0.335 in. what is the actual valve opening?

9. What is the purpose of a valve rotator?

10. Explain why some camshafts have tapered lobes.

Fill-in-the-Blanks

1. The camshaft is used to control valve _____, the _____ of opening and closing, and how long the valve will be open.

2. The _____ of the lobe determines the amount of valve lift or opening.

3. _____ is the length of time, expressed in degrees of crankshaft rotation, the valve is open.

4. The camshaft lobes generally have a very slight taper. This taper, along with the _____ of the lifter and the lobe, provides for lifter rotation.

5. Lifters are mechanical or hydraulic connections between the _____ and the valves.

6. OHV engines with the camshaft located in the block use _____ to transfer motion from the lifter to the rocker arms.

7. The design of the rocker arm is such that the side to the valve stem is usually _____ than the side to the pushrod.

8. When the valve is opened 50 percent of its travel, the rocker arm-to-stem contact should be _____ on the valve stem.

9. Valve springs can have a _____ to reduce valve spring temperatures.

10. _____ lock the valve spring retainer on top of the valve spring.

Multiple Choice

1. Each of the following is part of a typical valvetrain *except*:
   A. Lifters   C. Ports
   B. Pushrods   D. Rocker arms

2. Camshaft _____ is how long the valve stays open.
   A. Lift   C. Ramp
   B. Lobe center   D. Duration

3. The lobe separation angle defines.
   A. The height of the lobe.
   B. The lift of the lobe.
   C. The amount of overlap.
   D. The valvetrain clearance.

4. *Technician A* says that overlap improves volumetric efficiency at higher rpm.
   Technician B says that overlap reduces HC emissions.
   Who is correct?
   A. A only   C. Both A and B
   B. B only   D. Neither A nor B

5. Hydraulic lifters can _____ at higher rpm causing _____.

   A. Collapse, clatter

   B. Pump-up, valve float

   C. Leakdown, low power

   D. Lose oil pressure, reduced valve lift

6. Mechanical lifters require.

   A. a specified clearance.

   B. periodic adjustment.

   C. Both A and B

   D. Neither A nor B

7. A rocker arm is 0.45 in. (11.43 mm) from the pushrod to the pivot and 0.66 in. (16.76 mm) from the pivot to the valve tip. The camshaft lift is 0.36 in. (9.14 mm). How far will the valve open?

   A. 0.25 in. (6.35 mm)

   B. 0.81 in. (20.57 mm)

   C. 0.36 in. (9.14 mm)

   D. 0.53 in. (13.46 mm)

8. Which valve spring runs coolest?

   A. A coated spring

   B. A variable-rate spring

   C. A double-coil spring

   D. A quenched-coil spring

9. *Technician A* says that a weak spring can cause valve float.

   Technician B says that a weak valve spring can cause the valve to burn.

   Who is correct?

   A. A only   C. Both A and B

   B. B only   D. Neither A nor B

10. *Technician A* says that advanced camshaft timing will increase power at high rpm.

   Technician B says that retarded camshaft timing will increase power at low rpm.

   Who is correct?

   A. A only   C. Both A and B

   B. B only   D. Neither A nor B

CHAPTER 11
TIMING MECHANISMS

Upon completion and review of this chapter, you should understand and be able to describe:

- The function of the valve timing system.
- The operation of a timing chain system.
- The function and advantages of a timing belt system.
- The uses of a gear-driven timing system.
- The difference between an interference engine and a freewheeling engine.
- The advantages of variable valve timing systems.
- The operation of a variable valve timing and lift system.
- The basic operation of displacement-on-demand.

Terms To Know

Atkinson-cycle engine	Interference engine	Valve timing mechanism
Cam-phasing	Solenoid	Variable cylinder displacement
Cam-shifting	Timing chain guide	Variable valve timing (VVT)
Distributor	Timing chain tensioner	system

INTRODUCTION

The front end of the engine consists of the **valve timing mechanism**. This is either a belt(s), a chain(s), or gears used to time the rotation of the camshaft to the crankshaft. Some older gasoline and diesel engines used gear sets to keep valve timing; modern engines typically use either chains (**Figure 11-1**) or belts (**Figure 11-2**) to drive the camshaft(s). Perfect valve timing is essential for proper engine performance. Good volumetric efficiency of the engine depends on valves opening and closing at the correct time. The timing mechanism may also drive other auxiliary shafts, balance shafts, and components (such as the water pump). This chapter will expose you to the various design configurations commonly used. We will also discuss the purposes, advantages, and disadvantages of the traditional and variable valve timing and lift designs.

> The **valve timing mechanism** drives the camshaft in the correct timing with the crankshaft. It may use a belt, chain, or gears.

VALVE TIMING SYSTEMS

The timing mechanism is responsible for maintaining a perfect relationship between the rotation of the crankshaft and the camshaft, to open the valves at the right time. The intake valve must begin to open prior to the piston completing the exhaust stroke (**Figure 11-3**). Also, the valve must be closed at the proper time to prevent loss of compression. If the exhaust valve is not opened and closed at the precise time, the engine will not be able to breathe properly and draw in a fresh air-fuel charge.

Valve timing is a function of the camshaft, which is driven off the crankshaft. The camshaft rotates at half the speed of the crankshaft to allow the valves to open only for the desired length of time. To complete one full engine cycle, the camshaft will rotate once

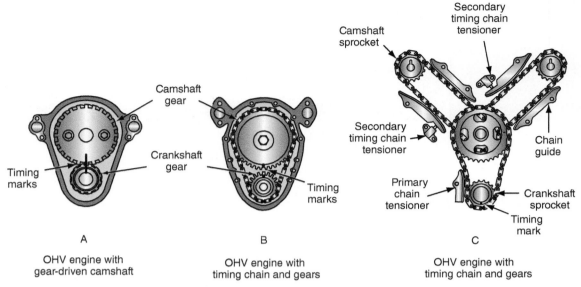

Figure 11-1 Some timing mechanisms use chains or gears.

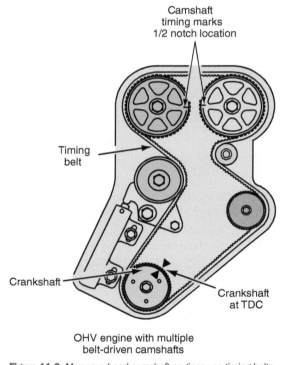

Figure 11-2 Many overhead camshaft engines use timing belts.

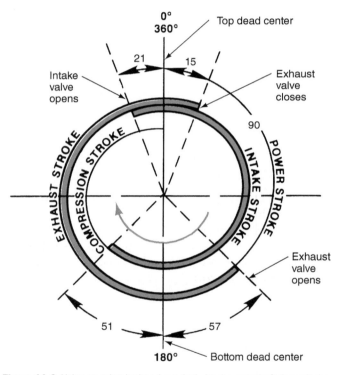

Figure 11-3 Valve opening is timed precisely by the camshaft through the valve timing mechanism.

and the crankshaft will complete two revolutions. The camshaft sprockets will be twice the size of the crankshaft sprockets unless an intermediate shaft is used. The valve timing system must provide perfect correlation between the camshaft and crankshaft. If the cam timing is off by one tooth, the engine will likely run very poorly; emissions will increase, and power and fuel economy will decrease. Variable valve timing and lift systems must also start with the correct alignment between the crank and the cam to allow proper variations on valve timing. Some of these systems sometimes use a secondary timing chain that can allow for changes in a camshaft's timing (**Figure 11-4**).

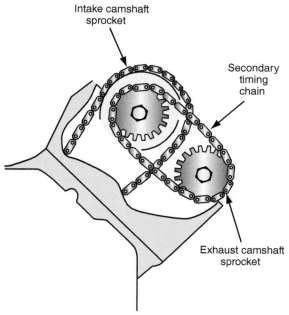

Intake camshaft
sprocket

Secondary
timing
chain

Exhaust camshaft
sprocket

Figure 11-4 A secondary chain may be used for a camshaft, particularly when the engine uses variable valve timing.

The engine may be a free-wheeling or interference type. A free-wheeling or noninterference engine means that when the timing mechanism fails, the valves will usually not hit the pistons. Even some free-wheeling engines can allow the valves to contact the pistons if the engine rpm is high enough when the mechanism breaks. On an **interference engine**, it is very likely that when the timing mechanism breaks, the valves will contact the pistons (**Figure 11-5**). At best, valves will need to be replaced. At worst, valves will float at their maximum lift and then be driven into the pistons and the whole engine will need to be overhauled or replaced. Many timing belts require replacement at regular maintenance intervals. Timing gears and chains are generally replaced during an engine overhaul or when they begin to make noise or show signs of wear. When a timing mechanism breaks, engine damage can be severe. You can help your customers by explaining the importance of proper timing system maintenance, including routine timing belt replacement.

Every valve timing system will have alignment marks on the sprockets or gears to make sure you can properly time the engine (**Figure 11-6**). Camshaft or valve timing is usually established with cylinder number 1 at TDC on the compression stroke.

In an **interference engine**, the valves hit the top of the pistons when they are open and the piston is at TDC. This can occur when the timing mechanism breaks.

Figure 11-5 Every valve that came out of this engine was bent after the timing belt snapped.

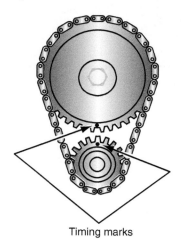

Timing marks

Figure 11-6 Timing marks on the sprockets allow proper alignment during service.

CHAIN-DRIVEN SYSTEMS

Chain-driven systems can be used either on cam-in-the-block or on overhead-cam engines. The timing chain is a strong, long-lasting way of linking the cam to the crank. A flat-link silent chain makes much less noise than a roller-style chain. The chain system is still noisy, however, which is objectionable to some consumers. It is also more expensive for the manufacturers to fit a chain rather than a belt.

When a chain is used, steel, fiber composite, or composite plastic sprockets may be used to rotate with the chain. To drive an overhead camshaft(s), one or more **timing chain guides** and a **timing chain tensioner** must be used. The guides hold the chain against a synthetic rubber, nylon, or Teflon face. The timing chain tensioner holds proper tension on the guide to take up slack as the chain and guides wear. The tensioner is usually a spring-loaded plunger that may be fed with oil pressure to take up chain slack. The tensioner also has a spring-loaded ratcheting system. When the chain or guide is worn enough, the spring will push the plunger out one more step on the ratchet. This also prevents excessive noise at start-up, before oil pressure is held behind the plunger (**Figure 11-7**). Timing chains receive splash oiling often with a slinger placed at an oil passage. Tensioners are items that wear out with time. A tensioner is usually replaced when a timing belt or chain is replaced on regular service intervals.

Engines that use balance shafts may use two chains: one will drive the camshafts and the other will drive the balance shaft (**Figure 11-8**). Note the guides and tensioners as well. This engine has dual balance shafts. You can see that they will spin at twice the speed of the crankshaft, which is typical of balance shafts. You can see this by comparing the size of the balance shaft sprockets to that of the crankshaft sprocket.

Chains are generally replaced during an engine overhaul or if they begin to make excessive noise. As they wear, you can often hear the chain slap during a quick deceleration.

A **timing chain guide** is a length of metal with a soft face to hold the chain in position as it rotates.

A **timing chain tensioner** maintains tension on the chain or belt to prevent slapping or skipping a tooth.

Shop Manual
Chapter 11, page 496

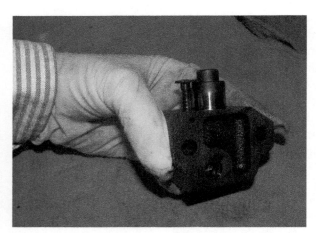

Figure 11-7 The timing chain tensioner uses spring pressure and oil pressure to maintain the correct tension on the chain.

Primary chain   Balance shaft chain

Tensioner   Guide

Figure 11-8 This DOHC engine shows the tensioner, guides, and primary and balance shaft chains.

AUTHOR'S NOTE I heard the telltale rattle of my timing chain hitting the cover during deceleration for several weeks before I took a Saturday and replaced the timing chain, guides, sprockets, and tensioner. My chain and sprockets were very worn; I was lucky they didn't cause me to walk home one day.

BELT-DRIVEN SYSTEMS

A timing belt is frequently used on overhead camshaft engines. The timing belt is strong and very quiet. It is much more cost-effective for the manufacturer than gears or chains. As belt materials have improved, belt replacement intervals have increased. This is good for the consumer because timing belt replacement can be a costly maintenance item. Some manufacturers do not even specify a replacement interval; they simply require the technician to inspect the belt periodically.

The timing belt runs from the crank sprocket to the cam sprocket(s) without a guide. Instead, an idler pulley is used to guide the belt and provide a tension adjustment. Most newer timing belt systems use an automatic tensioner that self-adjusts; others require periodic adjustment (**Figure 11-9**). These tensioners used with timing belt systems have chambers on both sides of a piston that is filled with silicone oil. They are not fed oil externally. If belt tension increases, the piston moves downward. The piston movement causes an increase in hydraulic pressure in the lower chamber. Since the check ball is closed, the oil must seep past the piston and cylinder wall to go into the reservoir chamber. As oil seeps into the reservoir, the tension piston moves slowly down the cylinder until the load from the belt balances with spring tension (**Figure 11-10**). Alternately, when the belt tension is increased, spring tension forces the piston to move up in the cylinder and take up the slack.

The timing belt may also drive other components. On older engines, it frequently drove the auxiliary shaft that turned the **distributor** (**Figure 11-11**). Other manufacturers use the timing belt to drive the water pump. When servicing or replacing

The **distributor** is usually driven by the camshaft. It uses the timing of the valves to control the timing of the spark ignition.

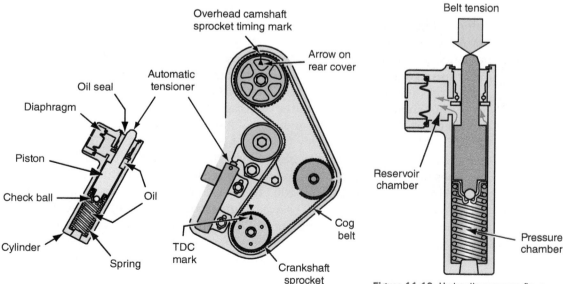

Figure 11-9 An automatic adjusting tensioner may be used to maintain proper belt tension.

Figure 11-10 Hydraulic pressure flows through the tensioner when belt tension increases.

Figure 11-11 A timing belt may also be used to drive the distributor shaft.

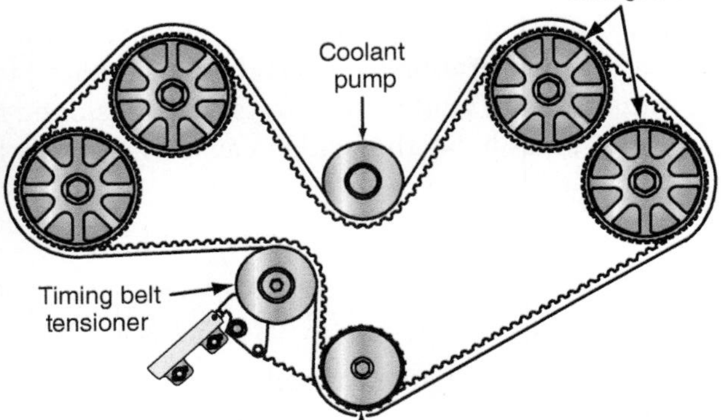

Figure 11-12 This DOHC timing belt drives all four camshafts.

the timing belt, it is recommended that the water pump be replaced at the same time. This type of setup makes water pump removal easier when the timing belt and other components are already removed. You will certainly be saving the customer a headache and money by taking care of it now. In some dual overhead camshaft (DOHC) V6 engines, a single timing belt may be used to drive the four camshafts and idler pulleys (**Figure 11-12**). In other DOHC V6 engines, a timing chain surrounds the crankshaft sprocket and intermediate shaft sprocket and the chain drives the intermediate shaft sprocket. A front cover is mounted over the chain drive, and an outer sprocket is mounted on the intermediate shaft on the outside of the front cover. This allows for replacement of the belt without disassembling the front chain cover. A timing belt surrounds the outer intermediate shaft sprocket to drive the overhead camshafts on both sides of the engine. The back side of the belt contacts two idler pulleys and a tensioner pulley (**Figure 11-13**). A timing belt may also be used on engines with a balance shaft(s). Not to be confused with a "wet-belt," a standard timing belt does not need lubrication; in fact, one reason for premature failure is oil or coolant leaking onto the belt. From 2007 to 2011, Ford of Europe used "wet-belts" in place of timing chains on some DLD series engines. The advantage to the wet-belt system is reduced timing mechanism friction, bringing about an increase in fuel economy while lowering emissions with quieter operation. However, due to cases of early failure, it is not an uncommon practice for service technicians to replace the wet-belt and its components with the older timing chain kit (cartridge) at the 125,000-mile or 10-year replacement interval.

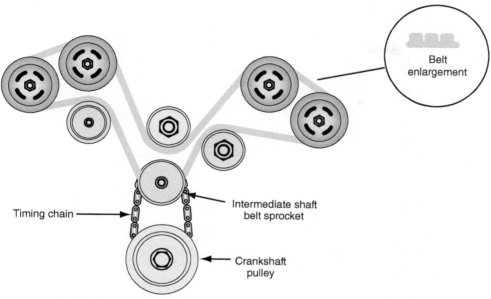

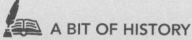

Figure 11-13 A DOHC engine that uses a timing chain to drive an intermediate shaft. A belt off the intermediate shaft sprocket drives the camshafts.

GEAR-DRIVEN SYSTEMS

Gear-driven systems are not as popular as they used to be, in part because they are used only on camshaft-in-the-block engines (**Figure 11-14**). They are a very reliable and precise system, although a little noisy. Because there is no chain or belt slack, valve timing is always very close to perfect. The gears are very durable, as long as they receive adequate lubrication. Many diesel engines use timing gears because the valve timing is very critical and it adds to the idea of the long-lasting engine (**Figure 11-15**).

A BIT OF HISTORY

Chrysler's first V-style engine was its legendary 331 c.i.d. "Hemi," introduced in 1951. It had a compression ratio of 7.5:1, with an overhead valve configuration. It put out 180 hp. Though today's new "Hemi" engine is slightly larger, at 348 c.i.d., it puts out 340 hp.

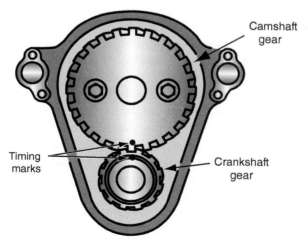

Figure 11-14 The crankshaft gear drives the camshaft directly when gears are used.

Figure 11-15 This 7.3-liter diesel engine uses timing gears.

VARIABLE VALVE TIMING SYSTEMS

In years past, manufacturers were able to design a camshaft profile that gave the engine certain horsepower and torque characteristics at specific engine speeds. One of the most significant changes to engine technology in recent years has been the ability to vary the valvetrain geometry during engine operation. The term **variable valve timing (VVT) system** is used to describe a vast array of different valvetrain changing type systems. These systems have added components that allow the intake or exhaust valves (or both) to change the lift, duration, or timing during engine operation.

> A **variable valve timing (VVT) system** refers to an engine that uses an adjustable valve train to vary the amount and timing of an air-fuel mixture entering or leaving a cylinder.

Variable Valve Timing (Cam-Phasing)

Variable valve timing systems have either mechanical, hydraulic, or electrical controls, or even a combination. Most modern engines use electrical components to control or monitor the VVT system. The timing of the valves is a critical component of the engine's operation. Changing valve timing is called **cam-phasing** (**Figure 11-16**). Of the different VVT systems, the cam phasing *exclusive* systems are the most common due to being the least expensive to manufacture and the simplest in design.

With a traditional valvetrain an engine may have a lot of power and run very smoothly at a higher speed but then run very rough and have low horsepower at idle. In an ideal situation, at low engine speeds, the engine should have a lot of torque, good efficiency, and low emissions. At high speeds, the engine should have more power.

At high engine speeds, the engine requires more air to produce more power. A cam-phasing system can retard the valve timing at higher rpms to keep the valves open later in the stroke. This is because the stroke is occurring so quickly and it still takes significant time to get the stream of air into the cylinder. Cam-phasing systems can also lower the cold-start emissions and overall emissions of a vehicle. By reducing valve overlap at idle, hydrocarbon emissions are reduced. Less of the fresh air charge is allowed out the exhaust port. By increasing the period of overlap at higher rpm, the engine can breathe better. In some systems, during cruise conditions the exhaust valves are held open longer. This allows some exhaust gases to be held in the combustion chamber to mix with the incoming intake charge. By adding some inert gases to the charge, the heat of combustion is reduced and oxides of nitrogen (NO_x) emissions

Figure 11-16 Cam phasers change valve timing.

are decreased. The exhaust gas recirculation (EGR) valves used on many vehicles perform this same function. They actually add exhaust gases into the intake charge to dilute the mixture and reduce combustion temperatures. The EGR valves and their systems in general have had many problems over the years. The function of the variable valve timing used on Ford's 2.0-liter Zetec engine has allowed Ford to eliminate the EGR system and still meet the NO_x standards.

This type of technology allows manufacturers to use smaller displacement engines. Most manufacturers are now using VVT systems in many engine designs to get better fuel economy, lower emissions, and more performance. A few examples of these designs include Audi's AVL, BMW's Valvetronic, Chrysler's VVT, Ford's VCT, GM's DCVCP, Honda's VTEC, Hyundai, Kia, and Mitsubishi's CVVT, Mazda's SVT, Mercedes Camtronic, Nissan's VCT, Porsche's VarioCam, Subaru's AVCS, and Toyota's VVT.

GM VVT

GM's VVT system improves engine performance, fuel economy, and emissions by changing valve timing only. It also eliminates the need for an EGR and AIR systems. On some applications, it allows 90 percent of the engine's torque to be available between the wide range of 1,900 rpm and 5,600 rpm. The VVT system uses a camshaft position actuator for each of the intake and exhaust camshafts that bolt onto the end of the camshafts. The actuators lock the exhaust cam in its fully advanced position and the intake cam in its fully retarded position for start-up and idle. This minimizes overlap to create a very smooth idle. Off idle, the powertrain control module (PCM) controls the camshaft actuator solenoid valves to allow the position actuators to alter intake and exhaust timing by 25 camshaft degrees. The PCM looks at information about engine coolant temperature, engine oil temperature, intake airflow, throttle position, vehicle speed, and volumetric efficiency to determine the optimum positioning of the camshafts for any given set of conditions. The solenoid valve controls oil pressure to the position actuator to effect changes in camshaft timing. The camshaft position actuator has an outer housing that is driven by a timing chain. A rotor with fixed vanes lies inside the assembly and is attached to the camshaft. The solenoid valve, as commanded by the PCM, controls the direction and pressure of oil flow against the vanes. Increasing pressure against one side of the vanes will advance the camshaft variably until the desired position is achieved. Changing the oil pressure to the opposite side of the vanes will retard the camshaft. Whenever the engine is off idle, the PCM is continuously providing electrical pulses to the solenoid valves to modify camshaft timing or to hold it in the desired position. Under light engine loads, the PCM retards the valve timing to reduce overlap and maintain stable engine operation. When the engine is under a medium load, overlap is increased to improve fuel economy and emissions. To improve mid-range torque under a heavy load and at moderate or low rpm, intake valve timing is advanced. Under a heavy load at high rpm, intake valve closing is retarded to improve engine output (**Figure 11-17**).

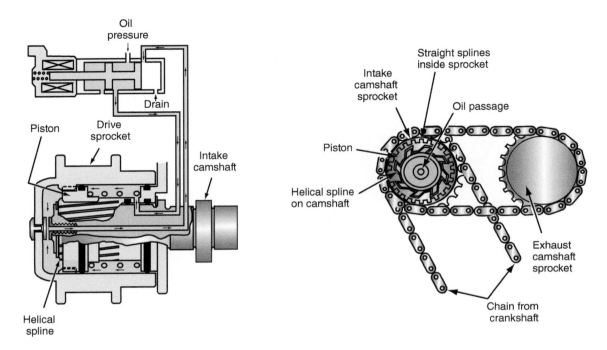

Figure 11-17 This cam-phasing actuator rotates the camshaft in relation to the crankshaft.

VARIABLE VALVE LIFT (Cam-Shifting)

Systems that can change valve lift are known as **cam-shifting** (sometimes called cam-switching) systems. Changes in valve lift can be used to increase power at high rpm, to vary engine displacement, or to control engine speed, thus allowing the removal of the throttle plate for increased intake airflow. Manufacturers accomplish this in different ways. One way is by using camshafts with additional lobes of different height that can switch back and forth during engine rpm and load changes. Changes in valve lift can also be achieved by changing the range of rocker arm or pushrod movement. This is done using hydraulic pressure or by an electrically operated actuator.

VARIABLE VALVE TIMING AND LIFT SYSTEMS

Anxious to reduce the limitations caused by compromises in camshaft design, many manufacturers are combining variable valve timing and variable valve lift systems. Used together, these technologies reduce inherent compromises. The systems may be designed to improve fuel economy and performance while reducing emissions.

For example, increasing valve lift and retarding valve timing can increase the output of an engine significantly at higher rpm when volumetric efficiency normally decreases. This is when the engine torque traditionally falls off due to the inability to fill the chambers. There are many variations in system design. More and more manufacturers are adding variable valve timing and/or valve lift systems to their vehicles every year. Honda has been using a variable timing and lift system since 1989.

Honda VTEC

Honda's popular variable timing and lift electronic control (VTEC) system modifies both valve timing and valve lift. Honda took the lead in production variable valve timing with its VTEC system, designed to improve performance at higher rpm. It increased the engine output over a single cam non-VTEC engine by 25 percent. In the VTEC system, each pair

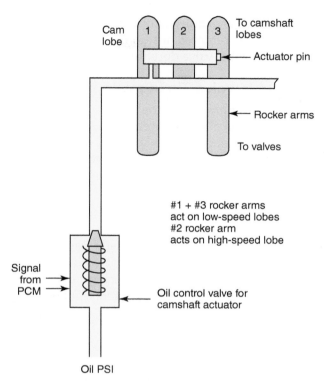

Figure 11-18 When a predetermined rpm and other conditions are met, the PCM signals the cam actuating control valve. This valve allows oil pressure to push a pin through the rocker arms, activating the high-speed rocker arm and cam lobe.

of valves has three cam lobes and three rocker arms. During low-speed operation, the low-speed rockers push directly on the two regular-speed lobes. At high rpm, when the PCM determines that the third lobe will improve performance, it signals an oil control valve to engage the third rocker to the high-speed lobe. A pin slides out of the low-speed rockers and into the high-speed rocker to engage it to use the camshaft lobe with more aggressive timing, lift, and duration. On one Acura model, the high-speed lobe has an advertised duration of 290 degrees; that is a race-quality camshaft. **Figure 11-18** shows a simplified diagram of the pin actuating system. Honda's i-VTEC was introduced in 2001 and uses a more powerful PCM to control the camshaft. This system uses an adjustable oil-driven cam gear (also called a cam-phasing actuator) that actually closes the intake valve well into the compression stroke when the PCM determines that it should be in "economy mode." This action is similar to what happens in the **Atkinson-cycle engine**. Later, Honda introduced the Advanced VTEC, which uses an even faster PCM. The intake valves have less lift at low engine speeds, which gives the engine an estimated 13 percent increase in fuel economy and a reduction in emissions.

Toyota VVT

Toyota is using a VVT system that changes the phase angle of the intake camshaft relative to the crankshaft to advance or retard the camshaft. A hydraulic actuator (called a variator), controlled by the PCM sits on the end of the camshaft to effect timing changes. The camshaft timing can be varied continuously up to 60 degrees. The system also employs a variable lift and duration system by using two different camshaft lobes for each set of valves. One rocker arm opens both intake valves. During low- to moderate-speed operation, the "high-speed" cam lobes do not contact the rocker arms. At higher speeds, a sliding pin is pushed by hydraulic pressure to take up the space between the higher-duration cam lobes. The increased rocker arm movement also increases valve lift. When the system is off, the variator defaults to the retarded position. The VVT-I system operates similarly to VVT but adds greater control of valve duration through camshaft rotation.

Dual VVT-i (Dual Variable Valve Timing with Intelligence)

Dual VVT-I is the most commonly used variable valve timing and lift system used by Toyota today. The Dual VVT-I system operates similarly to VVT-i but includes control of the exhaust camshaft. The advantages to this are lower emissions through faster catalytic converter warm-up and increasing internal exhaust gas recirculation (internal EGR), better startability, increased fuel economy and performance, and smoother-running at lower rpms.

Fiat's MultiAir

Fiat uses hydraulic oil pressure controlled by a normally closed, electrically operated solenoid (**Figure 11-19**). The solenoid is located in between the camshaft and the intake valves. When the solenoid is de-energized, the oil inside it causes it to act as a solid lifter. As the solenoid is energized, it prevents the intake valve from opening due to the oil being bled off. When desired, the MultiAir system has the ability to open the valves only part way.

Variable Cylinder Displacement

Variable cylinder displacement systems allow the engine under certain conditions to operate on fewer cylinders than it is equipped with, mainly to increase fuel economy.

Many manufacturers are now making engines that, when conditions permit, can "cancel" multiple cylinders to provide better fuel economy. These **variable cylinder displacement** systems are known as displacement-on-demand, variable displacement, variable cylinder management, or active fuel management engines. These systems are currently in some GM Active Fuel Management System (AFM), known previously as displacement-on-demand (DOD), Honda Variable Cylinder Management System (VCM), and Chrysler Multi Displacement System (MDS) vehicles.

GM's Active Fuel Management System

The AFM system is designed to improve fuel economy. This type of system is not new technology and has been tried by many manufacturers; it is a new configuration of an older system that did not work well. The 181 Cadillac 8-6-4 used a similar system to deactivate cylinders for fuel economy. Unfortunately, the car never succeeded with consumers because the technology was too far ahead of its time. Semiconductors in vehicles were still in their infant stage, and there were many consumer complaints of engines stalling and

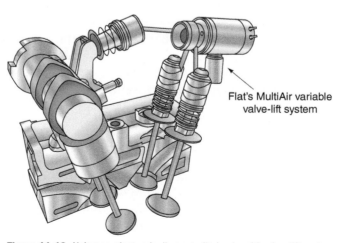

Fiat's MultiAir variable valve-lift system

Figure 11-19 Using an electronically controlled solenoid valve, lift and duration can be controlled.

rough running when the computer automatically turned cylinders off. Modern computer speed and complexity make the new versions of displacement-on-demand barely noticeable to the consumer. The system allows for maximum power when the consumer demands acceleration and a "smaller" engine when the driver is cruising. The engines use special hydraulic valve lifters to stop valve opening to selected cylinders during cruising and deceleration. The PCM activates a solenoid that allows oil pressure to unlock a pin in the valve lifter. Then rather than opening the valve, it collapses as the camshaft tries to push it up. The PCM also stops delivering fuel to the four cylinders during cylinder deactivation. Drivers say the transition is seamless; they do not even feel when the engine changes from four to eight or back again. Manufacturers claim up to 25 percent fuel economy increases. One V8 engine tested achieved 30 miles per gallon during highway driving.

The lifter used in an engine with an active fuel management system (also called DOD) is slightly different (**Figure 11-20**). When the vehicle is cruising down the highway, it does not require a lot of horsepower. A larger V8 engine can be reduced to generating power from only four of the cylinders; this is called cylinder deactivation. There are many examples from many manufacturers on how variable displacement works, but many of them use a method of controlling the oil flow in and out of the lifter. This, in turn, controls the action of the lifter and the opening of the valves. This is part of a strategy that the computer uses to deactivate a cylinder.

Cylinder deactivation components that are common on a V8 engine include a valve lifter oil manifold assembly that is mounted just below the intake manifold and above the cylinder block, oil control solenoids mounted in the manifold assembly, and an oil pressure relief valve that is located in the oil pan. When requested by the engine computer, one or many of the solenoids will be energized.

When a solenoid is energized it opens a pathway, thus allowing pressurized oil to flow to the lifter bores through a separate vertical passage. There are two passages per cylinder: one for intake and one for exhaust valves. A bleed passage leads excess oil pressure straight to the oil sump (**Figure 11-21**).

The lifter of a cylinder deactivation engine has a small pin inside the center. The lifter has many small parts, but the lifter itself is separated into two main areas: the outside of the lifter and the pin housing (inner part of the lifter where the pushrod sits). When the

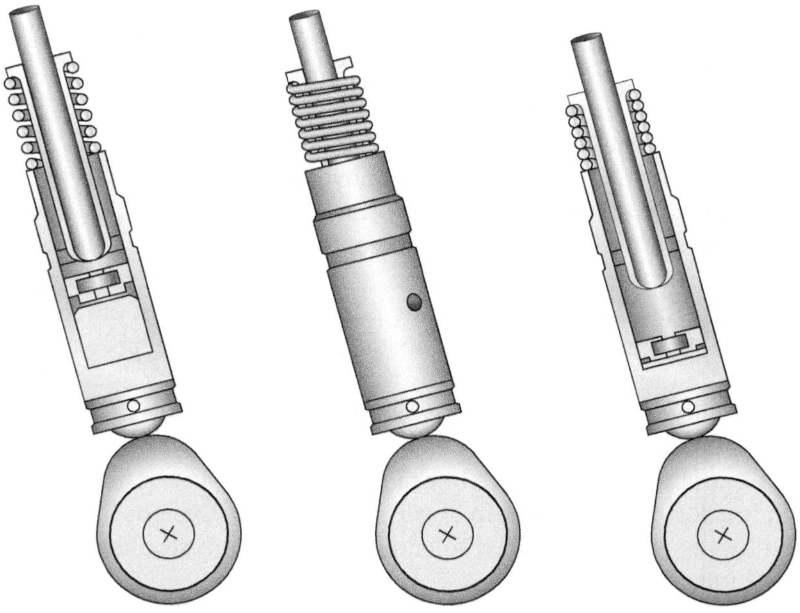

> A **solenoid** is an electromechanical device. It is typically operated by the PCM through an electrical signal and mechanically controls flow of air, oil, fuel, or other substances. A fuel injector is a type of solenoid.

Figure 11-20 These are the parts of a variable displacement (active fuel management) valve lifter.

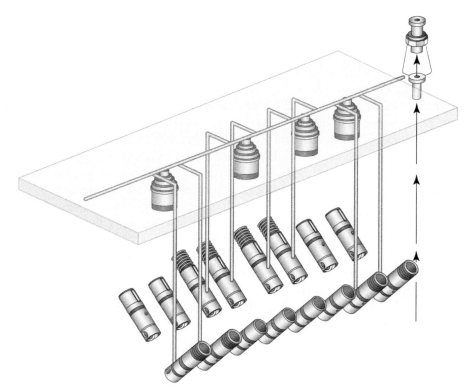

Figure 11-21 This is the variable displacement (active fuel management) valve lifter system.

solenoid is activated, pressurized oil pushes a lock pin in the pin housing. This locks the lifter in the valve closed position. The outside of the lifter body continues to move with the camshaft. The pin housing and the lifter body are now independent of each other, in terms of movement. This allows the pressure from the valve spring to force the pushrod down into the bore of the lifter and follow the contour of the camshaft, thus keeping the valve closed. This is similar to what happens when a valve lifter in a normal engine completely collapses and cannot hold its own pressure.

When cylinder deactivation is not needed, the solenoid is electronically deactivated by the computer, and oil flow to the lifter is no longer pressurized. This causes the pin to release and lock to the outer housing, because of a spring, and the inner pin housing and the outer part of the lifter now move as one component. This allows the pushrod to move and thus the valve to open and close with the movement of the camshaft.

When working with any engine, it is important to refer to technical service bulletins (TSBs) when there is a problem. One common complaint and TSB related to these types of lifters is related to the engine oil. If the engine oil is not maintained properly or the incorrect viscosity of oil is used problems with the lifter pin locking and unlocking can result, causing the check engine light and a diagnostic trouble code to appear;. There is also a small oil filter inside the manifold assembly that requires periodic maintenance. Having dirty oil, aerated oil, low oil level, incorrect oil viscosity, or worn lifters can sometimes cause lifter noise upon start-up.

The number and types of systems in use increases each year; the idea has proven to be effective in improving power and in lowering emissions. Designs in the works include valves opened by electrically operated solenoids. This system would offer infinitely variable valve lift, duration, and timing.

SUMMARY

- The front end of the engine consists of the valve timing mechanism.
- Proper valve timing is essential for proper engine performance.
- Valve timing systems may use a belt, chain, or gears to link the camshaft to the crankshaft.
- Marks are provided on the sprockets, gears, covers, block, or head to properly time the engine, usually at TDC firing cylinder number 1.
- A timing chain is a long-lasting drive mechanism that can be used on cam-in-the-block or overhead-cam engines.
- Timing chains are not generally replaced as a regular maintenance service.
- Dual timing chains may be used to drive camshafts individually on variable valve timing engines or on engines with balance shafts.

- Timing belts are a very popular and cost-effective way to drive overhead camshafts.
- Timing belts usually require periodic replacement as a maintenance service.
- Timing gears are sometimes used in heavy-duty diesel engines.
- Timing gears are very durable and provide precise control of timing.
- Manufacturers are increasingly using variable valve timing and/or lift systems to overcome the limitations of the compromise of one camshaft for all engine speeds and loads.
- Many manufacturers are now making variable cylinder displacement engines that can "cancel" multiple cylinders to provide better fuel economy.

REVIEW QUESTIONS

Short-Answer Essays

1. Explain the purpose of precision valve timing.

2. List the types of timing mechanisms.

3. Why do modern passenger car engine designs tend to use timing belts or chains rather than gears?

4. Describe the difference between a free-wheeling and an interference engine.

5. Explain the maintenance requirements of the different styles of timing mechanisms.

6. Describe the operation of a timing chain tensioner.

7. Explain the advantages of variable valve timing.

8. Briefly explain the operation of Honda's VTEC system.

9. Describe how a variable displacement (or displacement-on-demand) lifter works to deactivate a cylinder.

10. Briefly explain the operation of Fiat's MultiAir system.

Fill-in-the-Blanks

1. Valve timing is a function of the _____.

2. The camshaft rotates _____ as fast as the crankshaft.

3. Proper valve timing is established with cylinder number 1 at _____ of the _____ stroke.

4. Timing _____ mechanisms may be used on OHV or OHC engines.

5. The timing chain _____ holds the chain in proper alignment.

6. The _____ takes up slack on a timing chain.

7. Balance shafts spin at _____ the speed of the crankshaft.

8. A worn timing chain will make the most noise during engine start-up and _____.

9. By reducing the amount of overlap at idle, a VVT system can reduce _____ emissions.

10. Some VVT systems eliminate the need for the _____ emissions control system.

Multiple Choice

1. Valve timing affects each of the following *except*:
   A. Engine torque   C. Emissions
   B. Spark timing   D. Fuel economy

2. Each of the following is a typical type of valve timing mechanism *except*:
   A. Chain
   B. Belt
   C. Gear
   D. Band

3. *Technician A* says that timing belts are quieter than timing chains.
   Technician B says that timing chains generally last longer than timing belts.
   Who is correct?
   A. A only
   B. B only
   C. Both A and B
   D. Neither A nor B

4. An automatic timing belt tensioner:
   A. receives oil under pressure to tension the belt.
   B. has triple-coil springs to ensure proper tension.
   C. should be adjusted every 60,000 miles.
   D. is filled with silicone oil.

5. Which engines use an auxiliary shaft chain and a timing belt driven off of the auxiliary shaft sprocket?
   A. Eight-cylinder OHV engine
   B. Four-cylinder DOHC engine
   C. V6 DOHC engine
   D. Diesel engines

6. The quietest timing mechanism is the.
   A. belt.
   B. chain.
   C. band.
   D. gears.

7. A VVT system may be able to eliminate the system:
   A. Spark timing
   B. Positive crankcase ventilation
   C. Exhaust gas recirculation
   D. Valvetrain

8. The Honda VTEC system:
   A. modifies valve lift.
   B. changes valve timing.
   C. changes valve duration.
   D. All of the above.

9. Reducing valve overlap at idle reduces:
   A. hydrocarbon emissions.
   B. oxides of nitrogen emissions.
   C. carbon monoxide emissions.
   D. All of the above.

10. *Technician A* says that a VVT system may increase the overlap at higher rpm.
    Technician B says that a VVT system may retard the intake timing under heavy-load, high-rpm conditions to improve power.
    Who is correct?
    A. A only
    B. B only
    C. Both A and B
    D. Neither A nor B

CHAPTER 12

ENGINE BLOCK CONSTRUCTION

Upon completion and review of this chapter, you should understand and be able to describe:

- The purpose of the cylinder block.
- The construction and components of the cylinder block.
- The purpose and machining processes of the cylinder bore.
- The purpose and machining process of the main bore.
- Short block, long block, and crate engine and their uses.
- The components of the crankshaft.
- The relationship of crankshaft throw to stroke and firing impulses.
- The forces applied to the crankshaft.
- The manufacturing processes and materials used in crankshaft design.
- The purpose of the harmonic balancer.
- The purpose of the flywheel.

Terms To Know

Bearing inserts
Bedplate
Connecting rod–bearing journals
Core plugs
Crankshaft counterweights
Crankshaft throw

Cylinder block
Harmonic balancer
Honing
Long block
Main-bearing journals
Main bore

Main caps
Oil galleries
Short block
Splayed crankshaft
Torque converter
Torsional vibration

INTRODUCTION

The cylinder block and the components within it make up the bottom end of the engine. This is also called a **short block**. A **long block** is essentially a short block with cylinder head installed (**Figure 12-1**). Most crate engines can be ordered as short or long blocks. Engine materials and machining have changed significantly in the past 20 years. Clearances have decreased, ring and piston choices abound, and block reconditioning is often limited. A silicone-impregnated, cast aluminum alloy block with integral bores cannot be refinished, for example. The function and purpose of the flywheel and harmonic balancer are also covered in this chapter because these components affect the function of the block assembly. You must have access to thorough service information to properly service a modern engine (**Figure 12-2**). The service information will provide you with what you need to know about the materials used for the block, machining processes allowed for repairs, and proper repair procedures.

A **short block** is a block assembly without the cylinder heads and some other components.

A **long block** is an engine block with the cylinder head(s) and other components mounted.

Figure 12-1 This is a short block with a cutaway cylinder head unit.

AUTHOR'S NOTE You may be able to improve the performance and durability of an older engine by using newly designed replacement parts with improved engineering, materials, and machining.

GENERAL MOTORS ENGINES
3.0L, 3.8L, and 3.8L "3800" V6 (Continued)

Engine Specifications (Continued)

Crankshaft Main and Connecting Rod Bearings

Engine	Main Bearings				Connecting Rod Bearings		
	Journal Diam. In. (mm)	Clearance In. (mm)	Thrust Bearing	Crankshaft End Play In. (mm)	Journal Diam. In. (mm)	Clearance In. (mm)	Side Play In. (mm)
All Models	[1] 2.4988–2.4998 (63.469–63.494)	.0003–.0018 (.008–.005)	2	.003–.009 (.08–.23)	[1] 2.2457–2.2499 (57.117–57.147)	.0003–.0028 (.008–.071)	.003–.015 (.076–.38)

[1] Maximum taper is .0003 in. (.008 mm).

Pistons, Pins, and Rings

Engine	Pistons	Pins		Rings		
	Clearance In. (mm)	Piston Fit In. (mm)	Rod Fit In. (mm)	Ring No.	End Gap In. (mm)	Side Clearance In. (mm)
All Models	[1] .0013–.0035 (.033–.089)	.0004–.0007 (.010–.018)	.00075–.00125 (.019–.032)	1	.010–.020 (.254–.508)	.003–.005 (.08–.13)
				2	.010–.020 (.254–.508)	.003–.005 (.08–.13)
				3	.015–.035 (.381–.889)	.0035 (.09)

[1] Measured at bottom of piston skirt. Clearance for 3.8L turbo is .001–.003 in. (.03–.08 mm).

Figure 12-2 This service information includes critical measurement and torque specifications. You cannot service an engine properly without an engine's specific information.

Valve Springs

Engine	Free Length In. (mm)	Pressure Lbs. @ In. (Kg @ mm)	
		Valve Closed	Valve Open
3.8L (VIN C)	2.03 51.6	100–110@1.73 (45–49@44)	214–136@1.30 (97–61@33)
3.0L & 3.8L (VIN 3)	2.03 51.6	85–95@1.73 (39–42@44)	175–195@1.34 (79.1–88.2@34.04)
3.8L (VIN 7)	2.03 (51.6)	74–82@173 (33–37@44)	175–195@1.34 (79.1–88.2@34.04)

Camshaft

Engine	Journal Diam. In. (mm)	Clearance In. (mm)	Lobe Lift In. (mm)
3.0L	1.785–1.786 (45.34–45.36)	.0005–.0025 (.013–.064)	Int. .210 (5.334) Exh. .240 (6.096)
3.8L (VIN C)	1.785–1.786 (45.34–45.36)	.0005–.0025 (.013–.064)	[1].272 (6.909)
3.8L (VIN 3)	1.785–1.786 (45.34–45.36)	.0005–.0025 (.013–.064)	[1].245 (6.223)
3.8L (VIN 7)	1.785–1.786 (45.34–45.36)	.0005–.0025 (.013–.064)	. . . .

[1] Specification applies to both intake and exhaust.

Caution: Following specifications apply only to 3.0L (VIN L) 3.8L (VIN 3), 3.8L (VIN 7) and 3.8 "3800" (VIN C) engines.

TIGHTENING SPECIFICATIONS

Application	Ft. Lbs. (N.m)
Camshaft Sprocket Bolts	20 (27)
Balance Shaft Retainer Bolts	27 (37)
Balance Shaft Gear Bolt	45 (61)
Connecting Rod Bolts	45 (61)
Cylinder Head Bolts	[1]60 (81)
Exhaust Manifold Bolts	37 (50)
Flywheel-to-Crankshaft Bolts	60 (81)
Front Engine Cover Bolts	22 (30)
Harmonic Balancer Bolt	219 (298)
Intake Manifold Bolts	2
Main Bearing Cap Bolts	100 (136)
Oil Pan Bolts	14 (19)
Outlet Exhaust Elbow-to-Turbo Housing	13 (17)
Pulley-to-Harmonic Balancer Bolts	20 (27)
Outlet Exhaust Right Side Exhaust Manifold-to-Turbo Housing	20 (27)
Rocker Arm Pedestal Bolts	37 (51)
Timing Chain Damper Bolt	14 (19)
Water Pump Bolts	13 (18)

[1]Maximum torque is given. Follow specified procedure and sequence.
[2]Tighten bolts to 80 in. lbs. (9 N.m).

Figure 12-2 Continued

BLOCK CONSTRUCTION

The **cylinder block** is the foundation of the engine. It houses all the rotating and reciprocating parts of the bottom end and provides lubrication and cooling passages (**Figure 12-3**). The engine block is cast; hot, liquid metal is poured into a mold and allowed to cool and solidify. This forms the structure of the block. Sand is used inside the mold to provide water jackets around the bores. Holes are drilled in the outside of the block to remove the sand after casting. These holes are then sealed with core plugs.

Blocks have traditionally been made of gray cast iron because of its strength and resistance to warpage. After it is cast, it is easily machined to provide rigid bores for the pistons, crankshaft, and lifters (and camshaft, if applicable). The block is also machined on the top, bottom, and ends to provide good sealing surfaces for the head, oil pan, transaxle, and timing cover (**Figure 12-4**).

Many newer designs use a cast aluminum alloy for the block. This offers a significant weight reduction that helps manufacturers meet fuel economy standards without sacrificing performance. Cast aluminum alloy blocks are more prone to cracking and warping problems when they are overheated, however. The cylinder wall may have replaceable or cast-in-place steel liners. When they are cast in place, they are called dry liners because no coolant hits the surface directly. Replaceable liners are called wet sleeves; coolant circulates all around the sleeve. The sleeve is sealed by an O-ring on the bottom and the head gasket and head on the top. Some aluminum blocks are fitted with cast-iron cylinder heads and/or thickly cast oil pans to provide added rigidity to the block. Other newer engines are all aluminum.

In some cases, a silicone-impregnated aluminum alloy is used in the casting. After the bores are machined, a special honing process removes the aluminum and uncovers a layer of hard silicone to form the thin wear wall of the bore. These cylinders cannot be bored oversize unless a sleeve is installed.

The **cylinder block** is the main structure of the engine. Most of the other engine components are attached to the block.

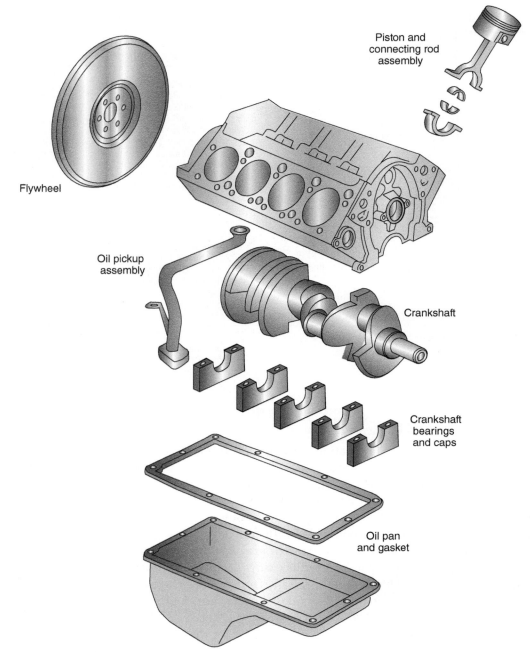

Piston and
connecting rod
assembly

Flywheel

Oil pickup
assembly

Crankshaft

Crankshaft
bearings
and caps

Oil pan
and gasket

Figure 12-3 Cylinder block and related components.

Other aluminum blocks use a special lost-foam casting process. The foam is vaporized in the casting process. The casting molds are made in a foam pattern that allows more complex shapes. These aluminum blocks (and heads) almost look as though they are made of silver Styrofoam (**Figure 12-5**). The manufacturers claim increased reliability and accuracy with lower machining costs.

Some manufacturers are using a two-piece block design. This design has a **bedplate** or a one-piece main-bearing cap (**Figure 12-6**). It provides a stronger lower end because it ties all of the main caps together to improve block stiffness and reduce engine vibration. This design is often called modular. "Modular" actually refers to the ability of the production plant to easily build the same design engine in different configurations (**Figure 12-7**).

The **bedplate** is the lower part of a two-piece engine block. The bedplate is also called the main girdle.

Figure 12-4 A bare V8 block.

Figure 12-5 This Saturn engine uses the lost-foam casting process.

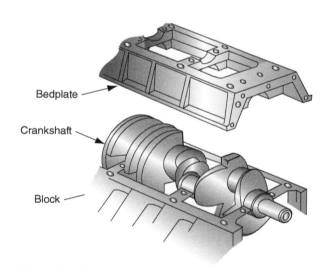

Figure 12-6 A cylinder block with a bedplate.

Lubrication and Cooling

A cylinder block contains a series of oil passages, called **oil galleries**, that allow oil to be pumped through the block and crankshaft to reach the main bearings, and to the camshaft and up into the cylinder head. The oil lubricates, cools, seals, and cleans engine components (**Figure 12-8**).

Some of the heat generated by the engine must be absorbed by the block and cylinder heads. This heat is wasted power and must be removed before it damages the engine. The engine has water jackets throughout the engine to help cool the engine. The area around the cylinder bores has coolant circulating and coolant flowing into the cylinder head to cool the area around the valve seats. Coolant circulates through the block jackets and coolant passages to transfer heat away from these hot areas into the rest of the cooling system.

Oil galleries are drilled passages throughout the engine that carry oil to key areas that need lubrication.

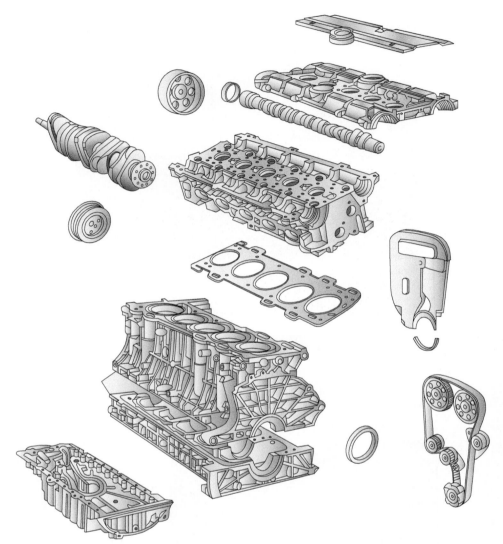

Figure 12-7 A two-piece block and head.

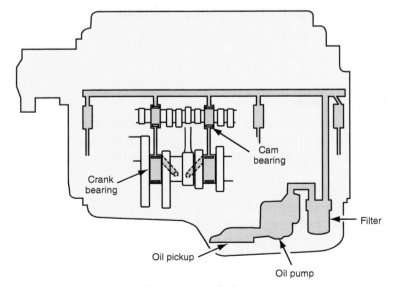

Figure 12-8 Cylinder block oil system and galleries.

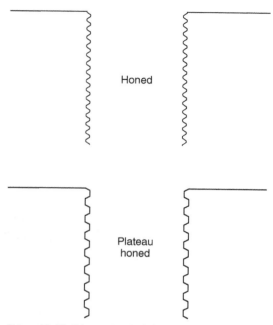

Honed

Plateau
honed

Figure 12-10 This exaggerated view shows the rough edges of honing, which are then cut down by plateau honing. This process holds oil for the rings but causes less wear to the rings than regular honing.

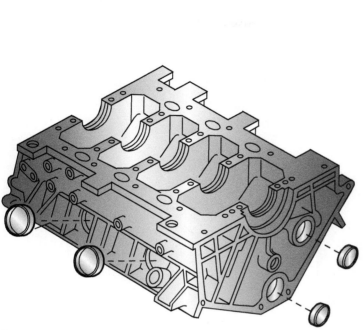

Figure 12-9 Cylinder block core and gallery plugs.

Core Plugs

All cast cylinder blocks use **core plugs**. These may also be called expansion plugs or freeze plugs. During the manufacturing process, sand cores are used. These cores are partly broken and dissolved when the hot metal is poured into the mold; however, holes must be placed in the block to get the sand out after the block is cast. The core holes are machined, and core plugs are placed into them to seal them (**Figure 12-9**).

Cylinder Bores

The cylinder bores may be either integral or sleeved. Integral bores are machined out of the casting to fit the pistons. They are bored first and then honed to size to provide the proper surface finish for the rings. **Honing** a cylinder uses special stones to make small scratches on the surface. This removes the glaze on the surface. It also allows oil to sit in the small scratch lines and help seal the piston rings. Bores are plateau honed from the factory to provide good cylinder wall lubrication with minimal wear. A rough stone hone is used to provide a crosshatch pattern in the cylinder. The crosshatch scratches of 50 to 60 degrees on the cylinder wall hold oil to lubricate the pistons and rings. The cylinder is then finish honed with a soft stone to cut the rough edges off of the grooves in the cylinder wall. The result is a smooth plateau for the rings and pistons to contact (**Figure 12-10**).

Clearances between the pistons and cylinder bores are commonly very small on modern engines, as little as 0.0005 in. the machining processes are precise. Many aluminum blocks will use cast-iron sleeves either cast or dropped into the block. The block is supported at the top and bottom during the machining processes to prevent twisting. The lower portion of the block has webbing to help keep the cylinders straight during engine operation. The cylinder head helps secure the top half of the block. The top of the block must be very smooth so that the cylinder head can seal it. The base or bottom of the block is also machined to allow for proper sealing of the oil pan.

Core plugs are used to seal the holes machined into the block to remove the sand after casting. Core plugs are often called freeze plugs. Occasionally, when coolant is not adequately protected with antifreeze, it freezes in the block; a core plug will pop out and prevent the block from cracking. People are not always that lucky, however, and that is not the purpose of core plugs.

Honing a cylinder is a machining process that prepares the surface for the installation of new piston rings.

Main caps with steel inserts

Girdle assembly

Figure 12-11 This all-aluminum, two-piece block engine uses a main girdle; all the main caps are formed in one casting and bored out to fit the crankshaft.

Figure 12-12 Bearings are fit into the main bores to protect the crankshaft. This is an example of a two-bolt main block.

Main Bore

The **main bore** is a straight hole drilled through the block to support the crankshaft.

The **main caps** form the other half of the main bore and bolt onto the block to form round holes for the crankshaft bearings and journals.

The **main bore** holds the crankshaft. The bore is supported by the webbing in the lower part of the block casting. Some blocks use a cast-iron girdle that bolts onto the **main caps** to increase rigidity of the bore (**Figure 12-11**). Main-bearing caps are cast separately and then machined to bolt on the bottom of the block. The main bore is then line- or align-bored to form the round hole between the bottom of the block and the main caps. The main caps are bolted in place, while a long boring bar passes through the end of the block. Because the boring bar will never turn exactly true, particularly toward the end of its reach, the bores will only be round if the same cap is installed in the same direction for each bore. The main bores must be in perfect alignment with each other in a straight line. If the main bores are misaligned, rapid and uneven main bearing wear will occur.

The main caps are held in place by two, four, or six bolts. A two-bolt main cap is the most common (**Figure 12-12**), but many modern lightweight and performance engines use a four-bolt main cap. The extra bolts may be next to each other (parallel), or they may be bolted in from the sides (cross bolted). This design holds the crankshaft better and allows for less flexing when under higher stresses (**Figure 12-13**). Some engines use a six-bolt main cap.

Bearing inserts are soft-faced bearings used to protect the crankshaft main and rod journals as well as the camshaft journals.

Main **bearing inserts** snap into the halves of the main caps. They provide a soft surface to support, lubricate, and protect the crankshaft. They also prevent the main bore from wear; the crankshaft does not contact the main bores (**Figure 12-12**).

CRANKSHAFT

The crankshaft withstands tons of force when the piston pushes the connecting rod down on the power stroke. It converts that reciprocating motion of the piston into useful rotary motion. Crankshafts may be made of cast iron, nodular iron, or forged steel. Cast-iron cranks are quite

Figure 12-13 This is a four-bolt main bearing cap.

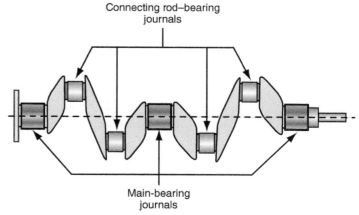

Figure 12-14 The rod-bearing journals are offset from the centerline of the crank. The length of the offset times two is the engine's stroke.

common on production vehicles. Forged crankshafts are stronger and are often found on high-performance vehicles, engines with forced induction systems, and diesel engines.

The crankshaft is bolted into the main bore on its **main-bearing journals**. These are machined on the centerline of the crankshaft and support the weight and forces applied to the crankshaft. Crankshaft journals are machined and polished smooth to minimize bearing wear and friction. The journals actually ride on a thin film of oil between the bearing and the journal.

The **connecting rod–bearing journals** or pins are offset from the centerline (**Figure 12-14**). The connecting rod clamps around these journals with a bearing insert in between to protect the crankshaft and the connecting rod. The offset places weight and pressure off the center of the crankshaft and provides leverage for the piston assembly to turn the crankshaft.

The distance the connecting rod journals are offset from the centerline of the crankshaft is called the **crankshaft throw**. The throw determines the stroke of the engine. Rod journal offset times two equals the stroke. The throws are set so that pistons in opposite banks move in opposite directions to minimize engine vibration. This design cancels some of the forces applied to the crankshaft.

Crankshafts have many different configurations, depending on the engine. They will usually be ground so that the cylinders fire in equal intervals during one full cycle. A four-cylinder engine produces a power stroke every 180 degrees; its rod journals are 180 degrees off of each other. A typical four-cylinder engine with a firing order of 1–3–4–2 will fire cylinder number 1 at 180 degrees of crank rotation, number 3 at 360 degrees, number 4 at 540 degrees, and complete the full cycle at 720 degrees with the firing of cylinder number 2. This puts cylinders number 1 and number 4 and cylinders number 3 and number 2 positioned at the same point in the cylinder during rotation (**Figure 12-15**). The connecting rod journals from an end view form a straight line above and below the main journals.

Just like the four-cylinder engine, the flat-plane V8 crankshaft (**Figure 12-16**) produces a power stroke every 180 degrees, making it an even fire engine since there are no cylinders firing in between the pistons being at TDC (top dead center) and BDT (bottom dead center). Flat-plane crankshafts have less mass, making them lighter and allowing them to rev faster and higher. They have more torsional vibration and can be fired alternating banks for smoother running.

A V6 engine typically produces a power stroke every 120 degrees of crank rotation. This is called an even-firing V6. The journals on the crankshaft are **splayed**. In this engine, on one throw, two journals are offset and separated by 30 degrees in this case (**Figure 12-17**). Other V6 engines use one throw with two equal journal surfaces. This means that two rods

Main-bearing journals are machined along the centerline of the crankshaft and support the weight and forces applied to the crankshaft.

Connecting rod–bearing journals are offset from the crankshaft centerline to provide leverage for the piston to turn the crankshaft.

Crankshaft throw is the measured distance between the centerline of the rod journal and the centerline of the crankshaft main journal. Crankshaft throw multiplied by two equals the piston stroke.

A **splayed crankshaft** is a crankshaft in which the connecting rod journals of two adjacent cylinders are offset on the same throw.

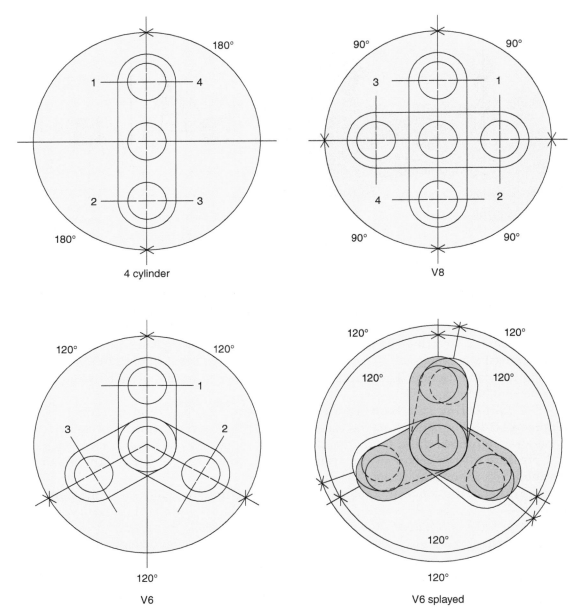

Figure 12-15 Typical connecting rod throw offsets.

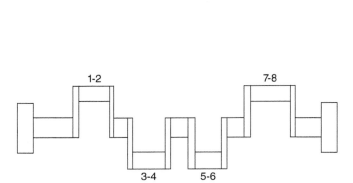

Figure 12-16 A flat-plane crankshaft.

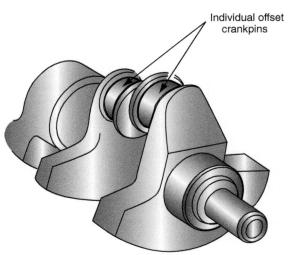

Figure 12-17 A splayed crankshaft.

Figure 12-18 Two rods share one journal on this V8 crankshaft. Notice the oil holes to feed each of the rod bearings.

will rotate together. The throws are machined at 120-degree intervals around the crankshaft. This produces an odd-firing V6. cylinders fire at 90 degrees, then after 150 degrees of rotation, then again after 90 degrees, and on through the two full revolutions of the crankshaft that create a cycle.

An even-firing V8 crankshaft has four throws each with two journals on them (**Figure 12-18**). The crankshaft produces power strokes every 90 degrees of crank rotation. This is what makes a V8 such a smooth-running engine; the crankshaft is rotated by a connecting rod every 90 degrees. This leaves little time for the crankshaft to speed up or slow down.

Crankshaft Oil Holes

The crankshaft is drilled to provide oil passages that feed each rod and main journal. The oil is held under pressure by the tight fit between the journal and the bearing. The bearing has an oil hole in it that lines up with the drilled hole in the crankshaft.

Crankshaft Fillets

The fillet on the journal is used to increase the strength of the journal (**Figure 12-19**). This joint area is subject to high stress and is a common location for cracks. Because sharp corners are weak, manufacturers call for a curved fillet of a specific radius. The fillet is increased for applications requiring severe duty. Most fillets are formed by grinding the joint. Some turbocharged or high-performance engines use a process of rolling the fillet. This is done by rolling hardened steel balls under pressure on the joint. This causes the metal at the joint to become compacted and tightens the grain structure, thus

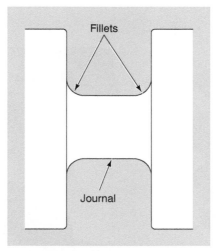

Figure 12-19 The fillet strengthens the crankshaft journal.

strengthening the crankshaft by as much as 30 to 40 percent. The fillet is not usually a wear area unless the bearing is worn, but the fillet is an area where the crankshaft can crack and should be carefully inspected. If the journal requires machining, the fillet should be reground to the proper radius. The crankshaft is also subject to lateral forces. These come from clutch engagement and automatic transmission loads. The lateral movement of the crankshaft is controlled by a thrust bearing and a thrust surface on the crankshaft. The thrust bearing is usually attached vertically to the sides of one main bearing.

Crankshaft Counterweights

Crankshaft counter-weights are positioned opposite the connecting rod journals and balance the crankshaft.

The crankshaft is cast or forged with large **counterweights** that oppose the rod journals. The counterweight overcomes the total weight of the rod-bearing journal, bearing, connecting rod, piston pin, and piston. These help support the crankshaft and even out the speed changes as the crank rotates through firing impulses. Crankshafts that use (only) their counterweights for balance are "internally" balanced. Externally balanced crankshafts depend on the harmonic balancer and flywheel or flexplate (in addition to the counter weights) to absorb vibration. You will often see holes drilled in the counterweights. These are drilled when the crankshaft is balanced as it spins with rods and pistons attached. Holes also lighten the crank; you may see lightening holes on the sides of the throws as well.

Crankshaft Construction

Many manufacturers construct their crankshafts from gray cast iron. This construction is strong enough for most applications. Some engine applications require additional strength above what gray cast iron can provide; in these instances, nodular iron is often used. A cast-iron crankshaft is identifiable by its straight cast mold parting line (**Figure 12-20**).

In other applications, such as some turbocharged engines, a forged medium-carbon steel crankshaft (carbon content between 0.30 and 0.60 percent) is used. High-performance engines may use a crankshaft using chromium and molybdenum (sometimes called chrome-moly) alloys to increase its strength. The forging process condenses the grain of the steel and increases its strength. Because of the density of the steel, these crankshafts weigh more than a cast-iron crankshaft. To lighten the weight, some manufacturers hollow the rod-bearing journals.

Forged crankshafts designed for use in V8 and in-line, six-cylinder engines must be twisted to achieve the desired throw offset. This process is done immediately after the crankshaft is forged, while the meal is still hot. A forged crankshaft can be identified by the staggered die parting lines because of this twisting (**Figure 12-21**).

Case-hardening crankshafts protects the journals from wear and fractures. Many manufacturers use a case-hardening process called ion nitriding. The component is placed

Figure 12-20 Casting lines on a cast-iron crankshaft.

Figure 12-21 Casting lines on a forged crankshaft.

into a pressurized chamber filled with hydrogen and nitrogen gases. An electrical current is then applied through the component. This changes the molecular structure of the metal and allows the induction of the gases into the surface area of the journal.

 A BIT OF HISTORY

Bentley Motors was founded in 1919, and within a short span of time, the Bentley automobile became a refined and complex "silent sports car." In 1931, Bentley acquired Rolls Royce, and thereafter the Bentley automobile became associated with it. Today, with a $500-million investment, Bentley Motors is focusing on designing and building automobiles of world-class standards. All Bentley automobiles go through a 154-point inspection designed by Bentley engineers in the factory.

CAMSHAFT

The camshaft may be fitted in the block (**Figure 12-22**). It rides on bearings placed in the camshaft bore. A thrust plate typically limits the camshaft end play. Refer to Chapter 10 of this manual for a review of the full characteristics and functions of the camshaft.

LIFTER BORES

When a block contains the camshaft, it also has lifter bores drilled above the camshaft lobes. These bores are slightly off center from the centerline of the camshaft lobe so that the convex face of the lifter turns the lifter in its bore. Lifter bores must be clean and free of coring to provide a smooth surface for the lifters to ride on.

Shop Manual
Chapter 12, page 537

HARMONIC BALANCING

A two-piece **harmonic balancer** is used on the front end of the crankshaft. It is used to absorb the **torsional vibrations** of the crankshaft created as the crankshaft tries to speed up due to firing impulses on cylinders. At the same time, another cylinder is trying to slow the crankshaft down as it pushes against air pressure on the compression stroke. The inner piece is bolted to the crankshaft. An inertia ring is bonded with rubber to the inner ring. As the crankshaft winds and unwinds, the twisting of the inertia ring and the rubber ring absorb some of the vibrations (**Figure 12-23**). If the bonding fails, the two rings will separate and you will need to replace the harmonic balancer. A faulty harmonic balancer can cause engine vibration, resulting in a broken crankshaft.

Harmonic balancers are used to control and compensate for torsional vibrations.

Torsional vibration is the result of the crankshaft twisting and snapping back during each revolution.

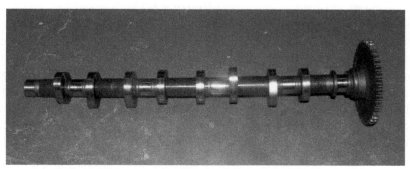

Figure 12-22 The camshaft may be housed in the block or in the head.

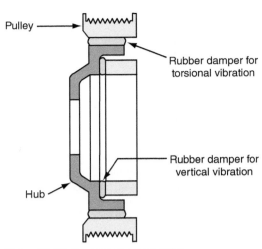

Figure 12-24 Harmonic balancer action.

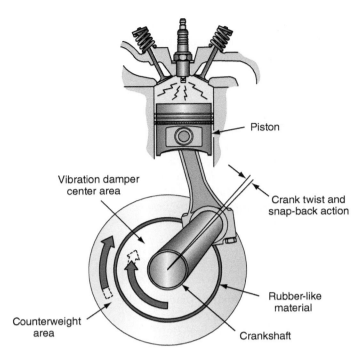

* Harmonic balancer scale exaggerated.

Figure 12-23 Dual-mode harmonic balancer.

Many of today's engines use dual-mode harmonic balancers (**Figure 12-24**). The first mode eliminates torsional vibration, and the other mode eliminates vertical vibration.

Many harmonic balancers are pressed onto the end of the crankshaft, and a square key fits into a matching keyway in the crankshaft and harmonic balancer to prevent rotation. A bolt and washer usually retain the harmonic balancer on the end of the crankshaft.

AUTHOR'S NOTE One day a customer had her vehicle towed in. The engine had stopped running after developing a very serious knocking noise from the bottom end. While disassembling the engine, I noticed that the harmonic balancer had separated badly. The result: the crankshaft had broken in half!

The **torque converter** mounts to the crankshaft through a flexplate. Its weight and fluid coupling do a fine job of maintaining a more constant crankshaft speed.

Crankshaft specifications, stampings, part numbers, and tolerances may be different if an engine is mated with the vehicle option of an automatic or a manual transmission.

FLYWHEEL

The flywheel is a large round disc of heavy mass attached to the rear end of the crankshaft to smooth the rotation of the crankshaft (**Figure 12-25**). It helps the crankshaft speed remain more constant between firing pulses by storing inertial energy. The outer ring, the ring gear, is pressed on to provide teeth for the starter pinion gear to turn the engine over. The machined center surface provides a surface for the clutch disc to grab onto to transmit the rotation of the crankshaft to the transmissions. Flywheels are used only on engines mounted to a manual transmission. An automatic transmission is mounted to the engine by a light flexplate that bolts to the **torque converter**. The torque converter is filled with transmission fluid and is as heavy as a flywheel and clutch assembly. The fluid coupling of the torque converter does an excellent job of helping maintain a constant crankshaft speed.

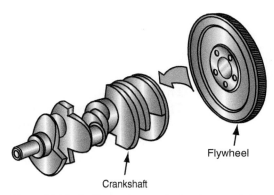

Flywheel

Crankshaft

Figure 12-25 The flywheel is attached to the rear of the crankshaft. A flywheel is used with a manual transmission. A flexplate is used with a manual transmission.

SHORT BLOCKS, LONG BLOCKS, AND CRATE ENGINES

Sometimes an engine block, the heads, or both are damaged badly enough that it becomes more cost-effective to replace the failed component(s) rather than machine or recondition it. New or rebuilt short or tall blocks are available for most engines. In addition, "crate" engines, usually supplied by an aftermarket company, are available fully assembled and rebuilt. The option of replacing a seriously damaged engine with a crate engine has become a very popular option in the past several years. Crate engines are remanufactured on assembly lines and are often cheaper than the cost of parts, machining, and labor to rebuild an engine in house (**Figure 12-26**).

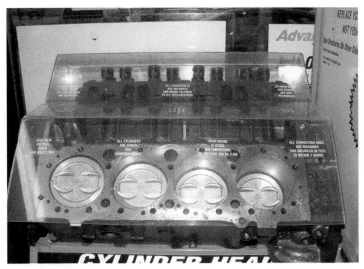

Figure 12-26 This remanufactured engine comes with bored cylinders, polished or new crankshafts, refurbished and balanced connecting rods, and new rings, pistons, and bearings.

SUMMARY

- The bottom end of the engine is made up of the block, crankshaft, pistons, and connecting rods.
- The cylinder block may be cast iron or aluminum.
- Cylinders are typically integral on a cast-iron block, as are steel sleeves on aluminum blocks.
- The block's main bore holds the crankshaft.
- The main bore is line-bored with the caps in place. Main caps must be installed in their original locations and in the same direction.

- Crankshafts may be made of cast iron, nodular iron, or forged steel; forged cranks are stronger.
- The crankshaft spins on its main journals.
- The connecting rod journals are offset from the centerline, which determines the stroke.
- The connecting rod transmits combustion forces to spin the crankshaft.
- The harmonic balancer bolts to the front of the crankshaft to reduce torsional vibration.
- The flywheel or torque converter helps maintain crankshaft speed between power pulses.

REVIEW QUESTIONS

Short-Answer Essays

1. Briefly explain how a cylinder block is formed.
2. Describe the advantages and disadvantages of an aluminum block.
3. What are the advantages of two-piece blocks?
4. Why must a main cap be installed in its original location and direction?
5. What is the purpose of the crankshaft?
6. How is the stroke of an engine determined?
7. Explain the purpose of the crankshaft fillet.
8. Explain why lifter bores are off center from the center of the camshaft lobe.
9. What is the purpose of a dual-mode harmonic balancer?
10. Explain the purpose of the flywheel.

Fill-in-the-Blanks

1. The _____ _____ or main girdle is the lower portion of a two-piece block.
2. _____ _____ _____ blocks are significantly lighter than cast-iron blocks.
3. The passages that allow oil to flow throughout the engine are called _____.
4. _____ _____ are used to seal holes drilled to remove sand from the block casting process.

5. Bores are _____ _____ at the factory to provide good cylinder wall lubrication with minimum wear.
6. The _____ _____ holds the crankshaft.
7. _____ crankshafts are stronger than _____ crankshafts.
8. An eight-cylinder engine produces a power stroke every _____ degrees of engine rotation.
9. The harmonic balancer reduces _____ _____.
10. A _____ engine is a fully assembled remanufactured engine.

Multiple Choice

1. Aluminum blocks often use.
   A. integral bores.
   B. steel housings.
   C. steel liners.
   D. cast-iron main girdles.
2. A harmonic balancer has separated. The response the technician should take is to
   A. reseal the bond with rubber Loctite.
   B. replace the inertia ring.
   C. replace the crankshaft; it is likely cracked.
   D. replace the whole harmonic balancer.

3. An engine bedplate.
   A. reduces vibrations.
   B. contains the main-bearing caps.
   C. adds strength to the block.
   D. All of the above.

4. On an even-firing V6 engine with a splayed crankshaft, the throws on the splayed connecting rod journals are offset.
   A. 10 degrees. C. 35 degrees.
   B. 30 degrees. D. 45 degrees.

5. The crankshaft is balanced by the _____ to overcome the weight of the rod journal, piston pin, connecting rod, and piston.
   A. Harmonic balancer
   B. Fillet
   C. Counterweights
   D. Flywheel

6. The lifter bores are offset from the camshaft lobe to.
   A. allow the camshaft to turn.
   B. spin the valve.
   C. turn the pushrod.
   D. rotate the lifter.

7. Each of the following is a crankshaft material, *except*:
   A. Gray cast iron
   B. Nodular iron
   C. Chrome-moly alloys
   D. Forged aluminum

8. A dual-mode harmonic balancer reduces.
   A. torsional and vertical vibrations.
   B. crankshaft end thrust and torsional vibration.
   C. speed fluctuations and torsional vibrations.
   D. vertical vibrations and crankshaft end thrust.

9. The purpose of core plugs is.
   A. to fully drain coolant during engine service.
   B. to prevent the block from freezing.
   C. to remove the sand after casting.
   D. to provide cooling jackets to remove excess heat.

10. While discussing V8 crankshafts,
    Technician A says that a crate engine built for a vehicle with an automatic transmission may have a different crankshaft than the same one built for a manual transmission.
    Technician B says that a flat-plane crankshaft is heavier and takes up more space than a cross plane crankshaft.
    Who is correct?
    A. A only
    B. B only
    C. Both A and B
    D. Neither A nor B

CHAPTER 13

PISTONS, RINGS, CONNECTING RODS, AND BEARINGS

Upon completion and review of this chapter, you should understand and be able to describe:

- The function and design of engine bearings.
- The importance of proper bearing oil clearances.
- The design and function of pistons.
- The methods used in piston design to control heat expansion.

- The advantages of cast, forged, and hypereutectic cast pistons.
- The purpose of piston pin offset.
- The purposes of piston rings.
- The design and materials of piston rings.
- The design and function of the connecting rods.

Terms To Know

Balance shaft
Bearing crush
Bearing spread
Blow-by
Cam ground pistons
Compression rings

Connecting rod bearings
Locating tab
Major thrust surface
Minor thrust surface
Offset pin
Oil control rings

Oversize bearings
Tapered pistons
Thrust bearing
Undersize bearings

INTRODUCTION

The components within the engine block, the pistons, piston rings, connecting rods, and engine bearings, all play a critical role in transmitting the power of combustion from the top of the piston to the rod journal of the crankshaft to spin the engine. The piston is forced downward when combustion peaks just after TDC on the power stroke. The piston rings seal the small gap between the piston and the cylinder wall to use all the pressure of combustion. The connecting rods connect the piston to the crankshaft, delivering the reciprocating motion of the piston to the rotating crankshaft. Soft-coated engine bearings are held in the main, connecting rod, and camshaft bores to protect the crankshaft and camshaft. A thin film of oil is held between the bearing and the journal to reduce friction and minimize wear. Piston rings; main, connecting rod, and camshaft bearings; and timing chains are the main wear items in an engine block. Each of the bottom end components is carefully designed to match the characteristics of the engine. By knowing how these components are designed and should perform, you will be better able to evaluate their condition during engine repairs.

BEARINGS

Bearings (bearing inserts) are used to protect the main and connecting rod journals of the crankshaft and carry the rotational loads (**Figure 13-1**). A similar type of bearing is used on the camshaft when it is in the block. The camshaft bearings are one piece rather than two, but

Figure 13-1 This is one half of a set of new aluminum-coated main bearings.

their materials and characteristics are nearly the same as those used to protect the crankshaft. The engine bearings are designed to wear before the crankshaft. The bearings snap into the main caps and connecting rod caps and are then bolted around the crankshaft journals. The clearance between the bearing and the journal is critical. The clearance allows a thin film of oil to lubricate the bearing and the crankshaft. That oil also absorbs the pounding of the crank as the power stroke forces it downward in the main bores. The crankshaft should spin on a film of oil and not contact the bearing (**Figure 13-2**). The oil pump sends oil under pressure through the holes drilled in the crankshaft through an oil hole in the bearing into this small clearance. The correct bearing clearance allows a calibrated amount of oil to leak out past the side of the bearings so the oil is constantly refreshed. When the bearing clearances become too great, more oil leaks out and engine oil pressure decreases. This loss of oil pressure results in a decrease in the oil film strength. The crankshaft will knock in its bore and cause rapid wear and distortion of the bearings and crankshaft. If the engine is not repaired promptly, the damage to the crankshaft and bearings can become so severe that the bearings spin with the crankshaft in their bores. When bearings spin in their bores, they can destroy the main bores, connecting rod bores, or camshaft bores and even cause the engine to seize. You will usually replace an engine with this type of severe damage to the block and crankshaft.

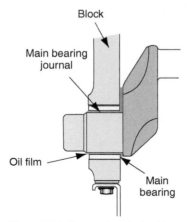

Figure 13-2 The crankshaft should ride on a film of oil roughly 0.001 inch thick. That is enough to reduce friction and protect the crankshaft.

Bearing Materials

Engine bearings are made in halves that snap into the main and connecting rod bores and caps. The back of the bearings is steel to provide strength to carry the tremendous loads impressed by the crankshaft. On top of the load bearing shell are layers of soft metal to protect the crankshaft.

Many different materials are used in the construction of bearings. The main concern of the manufacturers is to achieve a bearing construction that will have these desirable characteristics:

1. *Surface action.* The ability of the bearing to withstand metal-to-metal contact without being damaged.
2. *Embedability.* The ability of the bearing material to tolerate the presence of foreign material.
3. *Fatigue resistance.* The ability of the bearing to withstand the stress placed upon it.
4. *Corrosion resistance.* The ability of the bearing to withstand the corrosive acids produced within the engine.
5. *Conformability.* The ability of the bearing material to conform to small irregularities of the journal.

Given that no single material will have all of these characteristics, most bearings are constructed using multilayers of different materials. The backing is usually made of steel; then additional lining material is added. The most common materials used include the following:

1. Babbitt
2. Sintered copper-lead
3. Cast copper-lead
4. Aluminum

To assist the technician in the selection of bearings, most manufacturers use a part number suffix to identify the bearing lining material; for example, a Federal Mogul bearing with the part number 8-3400CP indicates the bearing is steel-backed copper-lead alloy with overplate. The following chart provides the codes used:

Part Number Suffix	Lining Material and Application
AF	Steel-backed aluminum alloy thrust washer
AP	Steel-backed aluminum alloy with overplate connecting rod or main bearings
AT	Solid aluminum alloy connecting rod or main bearings
B	Bronze-backed babbitt connecting rod or main bearings
BF	Steel-backed bronze or babbitt thrust washers
CA	Steel-backed copper alloy connecting rod or main bearings
CH	Steel-backed copper-lead alloy with overplate connecting rod and main bearings for performance applications
CP	Steel-backed copper-lead alloy with overplate connecting rod and main bearings
DR	Steel-backed babbitt, steel-backed copper-lead alloy, or solid aluminum camshaft bushing
F	Bronze-backed babbitt, bronze or aluminum thrust washers
RA	Steel-backed aluminum alloy connecting rod and main bearings
SA, SB, SBI, SH, SI, SO	Steel-backed babbitt connecting rod and main bearings
TM	Steel-backed copper alloy with thin overplate connecting rod and main bearings
W	Camshaft thrust washer

When choosing replacement bearings, you will usually use original equipment quality components or updated designs as recommended by the parts supplier.

Bearing Characteristics

Figure 13-3 provides common nomenclature used with insert-type bearings. Most main and connecting rod bearings are designed to have a certain amount of spread. **Bearing spread** is provided by manufacturing the bearing inserts so they have a slightly larger curvature compared to the curvature of the rod bearing or cap. This design ensures positive positioning of the bearing against the total bore area. When the bearing is installed into the bore, a light pressure is required to set the bearing (**Figure 13-4**).

In addition to bearing spread, most main and connecting rod bearings are designed to provide **bearing crush** (**Figure 13-5**). When the bearing half is installed in the connecting rod or cap, the bearing edges protrude slightly from the rod or cap surfaces. When the two halves are assembled, the crush exerts radial pressure on the bearing halves and forces them into the bore. Proper crush is important for heat transfer. If the bearing is not tight against the bore, it will act as a heat dam and not transfer the heat to the engine block. Bearing spread and bearing crush also help prevent the bearing from spinning in its bore as rotational friction acts on it.

> **Bearing spread** means the distance across the outside parting edges is larger than the diameter of the bore.

> **Bearing crush** refers to the extension of the bearing half beyond the seat.

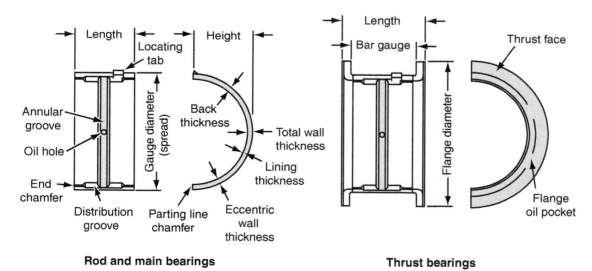

Rod and main bearings

Thrust bearings

Figure 13-3 Bearing nomenclature.

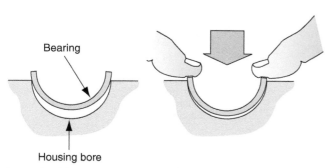

Figure 13-4 Designing the bearing with a certain amount of spread ensures positive positioning of the bearing against the total bore area.

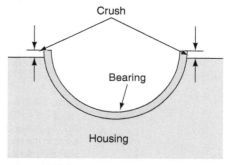

Figure 13-5 Bearing crush exerts radial pressure on the bearing halves and forces them securely into the bore.

The **locating tab** is a tab sticking out of bearing inserts that fits in a groove in the block, main bearing cap, connecting rod, or connecting rod cap to prevent insert rotation.

To ensure proper installation of the bearing in the bore, most manufacturers use a **locating tab** (**Figure 13-6**). A protruding tab at the parting of the bearing halves is set into a slot in the housing bore. When properly installed, the tab will locate the bearing, preventing it from shifting sideways in the bore.

Main Bearings

Main bearings carry the load created by the movement of the crankshaft. Most of these bearings are insert design. The advantages of this design include ease of service, variety of materials, and controlled thickness.

Split bearings surround each main bearing journal. The bearings are supplied oil under pressure from the oil pump through oil passages drilled in the engine block and crankshaft (**Figure 13-7**). The crankshaft does not rotate directly on the main bearings; instead, it rides on a thin film of oil trapped between the bearing and the crankshaft. If the journals are worn and become out-of-round, tapered, or scored, the oil film is not formed properly. This will result in direct contact between the crankshaft and bearing and will eventually damage the bearing, crankshaft, or both. Soft materials are used to construct the bearings in an attempt to limit wear of the crankshaft.

Oil Grooves. Providing an adequate oil supply to all parts of the bearing surface, particularly in the load area, is an absolute necessity. In many cases, this is accomplished by the oil flow through the bearing oil clearance. In other cases, however, engine operating conditions are such that this oil distribution method is inadequate. When this occurs, some type of oil groove must be added to the bearing. Some oil grooves are used to ensure an adequate supply of oil to adjacent engine parts by means of oil throw-off.

Oil Holes. Oil holes allow for oil flow through the engine block galleries and into the bearing oil clearance space. Connecting rod bearings receive oil from the main bearings by means of oil passages in the crankshaft. Oil holes are also used to meter the amount of oil supplied to other parts of the engine; for example, oil squirt holes in connecting rods are often used to spray oil onto the cylinder walls. When the bearing has an oil groove, the oil hole normally is in line with the groove. The size and location of oil holes are critical. Therefore, when installing bearings, you must make sure the oil holes in the block line up with holes in the bearings.

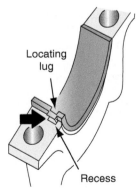

Figure 13-6 A protruding tab at the parting of the bearing halves is mated to a slot in the housing bore. The tab locks the bearing to prevent it from spinning or shifting sideways.

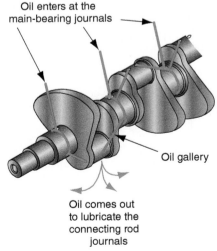

Figure 13-7 Internal passages in the crankshaft send lubrication to the main and connecting rod bearings.

Thrust Bearings

The crankshaft is fitted with one **thrust bearing** to limit end play within the block (**Figure 13-8**). There must be some clearance for the crankshaft to move horizontally (end to end), to prevent it from seizing as components heat up. The thrust bearing protects the crank from slamming against the block as it moves forward and backward in its bore. Under acceleration, the crankshaft will be forced toward the back of the block. A thrust bearing allows a thin film of oil between a soft bearing and the crankshaft to absorb this force. The thrust bearing is often one of the main bearings, with flanges on the sides to control crank end play (**Figure 13-9**). It may be positioned at an end of the crank or in the middle. Some thrust bearings are not part of a main bearing; they can be multipiece (**Figure 13-10**). In this case, they are called thrust washers. Thrust bearings will typically wear faster when the engine is attached to a manual transmission because the clutch friction disk is constantly being engaged and disengaged from the flywheel while accelerating. This extra stress places horizontal loads on the crankshaft that are not found when the engine is coupled with an automatic transmission. Some manufacturers will have a different specification for the crankshaft end play. You will measure crankshaft end play during engine disassembly and reassembly to be sure it is within specifications.

> The **thrust bearing** prevents the crankshaft from sliding back and forth by using flanges that rub on the side of the crankshaft journal.

> Crankshaft end play is the measure of how far the crankshaft can move lengthwise in the block.

Rod Bearings

Split bearings surround each connecting rod journal. Like the main bearings, the **connecting rod bearings** do not rotate directly on the crankshaft. They allow the rod journal to rotate on a thin film of oil trapped between the bearing and the crankshaft. Journals that are worn, out-of-round, tapered, or scored will not maintain this oil film, resulting in damage to the bearing, crankshaft, or both.

Most main and rod bearings are designed with a bearing crown to maintain close clearances at the top and bottom of the bearing. This is the area where most of the loads are applied. The crown also allows increased oil flow at the sides of the bearing (**Figure 13-11**). In addition, most bearings have a beveled edge to provide room for the journal fillet.

> **Connecting rod bearings** are insert-type bearings retained in the connecting rod and mounted between the large connecting rod bore and the crankshaft journal.

Bearing Oil Clearances

There must be a gap or clearance between the inside diameter of the bearing face and the outside diameter of the journal of the main, connecting rod, and camshaft (if insert bearings are used). This clearance allows engine oil pressure to hold a film of oil within the clearance and prevent too much oil from leaking out. When the clearance becomes too great, either from wear of the bearing or the crankshaft journal, engine oil pressure is reduced. The bearing oil clearances provide a calibrated "leak" of oil from between the bearing and the journal. The clearances must be within the specifications to provide good oil pressure, prevent knocking, and protect the journal from wear. Many modern

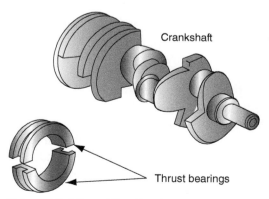

Figure 13-8 A flanged thrust bearing.

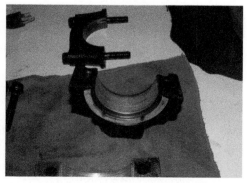

Figure 13-9 This thrust bearing has a soft aluminum-coated side flange to cushion crankshaft movement back and forth in the block.

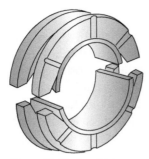

Figure 13-10 A multiple-piece thrust bearing.

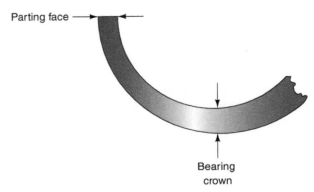

Figure 13-11 Bearing crown maintains close clearances at the top and bottom of the bearing, where most of the loads are applied.

oil bearing clearances fall within the range of 0.0008 inch (0.0203 mm) to 0.002 inch (0.0508 mm). These are just typical clearances; always check the manufacturer's specifications for the engine you are working on.

Bearings are a wear item. They are designed to wear and protect the crankshaft or camshaft. They do wear out after high mileage and exposure to contaminated oil. When the engine is disassembled for repairs, it is customary to replace the main, rod, and cam bearings.

When fitting new bearings, you will have to check the clearance to ensure that it is within the specified range. We will discuss this procedure in Chapter 13 of the Shop Manual.

Shop Manual
Chapter 13, page 588

Undersize and Oversize Bearings

Undersize bearings have the same outside diameter as standard bearings, but the bearing material is thicker to fit an undersize crankshaft journal.

When a crankshaft is worn lightly and polished to be usable, bearings that are 0.001 and 0.002 inch undersize are often available. If the crankshaft has been ground to repair serious journal wear, **undersize bearings** can be fitted. These bearings are generally available in 0.010, 0.020, and 0.030 inch undersize. Metric undersized bearings may include 0.050, 0.250, 0.500, and 0.750 mm. Undersize means that the inside diameter of the bearings is smaller, to fit the reduced diameter of the crankshaft journals.

Oversize bearings are thicker than standard to increase the outside diameter of the bearing to fit an oversize bearing bore. The inside diameter is the same as standard bearings.

Oversize bearings may be used when the block has been line bored to an oversized diameter. Oversize bearings are often available in 0.010, 0.020, 0.030, and 0.040 inch. Available metric oversized bearings are typically 0.250, 0.500, 0.750, and 1.000 mm. These bearings have a larger outside diameter than the standard bearing. The use of oversize and undersize bearings has decreased dramatically as component replacement has become more cost-effective than many complex machining operations. Some bearings are stamped with a code that allows the technician to check the size of the bearing that is currently in use. New bearings come stamped from the factory with the size correlation on it. Some engine blocks have a stamping on them that helps the technician indicate what size bearing came on the engine originally. If different sized bearings are installed in this type of engine, the technician should change this mark or remove it.

> **AUTHOR'S NOTE** It has been my experience that most connecting rod and main bearing failures are caused by contaminated engine oil or lack of lubrication. Often the cause of failure is clear by the sludge in the oil pan, the coolant-contaminated oil, or the lack of oil in the crankcase. Try to determine the cause of the failure if it is premature. Most bearings should last the life of the engine; these days, that may reach 150,000 to 200,000 miles (241,402 to 321,870 km). Be sure to discuss the importance of regular oil and filter changes with your customers.

A BIT OF HISTORY

Isaac Babbitt invented his bearing material in 1839 using a mixture of tin, lead, and antimony. This tin-based babbitt was later replaced with a lead-based material. When lead-based babbitt was used in the first automotive engines, it was melted and then poured into the block and connecting rods. After the babbitt cooled, the correct oil clearance was achieved by hand scraping the excess material. Final adjustment of the bearings was accomplished by using shims between the bearing caps. Due to the nature of these early bearings, they often smeared under heavy loads. To decrease the oil clearance as the bearing wore, the caps were refilled. This operation was considered owner maintenance, and babbitt material was carried in the vehicle to make the necessary roadside repairs.

CAMSHAFT AND BALANCE SHAFT BEARINGS

In an overhead valve (OHV) engine, the circular bearings are pressed into openings in the block (**Figure 13-12**). The camshaft journals are positioned in these bearings. The specified bearing clearance must be present between the camshaft journals and the bearings. Oil from the lubrication system is supplied through passages in the block to an opening in each camshaft bearing. The opening in the camshaft bearing must be properly aligned with the oil passage in the block. Excessive camshaft bearing clearance causes low oil pressure. During an engine rebuild, the camshaft bearings should be replaced. To make camshaft removal and replacement easier, the camshaft bearings on many engines are progressively larger from the back to the front of the block. Special tools are available to remove and replace the camshaft bearings. Balance shaft bearings are similar to camshaft bearings.

BALANCE SHAFTS

Many engine designs tend to have inherent vibrations. The in-line four-cylinder engine is a good example of this, but other engines use balance shafts for varying reasons, including customer demand for smoothness. Even some V-style engines use balance shafts. Some engine manufacturers use a balance shaft to counteract this tendency (**Figure 13-13**). Four-cylinder engine vibration occurs when the piston is moving down during its power stroke. When the engine completes one revolution, two vibrations occur. The **balance shaft** rotates at twice the speed of the crankshaft. This creates a force that counteracts crankshaft vibrations.

The **balance shaft** is a shaft driven by the crankshaft that is designed to reduce engine vibrations.

Shop Manual
Chapter 13, page 591

PISTONS

The piston, when assembled to the connecting rod, is designed to transmit the power produced in the combustion chamber to the crankshaft (**Figure 13-14**). The piston must be able to withstand severe operating conditions. Stress and expansion problems are

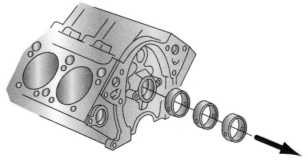

Figure 13-12 Camshaft bearings in the cylinder block.

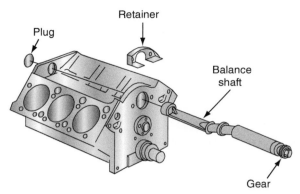

Figure 13-13 A balance shaft in a V6 engine.

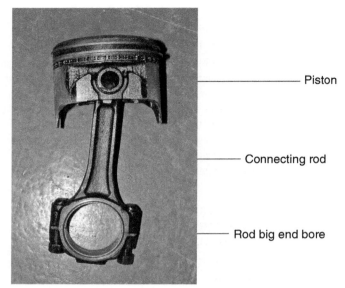

Figure 13-14 The forces of combustion push the piston down in the cylinder. The connecting rod bolts to the crankshaft and transmits this force.

compounded by the extreme temperatures to which the top of the piston is exposed. In addition, the rapid movement of the piston creates stress and high pressures. To control these stress conditions, most pistons are constructed from aluminum. The lightweight aluminum operates very efficiently in high-rpm engines.

The piston has several parts, including the following (**Figure 13-15**):

1. *Land*. Used to confine and support the piston rings in their grooves.
2. *Heat dam*. Used on some pistons to reduce the amount of heat flow to the top ring groove. A narrow groove is cut between the top land and the top of the piston. As the engine is run, carbon fills the groove to dam the transfer of heat.
3. *Piston head (or crown)*. The top of the piston against which the combustion gases push. Different head shapes are used to achieve the manufacturers' desired results (**Figure 13-16**). Most engines use flat-top piston heads. If the piston comes close to the valves, the recessed piston head is used to provide additional clearance. Different types of domed and wedged piston heads are used to increase compression ratios.
4. *Piston pins (or wrist pins)*. Connect the piston to the connecting rod. Three basic designs are used: piston pin anchored to the piston and floating in the connecting rod; piston pin anchored to the connecting rod and floating in the piston; and piston pin full floating in the piston and connecting rod.
5. *Skirt*. The area between the bottom of the piston and the lower ring groove and at a 90-degree angle to the piston pin. The skirt forms a bearing area in contact with the cylinder wall. To reduce the weight of the piston and connecting rod assembly, many manufacturers use a slipper skirt. The piston skirt surface is etched to allow it to trap oil (**Figure 13-17**). This helps lubricate the piston skirt as it moves within the cylinder and prevents scuffing of the skirt.
6. *Thrust face*. The portion of the piston skirt that carries the thrust load of the piston against the cylinder wall.
7. *Compression ring grooves*. The two upper ring grooves used to hold the compression rings.
8. *Oil ring groove*. The bottom ring groove used to hold the oil ring.
9. *Gas ports*. Some pistons have holes drilled near the top compression ring to allow combustion gases behind the ring to force it out against the cylinder wall to improve sealing (**Figure 13-18**).

Slipper skirts allow shorter connecting rods to be used. Part of the skirt is removed to provide additional clearance between the piston and the counterweights of the crankshaft. Without this recessed area, the piston would contact the crankshaft when the shorter connecting rods are used.

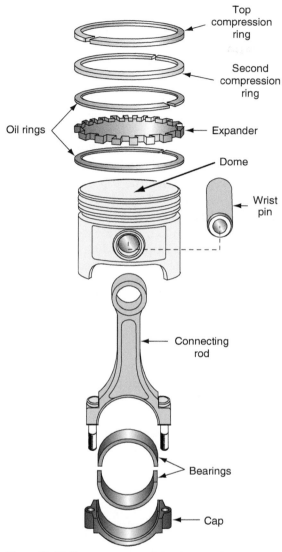

Figure 13-15 The components of the piston and connecting rod assembly.

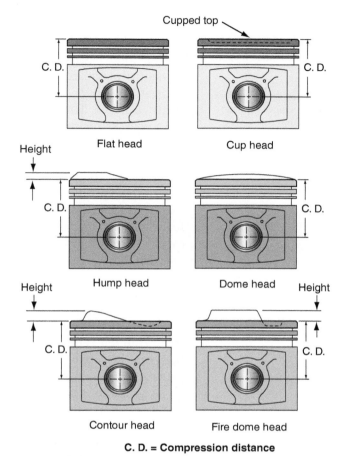

C. D. = Compression distance

Figure 13-16 Some of the common piston head designs used in automotive engines.

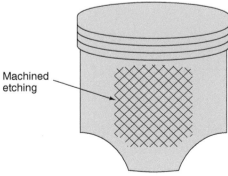

Figure 13-17 Some piston skirts have an etching machined into them to help trap oil to lubricate the skirt. Scuffing prevents oil from being trapped and causing increased friction and wear.

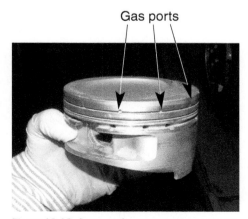

Gas ports

Figure 13-18 Some performance engines use gas-ported pistons to improve top ring sealing.

Thermal expansion of aluminum is about twice as much as iron.

A piston is not a perfect circle shape; it is more of an oval shape. **Cam ground pistons** are oval shaped so that they can expand to a more perfect circle shape when heated.

A **tapered piston** is narrower at the top of the piston where it will expand when heated to become a similar diameter.

Expansion Controls

When combustion occurs, high temperatures and pressures are applied to the top of the piston. Some of this heat is transmitted to the piston body, causing the piston to expand. To control piston expansion, most pistons are cam ground. **Cam ground pistons** are made to be an oval shape when the piston is cold. The larger diameter of the piston is across the thrust surfaces. The thrust surfaces are perpendicular to the piston pin (**Figure 13-19**). On a cold piston, the thrust surfaces have a close fit to the cylinder wall while the surface around the piston pin has greater clearance. As the piston warms, it will expand along the piston pin. As it expands, the shape of the piston becomes more round. Pistons are also tapered from the top to the bottom. The top of the piston runs much hotter than the skirt due to its exposure to combustion gases. On **tapered pistons**, the top of the piston is manufactured with a narrower diameter than the bottom area of the skirt, to allow room for the excess expansion. When the piston is fully warm, the top is designed to fit the cylinder perfectly (**Figure 13-20**). To measure piston diameter, use the manufacturer's guidelines; measuring at the wrong height could give you improper readings.

Many manufacturers design their cast-aluminum pistons with steel struts located in the skirt area (**Figure 13-21**). The strut works to control the amount and rate of expansion. The struts are cast into the piston during manufacturing. Since steel and aluminum expand at different rates for the same temperature, the struts keep the skirt from expanding too much (**Figure 13-22**). This allows the manufacturer to use tighter clearances between the piston and cylinder to prevent the piston from rocking in the cylinder and generating noise.

A graphite moly-disulfide coating is added to many modern piston skirts to reduce scuffing and increase durability (**Figure 13-23**). These lubrication coatings are helpful in reducing friction, to allow tighter piston-to-cylinder wall clearances. Many parts manufacturers offer replacement pistons with coatings. They are slightly more expensive but will increase the longevity of the piston and reduce cylinder wear.

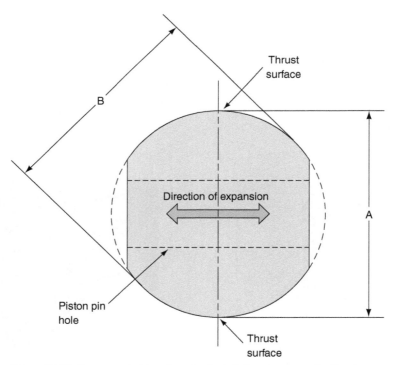

Figure 13-19 Cam ground pistons have the largest diameter across the thrust surfaces (A).

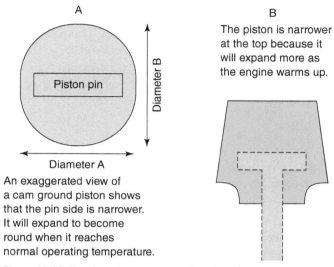

A

Piston pin

Diameter A

Diameter B

An exaggerated view of
a cam ground piston shows
that the pin side is narrower.
It will expand to become
round when it reaches
normal operating temperature.

B

The piston is narrower
at the top because it
will expand more as
the engine warms up.

Figure 13-20 The piston is cam ground (A) and tapered (B) to provide
good cylinder fit when running at normal operating temperature.

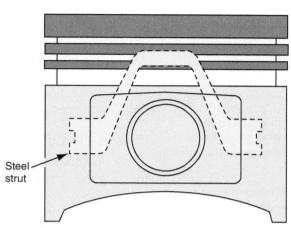

Steel
strut

Figure 13-21 Some manufacturers use a steel strut to help
control expansion and add strength to the piston.

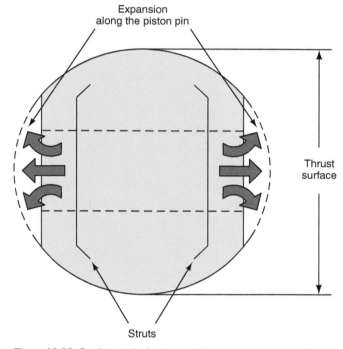

Expansion
along the piston pin

Thrust
surface

Struts

Figure 13-22 Steel struts in the piston help control piston expansion.

Lubrication
patch

Figure 13-23 Some pistons have a coating on the skirts to reduce friction and improve piston fit in the cylinder.

Piston offset is used to provide more effective downward force onto the crankshaft by increasing the leverage applied to the crankshaft.

The skirt that pushes against the cylinder wall during the power stroke is the **major thrust surface**. The skirt that pushes against the cylinder wall during the compression stroke is the **minor thrust surface**.

An **offset pin** is a piston pin that is not mounted on the vertical center of the piston.

The pin boss accepts the piston pin to attach the piston to the connecting rod.

Piston Pin

The piston is joined to the connecting rod by the wrist pin or piston pin. The piston pin bore is reinforced by a pin boss to support the huge forces transmitted to the connecting rod through the pin. The piston pin is often pressed into the piston on passenger car applications. Some higher-end engines and racing applications use full floating pins. A full floating pin is held in place by snap rings.

Piston Pin Offset

Most manufacturers offset the piston pin to reduce piston slap (**Figure 13-24**). The connecting rod is angled to different sides on the compression and power strokes, causing the piston to rock from one skirt to the other at TDC. During the compression stroke, not as much force is exerted on the piston skirt as during the power stroke. Offsetting the piston pin hole toward the major thrust surface about 0.062 inch (1.57 mm) reduces the tendency of the piston to slap the cylinder wall as it rocks from the minor to the **major thrust surface**. During the compression stroke, the connecting rod pushes the minor thrust surface against the cylinder wall (**Figure 13-25**). The **offset pin** causes more of the combustion pressure to be exerted on the larger half of the piston head (**Figure 13-26**). This uneven pressure application causes the piston to tilt so the top of the piston contacts the cylinder wall on the **minor thrust surface** and the bottom of the piston contacts the cylinder wall on the major thrust surface. When the piston begins its downward movement, its upper half slides into contact with the major thrust surface (**Figure 13-27**).

To assist in the installation of the piston, most manufacturers have a groove or other marking to indicate the side of the piston that faces the front of the block. On V-type engines, the pin offset for each bank is on opposite sides. This is because crankshaft rotation determines the major thrust surface side. If the pistons are installed in the wrong direction, it will cause excess friction and piston slap.

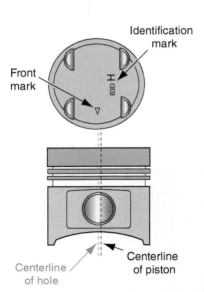

Figure 13-24 Some piston pins are offset to reduce piston slap. A forward marking on the piston ensures the offset will be in the correct orientation during reassembly.

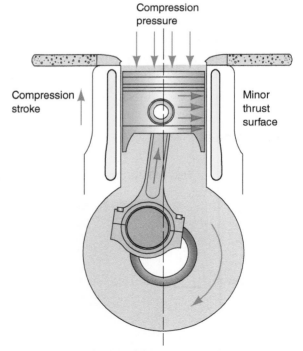

Figure 13-25 During the compression stroke, the connecting rod forces the piston against the minor thrust side of the cylinder wall.

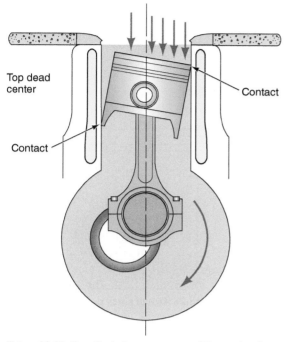

Figure 13-26 The offset pin causes more of the combustion pressure to be exerted on the larger half of the piston head, causing the piston to tilt and cushion the transition from the minor thrust side to the major thrust side. The illustration is exaggerated for clarity.

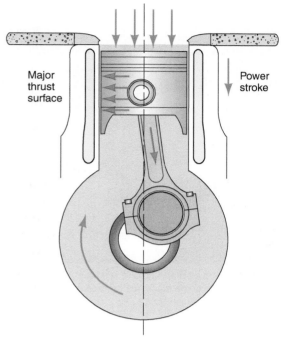

Figure 13-27 When the piston begins its downward movement, its upper half slides into contact with the major thrust surface.

AUTHOR'S NOTE I had a young man bring his newly restored, freshly rebuilt, and performance-modified engine in to diagnose a knocking noise. The noise was coming from the top of the block, and I suspected piston slap. After removing the cylinder head, I could see that the forward markings on the pistons were pointing toward the rear of the engine. After proper reassembly, the engine was quiet and powerful; he had done fine work aside from the piston assembly problem.

Piston Speed

As the piston moves up and down in the cylinder, it is constantly changing speeds. At top and bottom dead center, the speed of the piston is zero. It then immediately accelerates to maximum speed. This action places heavy loads on the piston pin and pin boss. Today, most pistons are constructed of aluminum instead of iron. Because of its mass, a piston made of iron will take more energy to accelerate and decelerate than one made of aluminum. This would accelerate wear on the piston pin, connecting rod, crankshaft, and bearings. It would also decrease the power output and efficiency.

PISTON DESIGNS AND CONSTRUCTION

If it is determined that the pistons require replacement, several piston designs are available from manufacturers and aftermarket suppliers. The technician must be capable of selecting the correct design for the engine application. This usually means replacing the piston

Figure 13-28 This piston has valve relief cutouts to prevent the valves from knocking the top of the piston. You can also see the forward marking dot on the right side of the piston head.

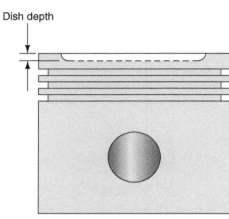

Figure 13-29 Dished piston design.

It is important to recognize that over 90 percent of technicians will work on restoring engines to their original operating condition. Very few technicians will actually find work modifying engines to improve performance. If you understand the basic principles of engine operating and service presented in this text, you will easily be able to transfer that knowledge to high-performance work if you find yourself in that often-coveted position.

with the same design as the one removed; however, if the engine is to be modified, piston selection is one aspect in increasing engine performance and durability.

It is possible to change the compression ratio of the engine by changing the head design of the piston. Most original equipment pistons have a flat head. These pistons may require valve reliefs machined into the head to prevent valve-to-piston contact (**Figure 13-28**). The depth of the relief is determined by camshaft timing, duration, and lift. A dished piston is designed to put most of the combustion chamber into the piston head and lower the compression ratio (**Figure 13-29**). When adding a turbocharger or supercharger to an engine, you may need to lower the compression ratio by installing dished pistons. A domed piston will fill the combustion area and increase the compression ratio (**Figure 13-30**). Other designs are variations of these three types. In addition to changing compression ratio, piston head design can increase the efficiency of the combustion process by creating a turbulence to improve the mixing of the air and fuel.

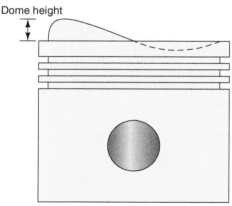

Figure 13-30 Domed piston designs can increase the compression ratio and add turbulence to the combustion chamber.

AUTHOR'S NOTE Changing the piston design can have a negative impact on emissions. For all street-application vehicles, it is illegal to raise the compression ratio of an engine by changing pistons or to increase emissions above the standard to which the vehicle was built. While modified parts are readily available and widely used, they come with a warning that they are for off-road use only. Making modifications to an engine that has to pass an emissions test can cause test failures and costly repairs. As a technician, you are responsible for maintaining the integrity of the emissions on a customer's vehicle.

Another consideration when replacing pistons is construction type. Weight and expansion rates of the piston are important aspects in today's high-rpm engines. The use of aluminum pistons reduces the weight of the assembly and the load on the crankshaft. Construction methods also affect the piston's expansion rate and amount. Most original equipment pistons are made of cast aluminum. These are very capable of handling the requirements of most original equipment engines. The advantage of cast pistons is their low expansion rate. The expansion rate is low because the grain structure of cast is not as dense as other manufacturing processes. This low expansion rate allows the engine manufacturer to tighten the piston clearance tolerances, helping to prevent piston slap and rattle when the engine is cold.

Forged pistons are stronger due to their dense grain structure. The tighter grain structure allows the piston to run about 20 percent cooler than a cast piston. Depending on the alloy used in the aluminum, the expansion rate may be greater than cast. This requires additional piston clearance and may result in cold engine piston rattle. Alloys have been developed in recent years that reduce the expansion rate of forged pistons. Consequently, it is important to follow the piston manufacturer's specifications for piston clearance.

A third alternative is hypereutectic cast pistons. Most cast pistons are constructed with about 9.5 percent silicon content, while forged pistons have between 0.10 and 10 percent, depending on the aluminum used. Silicone is added to help provide lubrication and increase strength. Hypereutectic pistons contain between 16 and 22 percent silicone. Aluminum can only dissolve about 12 percent silicone and hold it in suspension. The silicone is added when the aluminum is molten and dissolved. At the 12 percent level, the silicone remains dissolved when the piston cools. Normally, silicone added above this saturation point will not dissolve and settles at the bottom of the mold. Hypereutectic casting keeps the undissolved silicone dispersed throughout the piston by closely monitoring and controlling the heating and cooling rates during the manufacturing process.

Hypereutectic casting increases the temperature fatigue resistance of the piston. This allows the piston to be made thinner than most cast pistons, reducing the weight of the piston. The strength of hypereutectic pistons falls between cast and forged. The drawback to this piston construction is that silicone can reduce the piston's ability to conduct heat away from the combustion chamber. Although its temperature fatigue is increased, the piston runs hotter than cast and may cause detonation.

It is possible to tell if the piston is forged or cast by looking at the underside of the piston head. Cast pistons have mold dividing lines, while forged pistons do not have these parting lines. Gasoline direct injection (GDI) pistons have special domes or crowns to handle extremely lean fuel mixtures (**Figure 13-31**). Diesel pistons also have a different design because the combustion chamber is part of the piston (**Figure 13-32**).

Shop Manual
Chapter 13, page 594

Figure 13-31 GDI pistons are designed to handle leaner air-fuel mixtures and higher compression ratios.

Figure 13-32 Diesel pistons are shaped differently than gasoline pistons. A diesel piston uses the bowl in the center of the piston as part of the combustion chamber and ignition point.

AUTHOR'S NOTE GDI increases power and fuel economy while lowering emissions. Engine-related problems are mostly due to carbon buildup on the valves, piston rings, and fuel injector. Lack of maintenance or the use of incorrect oil can result in heavy valvetrain wear due to the friction of the high-pressure pump running off of the camshaft.

PISTON RINGS

Blow-by refers to compression pressures that escape past the piston.

For the engine to produce maximum power, compression and expansion gases must be sealed in the combustion chamber. To control **blow-by**, the piston is fitted with compression rings. The rings seal the piston to the cylinder wall, sealing in combustion pressures. In addition, oil rings are used to prevent oil from reaching the top of the piston. Oil in this area will result in blue smoke being expelled from the exhaust system.

Compressions rings are the top rings in a piston assembly that provide cylinder sealing during the compression and power strokes.

Compression rings are located on the two upper lands of the piston (**Figure 13-33**). The compression rings form a seal between the piston and the cylinder wall. To provide a more positive seal, many manufacturers design one or more of the compression rings with a taper (**Figure 13-34**). The taper provides a sharp, positive seal to the cylinder wall. This style of ring must be installed in the correct position. A dot or other marking is usually provided to indicate the top of the ring.

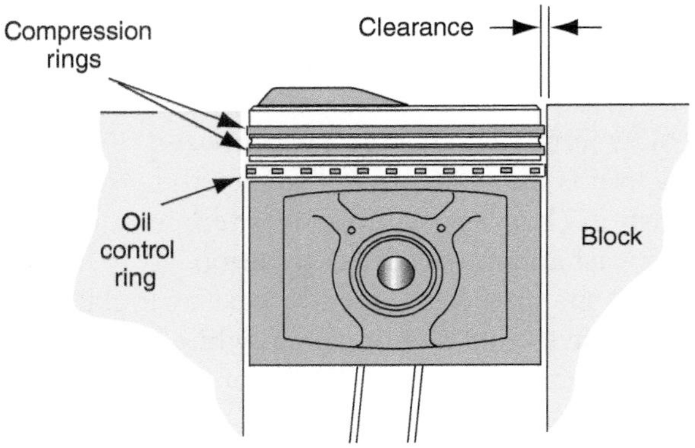

Ring seating is accomplished by the movement of the piston in the cylinder. If the cylinder is properly conditioned, this movement laps the rings against the wall.

Compression rings

Clearance

Oil control ring

Block

Figure 13-33 The top piston ring grooves are used to hold the compression rings. The rings seal against the cylinder wall.

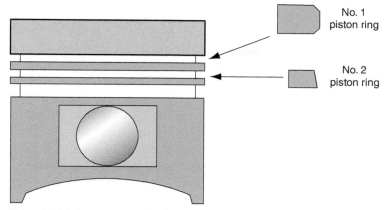

Figure 13-34 Some compression rings are designed with a chamfer or taper. These types of rings may be directional; always look for a top marking during installation.

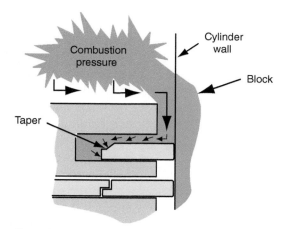

Figure 13-35 The rings are designed to use combustion pressures to exert additional outward forces of the rings against the cylinder walls.

In addition to tapering or chamfering the ring, the ring is designed to use combustion pressures to force the ring tighter against the cylinder wall and the bottom of the ring groove (**Figure 13-35**). The top ring provides primary sealing of the combustion pressures. The second ring is used to control the small amount of pressures that may escape past the top ring.

The most common materials for constructing compression rings are steel and cast iron. The face of the ring is usually coated to aid in the wear-in process. This coating can be graphite, phosphate, iron oxide, molybdenum, or chromium. The bottom piston ring groove houses the oil control ring. There are two common styles of oil control rings: the cast-iron oil ring and the segmented oil ring. Both styles of oil rings have slots to allow the oil from the cylinder walls to pass through. The ring groove also has slots to provide a passage back to the oil sump.

Many engine manufacturers now install low-tension piston rings. These rings are about as thin as an oil control ring rail. With cylinders in good condition they provide adequate combustion chamber sealing with reduced friction on the cylinder walls as the piston moves up and down. Because low-tension rings provide a reduced internal engine friction, they also provide a slight improvement in fuel economy. The downside is they are more susceptible to sticking and tension loss causing blow-by, oil consumption, and reduced compression.

Segmented **oil control rings** have three pieces (**Figure 13-36**). The upper and lower portions are thin metal rails that scrape the oil. The center section of the ring is constructed of spring steel and is called the expander or spacer. The movement of the piston causes the oil ring to scrape oil from the cylinder wall and return it to the sump.

Wear-in is the required time needed for the ring to conform to the shape of the cylinder bore.

Oil control rings are piston rings that control the amount of oil on the cylinder walls.

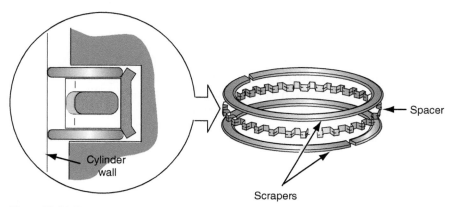

Figure 13-36 Typical three-piece oil control rings with two thin scrapers with a spacer in between.

Ring Materials

Rings are replaced whenever the engine is disassembled for repairs. Generally, you will use original equipment rings with an engine being built to original specifications. There are times, however, when you may choose an appropriate alternative. The following ring materials and their strengths and weaknesses are discussed to assist you in ring selection.

Rings have long been made from cast iron. These rings are a popular choice when overhauling an engine. The material is strong but soft. It can conform best to irregular cylinder surfaces and shapes. When re-ringing a worn engine without boring the cylinders, cast-iron rings are usually recommended. When selecting a replacement ring for a fully overhauled engine, be sure that the quality meets or exceeds the original equipment.

Other rings are made of steel or coated with chrome or moly. Many late-model engines are fitted with steel top rings for their increased durability. The top ring suffers the most; it is exposed to the combustion flame and the highest pressures. If an engine is originally equipped with a steel top ring, it should be replaced with the same steel ring. Many modern turbocharged, supercharged, high-output engines, and engines with the top ring positioned very close to the piston head require steel top rings.

Moly rings are cast rings with a groove cut in the face that is filled with molybdenum. Moly rings provide excellent scuff resistance and superior durability. The moly holds a little oil to lubricate the ring and reduce the friction against the wall. Moly rings are frequently used as original equipment rings. This is an excellent replacement choice for a freshly bored engine. They seal quickly against a properly finished cylinder wall.

Chrome-coated rings are another popular replacement option. They are stronger and can last 50 percent longer than plain cast-iron rings. They do not resist scuffing as well as moly rings, but they also do not absorb contaminants the way moly rings can. When an engine is operated in dusty or sandy conditions, consider chrome rings. They will not seal an out-of-round cylinder, however, so they should be used only in freshly bored and properly finished cylinders. Chrome-coated rings are very hard and can cause excessive cylinder wall wear unless the block is cast with additional amounts of nickel to increase its strength and resistance to abrasives.

Shop Manual
Chapter 13, page 599

Big end refers to the end of the connecting rod that attaches to the crankshaft. Small end refers to the end of the connecting rod that accepts the piston pin.

Oversize Rings

If the cylinder bore diameter is increased to correct for wear, the piston and rings must be replaced. The piston and rings are oversize to fit the new bore diameter. Piston rings are typically available in standard, 0.020, 0.030, 0.040, and 0.060 inch oversizes. If an odd size is needed, such as 0.010 or 0.050 inch oversize, use the ring one size smaller; for example, for 0.010 inch oversize, use a standard ring; for 0.050 inch oversize, use a 0.040 inch oversize. Metric oversizes are available in 0.50, 0.75, 1.0, and 1.5 mm. You cannot install oversized piston rings with a piston of standard size. An oversized piston must accompany it.

CONNECTING RODS

The connecting rod transmits the force of the combustion pressures applied to the top of the piston to the crankshaft (**Figure 13-37**). To provide the strength required to withstand these forces and still be light enough to prevent rough running conditions, the connecting rod is constructed from one of the following methods:

1. Forged from high-strength steel
2. Made of nodular steel
3. Made from cast iron
4. Made of sintered powdered metal

To increase its strength, the connecting rod is constructed in the form of an I (**Figure 13-38**). The small bore at the upper end of the rod is fitted to the piston pin. The piston pin can be

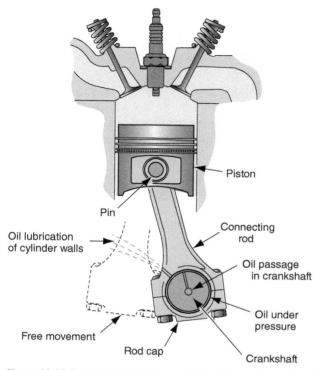

Figure 13-37 The connecting rod connects the piston to the crankshaft.

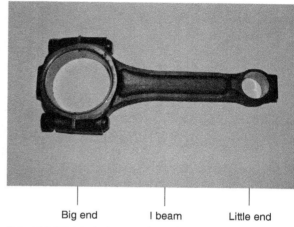

Figure 13-38 A typical connecting rod.

pressed into the piston and a free fit in the connecting rod. In this type of mounting, a bushing is installed in the small rod bore. The piston pin can also be pressed into the small rod bore and a free fit in the piston. The piston moves on the pin surface without the use of bushings.

The larger bore connects the rod to the crankshaft journal. The large bore is a two-piece assembly. The lower half of this assembly is called the rod cap. The rod cap and connecting rod are constructed as a unit and must remain as a matched set. During cylinder block disassembly, the connecting rods and caps should be marked to ensure proper match when installed. Most rod caps are machined from the connecting rods during the manufacturing process. Some manufacturers now use a new process of "breaking" the cap off of the connecting rod. This process uses connecting rods constructed from sintered powdered metal. The connecting rod is then shot peened to remove any flash and to increase surface hardness. After the rod bolt holes are drilled and tapped, the bolts are installed loosely. Next, the break area is laser scribed and the cap is fractured in a special fixture. This creates a rod and cap parting surface that is a perfect fit and creates a stronger joint (**Figure 13-39**).

Shop Manual
Chapter 13, page 600

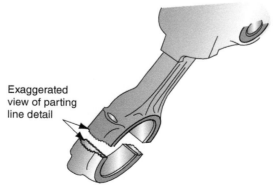

Figure 13-39 Fracturing, instead of machining, the connecting rod cap provides a perfect concentric fit when it is bolted back together.

SUMMARY

- Bearings are used to carry the loads created by rotational forces. Main bearings carry the load created by the movement of the crankshaft. Connecting rod bearings carry the load of the power transfer from the connecting rod to the crankshaft.
- Oil is supplied to the rod-bearing journals through holes in the crankshaft.
- Undersize bearings are used to maintain correct oil clearance when the crankshaft journal has been ground to provide a new journal surface.
- Oversize bearings are used if the bearing bore diameter has been increased.
- The piston, when assembled to the connecting rod, is designed to transmit the power produced in the combustion chamber to the crankshaft.

- The piston is manufactured with expansion controls to provide better cylinder fit when the engine is at normal operating temperature.
- Pistons are generally made of aluminum; they may be cast, forged, or hypereutectic cast.
- The piston pin is generally offset toward the major thrust side to reduce piston slap.
- To seal combustion pressures, the piston is fitted with compression rings. The rings seal the piston to the cylinder wall.
- Cast iron, molybdenum, chrome, and steel rings may be used, depending on the engine design and cylinder condition.
- The connecting rod links the piston to the crankshaft; it is typically formed in I-beam construction to provide the required strength to prevent bending.

REVIEW QUESTIONS

Short-Answer Essays

1. Describe the function of the engine bearings.
2. Explain the bearing characteristics, spread, and crush and why they are important.
3. Explain what problems occur when the bearing oil clearances are too large.
4. List the parts of the piston and their purpose.
5. Explain the advantages and disadvantages of cast, forged, and hypereutectic pistons.
6. Describe cam ground and tapered pistons, and explain how they enhance engine performance.
7. Explain why piston pins are offset in the pistons.
8. Describe the three rings and their purposes.
9. Explain the advantages and disadvantages of cast-iron, chrome, and moly rings.
10. Describe the design and purpose of the connecting rod.

Fill-in-the-Blanks

1. Typical bearing oil clearances on modern engines are between _____ inch and 0.0008 inch.

2. _____ _____ is important for heat transfer from the bearing to the block.
3. _____ _____ pistons are narrower across the pin side to allow for expansion.
4. Connecting rods are generally made with an _____ _____ shape construction to provide adequate strength.
5. The _____ _____ transmits the force of the combustion pressures applied to the top of the piston to the crankshaft.
6. _____ are used to carry the loads created by rotational forces. _____ _____ carry the load created by the horizontal movement of the crankshaft.
7. _____ bearings are used to maintain correct oil clearance when the crankshaft journal has been ground to provide a new journal surface. _____ bearings are used if the bearing bore diameter has been increased.
8. A _____ ensures proper installation of the bearing insert in the bore.
9. Steel struts help control piston _____.
10. Some connecting rod caps are _____ off the connecting rod during the manufacturing process to provide an improved cap-to-rod fit.

Multiple Choice

1. Undersize bearings
   A. have a smaller outside diameter.
   B. are used when the block has been line bored.
   C. have a thinner lining than standard bearings.
   D. are used when the crankshaft has been ground.

2. When bearing oil clearances are too large
   A. oil pressure will be reduced.
   B. shims are used to take up the clearance.
   C. the crankshaft will bend.
   D. All of the above.

3. Each of the following is a common type of piston, *except*:
   A. Forged aluminum
   B. Forged steel
   C. Cast aluminum
   D. Hypereutectic

4. The piston pin is generally offset to
   A. reduce bearing knock.
   B. protect the piston pin from combustion forces.
   C. reduce piston slap.
   D. All of the above.

5. Engine bearings may be made with each of the following as a liner *except*:
   A. Babbitt
   B. Aluminum
   C. Copper
   D. Cast iron

6. Rings are being chosen for an engine. Which of the following statements is true?
   A. If the engine has been freshly bored, cast-iron rings will provide the longest service life.
   B. Moly rings should be used on slightly worn cylinders for quick wear-in.
   C. Chrome rings will last a long time but may wear the cylinder walls.
   D. Many light-duty, lower-powered engines use steel rings.

7. Piston head designs can
   A. increase the compression ratio.
   B. increase turbulence.
   C. increase emissions.
   D. All of the above.

8. While discussing connecting rod bearings, *Technician A* says that the bearing curvature is slightly greater than the curvature of the connecting rod bore.
   Technician B says that when half of the bearing insert is installed in the rod cap, the bearing should protrude above the cap surfaces.
   Who is correct?
   A. A only
   B. B only
   C. Both A and B
   D. Neither A nor B

9. While discussing pistons, *Technician A* says that the larger diameter of a cam ground piston is across the piston pin bores.
   Technician B says that pistons are manufactured with a perfectly vertical skirt.
   Who is correct?
   A. A only
   B. B only
   C. Both A and B
   D. Neither A nor B

10. Common piston ring designs include
    A. a three-piece oil ring.
    B. a top compression ring with no chamfer or taper.
    C. a top compression ring with an expander behind the ring.
    D. an oil ring below the piston pin.

CHAPTER 14
ALTERNATIVE FUEL AND ADVANCED TECHNOLOGY VEHICLES

Upon completion and review of this chapter, you should understand and be able to describe:

- The reasons for manufacturers' research and development of alternate fuel and advanced technology vehicles.
- The common uses for alternatively fueled vehicles.
- The difference between dedicated and retrofitted alternate fuel vehicles.
- The basic operation of a propane-fueled vehicle.
- The benefits of E85.
- The basic operation of a flexible fuel vehicle.
- The benefits and limitations of compressed natural gas as an automotive fuel.

- The basic operation of a CNG-fueled vehicle.
- The advantages and disadvantages of electric automobiles.
- The benefits of hybrid electric vehicles.
- The difference between the various hybrid electric vehicle power system layouts.
- The basic operation of a hybrid electric vehicle.
- The basic operation of the Atkinson-cycle and Miller-cycle.
- Some of the benefits and limitations of fuel cell vehicles.

Terms To Know

Atkinson-cycle engine	E85	Parallel system
Closed loop system	Flexible fuel vehicle (FFV)	Power splitting device
Compressed natural gas (CNG)	HD-5 LPG	Pumping losses
Converter	Inverter	Regenerative braking
Dedicated alternative fuel vehicle	M85	Retrofitted
Dual parallel system	Miller-cycle engine	Series system
	Open loop system	SULEV

INTRODUCTION

Automobiles contribute greatly to North America's air pollution. In many urban areas, air pollution from vehicles is the primary source of pollution. According to the U.S. EPA, mobile sources contribute 51 percent of the carbon monoxide pollution. The governments of the United States and Canada and their respective environmental protection agencies implement legislation that requires vehicle manufacturers to continuously reduce the levels of emissions their vehicles emit. They also demand increases in fuel economy. In the United States, the Clean Air Act of 1990 produced a long-range set of milestones that require manufacturers to reach lower and lower levels of emissions with a greater percentage of their fleets. In an effort to reduce emissions, increase fuel economy, and comply with the environmental laws, manufacturers have invested millions of dollars into research and design of alternatively fueled and advanced technology vehicles. Many fleet vehicles run on propane, compressed natural gas (CNG), biodiesel, or E85. Recently, many fleets have purchased electric and plug-in hybrid vehicles. Fleets that have these vehicles are also installing charging stations. Many of these fleets use government incentives and grants to help purchase and maintain their fleets. When crude oil prices go high, more fleets and companies may choose alternative fuels for various reasons. There are several advanced technology hybrid vehicles now on the market that offer significantly reduced emissions and increased fuel economy.

The engines of the alternative fuel and hybrid vehicles on the market today are still very similar to the engines we have been discussing throughout the book. Changes will be noted when appropriate. The electric motors used in hybrid vehicles and their components are an entirely different type of power source. Their operation is discussed in a general fashion in this text, and safety precautions will be offered in the Shop Manual.

ALTERNATIVE FUEL VEHICLE USE

Many urban areas in the United States and in Canada have air quality that falls below safe levels on certain days or periods of the year. These areas are called nonattainment areas, meaning the quality of air regularly falls below the safe standard. Individual regulations vary by state and province, but many urban areas require that fleets run a percentage of their vehicles on alternative fuels. The Clean Air Act also introduced the Clean Fuel Fleet program in the United States. The act requires that in some nonattainment areas, certain fleets meet special fuel vehicle emissions standards. Transportation money from the federal government in the United States is also tied to a state's actions to reduce mobile source emissions. Many municipal, government, and educational fleets run on alternate fuels.

Alternative fuel vehicles may be **dedicated** or **retrofitted**. A dedicated propane vehicle, for example, is one that the manufacturer designed from the ground up to run on propane. Changes to the body, suspension, and fuel system were all designed to optimize the vehicle to run on propane. The vehicle's oxygen sensors, catalytic converters, and powertrain control modules have all been designed to run on the alternative fuel. Many alternative fuel fleet vehicles have been retrofitted to run on an alternative fuel. This means that a particular gas vehicle was chosen and then conversions were made to allow the vehicle to run on the alternative fuel. The dedicated fuel vehicles are more efficient, but they are not always available from the manufacturer to serve the needs of the consumer.

Dedicated alternative fuel vehicles were designed to run on an alternative fuel.

A vehicle that was originally designed to run on gasoline but has been changed to run on an alternative fuel is often called **retrofitted** or converted.

PROPANE VEHICLES

The first widely used alternative fuel vehicle has been the propane vehicle. Propane is liquefied petroleum gas (LPG) but often simply called propane or LPG. Automotive engine grade is **HD-5 LPG**. HD-5 consists of a minimum of 90 percent propane and a maximum of 5 percent propylene. Many fleets operate propane trucks and buses. In the United States, there are an estimated 350,000 trucks, cars, and industrial vehicles operating on propane. Propane's popularity as an alternative fuel is growing because retrofitters are now starting to produce both throttle body and multiport injection systems. There are also retrofit kits available for diesel engines. The propane fuel is blended with the diesel fuel.

Propane cannot start the engine alone, because of the high compression pressure in the engine. Therefore, the ratio of propane to diesel is lower during engine starting and continues to increase as the engine temperature rises. This continues to a point where

HD-5 LPG is a "consumer-grade" propane.

there is only enough diesel fuel being injected to cause ignition; then the propane is allowed to complete the combustion. A few manufacturers have offered dedicated propane vehicles or vehicles with factory conversion kits with modern fuel-injection systems in the past several years.

Vehicles running on propane produce lower hydrocarbon (HC) and carbon monoxide (CO) emissions. The reductions can be as high as 50 to 80 percent. These vehicles also produce up to 30 percent lower oxides of nitrogen (NO_x) emissions. Propane is also a sealed system, so it has no evaporative emissions. It is a clean-burning fuel with low particulate and sulfur emissions. This makes it an attractive fuel for fleet operators in urban locations who are required to use an alternate fuel. LPG is a by-product of natural gas and crude oil refining; 65 percent of propane comes from natural gas. This is advantageous to both the United States and Canada, which have the largest stores of natural gas in the world. Natural gas is plentiful and cheaper to refine.

Most propane supply companies run all their vehicles on propane. The United States has more than 10,000 refilling stations available, while Canada has about 5,000 refilling facilities.

Unfortunately, fuel consumption, by volume, is slightly higher with propane than with gasoline. Propane actually has more energy available in it, but it is not as dense, so by volume it has less available energy. Because it is roughly 30 percent less dense than gasoline, it is also difficult to carry as much fuel as would be necessary to maintain the range of a gas vehicle. Propane tanks are specially made sealed units because the fuel is stored at pressures that may be over 200 psi (**Figure 14-1**).

Propane vehicles run very similarly to gas-fueled vehicles. With a few modifications to the engine, virtually any vehicle can be converted to run on propane. Because of the size of the storage tanks, it is more common to convert trucks or larger vehicles. Vehicles can be equipped to run solely on propane or to be dual-fuel vehicles and able to run on either gasoline or propane. The tanks are fitted either in the trunk or under the vehicle (**Figure 14-2**).

Propane is stored in the tank(s) at temperatures roughly between 100°F and 200°F (55°C and 110°C). A conversion system includes a fuel controller, valves, actuators, electronics, and software modifications. The liquid fuel travels from the tank to a

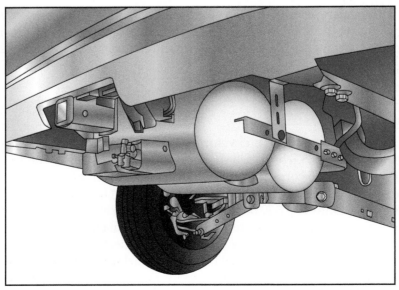

Figure 14-1 Liquid propane storage tanks may be mounted under the vehicle or in the trunk.

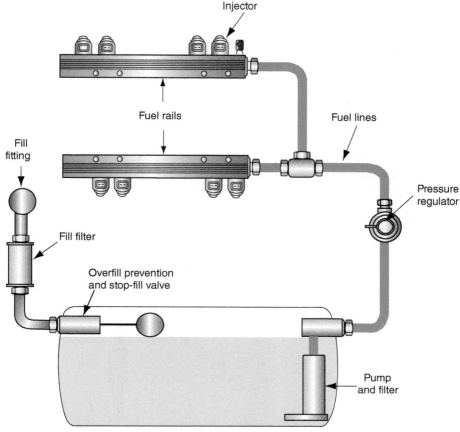

Figure 14-2 Fuel delivery components of the liquid propane injection system.

vaporizer/pressure regulator (also called a converter), which converts propane to a gaseous state. From the vaporizer, the fuel goes to the throttle body injector, the carburetor, or the fuel injectors for delivery to the intake manifold.

There are federal, and often state and local, tax incentives for converting vehicles to cleaner burning alternative fuels that help make up the cost of conversion. Other factors include the following:

- Original fuel economy of the vehicle
- Number of miles traveled per year
- Initial cost of conversion
- Engine life expectancy

Factory conversions of passenger cars and light-duty trucks can cost about $2,500 above the cost of the conventional vehicle. Aftermarket conversion kits may cost $1,500 to $3,000, depending on the amount of components needed and whether the system is an **open loop system** or a **closed loop system**. California requires warranties on conversion kit components, and Canada has regulations on component safety as well as installation and installers.

Dedicated propane fuel vehicles differ very little from gas vehicles aside from the fuel tank. A pump in the tank delivers liquid fuel at the correct pressure to the fuel injectors. Unused vaporized fuel returns to the tank. A single hose with two sections is used to deliver and return fuel (**Figure 14-3**).

When the system is a bifuel system, a fuel selector switch is used to inform the PCM which fuel is being used. The operator can change the fuel setting, or if the propane tank runs low enough, the PCM will automatically change to gasoline. This changes the fuel delivery strategy to optimize performance, emissions, and fuel economy (**Figure 14-4**).

An **open loop system** delivers fuel based on demand and has no way of detecting whether the amount sent was ideal in terms of performance, emissions, or fuel economy.

A **closed loop system** uses sensors (typically an oxygen sensor) that inform the PCM about the air-fuel ratio leaving the engine, so it can adjust accordingly.

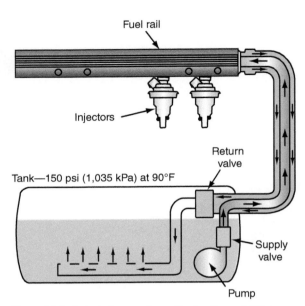

Figure 14-3 A single hose performs the functions of delivery and return.

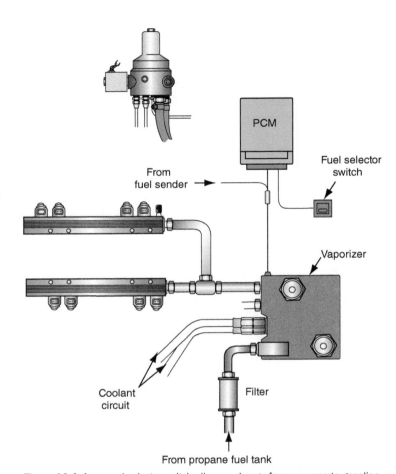

Figure 14-4 A manual selector switch allows a change from propane to gasoline.

The engines in propane vehicles may be identical to those in gasoline engines. Using propane reduces the wear on the engine because it is a cleaner fuel and forms less carbon and corrosive acids. Some fleet operators claim that propane engines last two to three times longer than their gasoline-powered counterparts. The National Propane Gas Association claims that spark plugs last 80,000 to 100,000 miles. The same mechanical failures and maintenance procedures exist; the occurrences are often extended.

E85 AND FLEXIBLE FUEL VEHICLES

E85 is an automotive fuel made of 85 percent ethanol and 15 percent petroleum.

A flexible fuel vehicle (FFV) is designed to be able to run on either gasoline or a methanol or ethanol blend, M85 or E85.

M85 is an automotive fuel that contains 85 percent methanol and 15 percent gasoline. It is not commonly used anymore.

E85 is made from 85 percent ethanol and 15 percent petroleum. Ethanol can be made from virtually any starch feedstock such as corn, sugarcane, or wheat. Most of the U.S. production of ethanol today comes from corn. A bushel of field corn can be converted into 2.7 gallons of fuel ethanol. Some ethanol producers are reaching even greater efficiencies. E85 is a domestically renewable energy source that helps reduce dependency on imported oil. This has a positive influence on the economy. It also helps reduce tailpipe and greenhouse gas emissions (carbon dioxide). E85 has the highest oxygen content of any fuel currently available. This makes it burn much more completely and cleaner than gasoline. EPA studies have shown that high-blend ethanol fuels can reduce harmful exhaust emissions by over 50 percent. E85 also reduces carbon dioxide (CO_2), the emission contributing to global warming, by nearly 30 percent.

Flexible fuel vehicles (FFVs) are designed to run on E85 or gasoline. M85 shares many of the attributes and benefits of E85. Older flexible fuel vehicles could run on E85, regular gasoline, or M85. **M85** uses a mix of methanol and gasoline. Methanol is sometimes used by race cars but is quickly being replaced by E85. The 2007 Indianapolis 500 race cars were all powered by ethanol. The use of M85 as an automotive fuel has decreased in popularity because of its toxicity, corrosiveness, and groundwater contamination issues. FFVs have been designed by the manufacturers to run on multiple fuels: They are not converted afterward. They typically use a fuel sensor that determines the level of alcohol in the fuel, and the PCM modifies its fuel delivery, ignition timing, and emissions strategies accordingly (**Figure 14-5**). The sensor can be located in the tank or in the fuel line after the tank (**Figure 14-6**). Instead of a special sensor, some manufacturers use wide-band oxygen sensors to determine the fuel content and adjust the fuel curve accordingly. The engine is essentially the same as a gasoline model, though different engine oil may be specified when running exclusively on E85. The fuel system components have been modified to resist the added corrosive effects of the alcohol in M85 and E85. A typical layout of an E85 vehicle is shown in **Figure 14-7**.

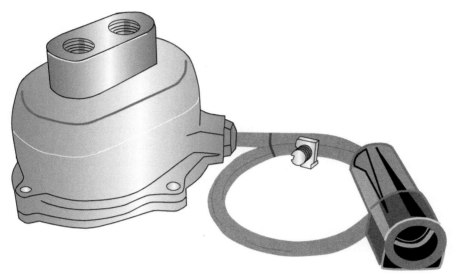

Figure 14-5 A flexible fuel sensor.

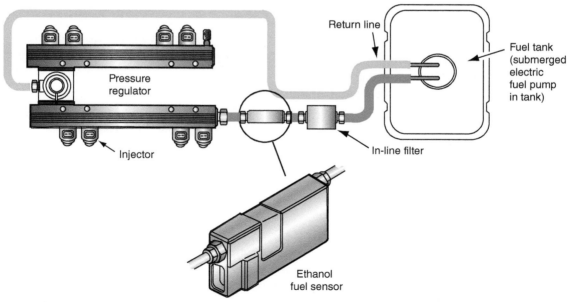

Figure 14-6 This E85 flexible fuel system mounts the fuel sensor in line between the tank and the fuel rail.

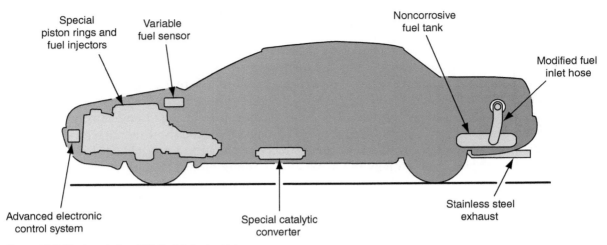

Figure 14-7 The layout of an E85 flexible fuel vehicle.

COMPRESSED NATURAL GAS VEHICLES

Compressed natural gas (CNG) is a promising alternative fuel for the future. It is a clean-burning fuel, is available in abundance domestically (both in the United States and Canada), and it is commercially available to end users at refilling stations. Natural gas is the cleanest burning fuel used today. CNG vehicles can emit 70 percent less CO, 89 percent less HC, and 87 percent less NO_x. CNG vehicles also produce significantly lower (20 percent) greenhouse gas emissions. Dedicated CNG vehicles produce little or no evaporative emissions during fueling or use. This is a huge advantage over conventional gas vehicles and gas filling stations. Gas vehicle evaporative emissions and fueling contribute over 50 percent of the vehicles' HC emissions.

Natural gas is minimally refined and widely available as a resource in the United States and Canada. Prices for CNG should be low once production is increased and refueling stations are abundant. Already many CNG refilling stations are available, which is another benefit of CNG as an alternative fuel choice (**Figure 14-8**).

CNG has an octane rating of between 110 and 130. This allows dedicated CNG vehicles to run higher compression ratios and advanced ignition timing. The potential power from a CNG vehicle is increased, even though CNG has a lower energy content than gasoline. Vehicles can be converted to CNG, or they can be dedicated CNG or bifuel vehicles. Several manufacturers produce CNG vehicles, including the following:

- Chevrolet Cavalier
- Dodge Caravan and B Van
- Ford Crown Victoria and Contour
- Select GM trucks
- Honda Civic GX
- Toyota Camry
- Volvo S70 and V70

Many of these models are offered as bifuel vehicles, which can run on CNG or unleaded gasoline.

Cylinder Safety

The layout of a typical CNG vehicle is shown in **Figure 14-9**. The CNG is stored in tanks under pressures between 3,000 and 3,600 pounds. Some of the tanks are going as high as 100 bar. These tanks are typically aluminum cylinders wrapped in fiberglass because they cost less and weigh less than comparable cylinders made of steel or composites. All tanks are sealed with epoxy paint and clear polyurethane to further protect them from environmental factors.

Figure 14-8 A compressed natural gas filling station.

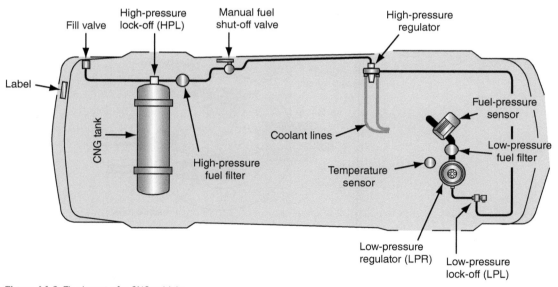

Figure 14-9 The layout of a CNG vehicle.

All cylinders are certified by the U.S. Department of Transportation and labeled with the psi capacity, serial number, manufacturer's name, and date of manufacture. The U.S. DOT requires that tanks be inspected every three years. The inspection must be performed by a qualified service person in accordance with the manufacturer's inspection process. Retest marking labels must be affixed to the tank near the original test date label and coated with epoxy. No repairs are allowed to the CNG cylinders. Cylinders must be removed from service 15 years after the date of manufacture.

In Canada, cylinders must be inspected and also hydrostatically tested every three years, as regulated by the Canadian Standards Association. A retest label must be placed near the original label and sealed with epoxy. Repairs are not allowed, and all cylinders must be replaced 15 years after the date of manufacture. A label may be attached to the fuel filler door stating the retest and expiration date.

Vehicle Safety

There are two safety devices designed to stop fuel flow if leaks or service need to be performed. The fuel control valve is a manually operated ball valve located on each tank (**Figure 14-10**). Full clockwise rotation turns the fuel flow off. The control valve also houses a pressure relief valve. The cylinder releases excess pressure when temperatures reach approximately 215°F (100°C). Remember, temperature increases as pressure increases. The relief valve will vent all the CNG from the cylinder. This does not pose a fire hazard, as CNG rises into the atmosphere rather than hangs at ground level like propane or gasoline. Once a relief valve has popped, it must be replaced.

There is also a quarter-turn manual shut-off valve (**Figure 14-11**). It is accessed from the outside of the vehicle and is labeled "Manual Shut-Off Valve" on a body panel. When the valve is parallel to the line, fuel can flow. When the handle is perpendicular to the fuel line, fuel flow is stopped from the tanks to the fuel delivery system.

CNG vehicles use high-pressure stainless steel seamless fuel tubes. Fittings on the high-pressure side of the system are double-ferrule, compression-type fittings made of special stainless steel (**Figure 14-12**). Fittings on the low-pressure side of the system use national pipe thread (NPT) and 45-degree flare fittings.

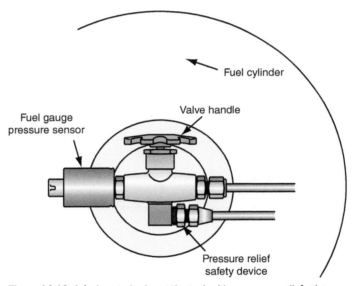

Figure 14-10 A fuel control valve at the tank with a pressure relief valve.

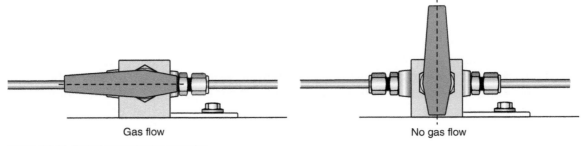

Gas flow   No gas flow

Figure 14-11 Typical manual valve operation.

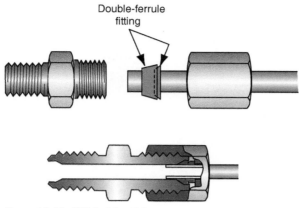

Figure 14-12 CNG fuel line fittings.

Vehicle Operation

Fuel flows through either a single dual-stage regulator or a primary and then a secondary regulator. The regulator reduces the pressure to between 90 psi and 140 psi. The regulator may contain a built-in pressure relief valve and fuel filter (**Figure 14-13**). The relief valve is on the low pressure side of the regulator and relieves pressure above roughly 200 psi, depending on the vehicle model. The pressure regulator has engine coolant running to it because as the pressure drops rapidly, so does the fuel temperature. The coolant prevents fuel from freezing (**Figure 14-14**). The PCM controls a high-pressure fuel shut-off solenoid and a low-pressure fuel solenoid. The solenoid is an on/off valve that controls fuel flow. The PCM operates a relay to turn the solenoid on and off. The PCM also uses a fuel low-pressure sensor to help with its injector timing strategy (**Figure 14-15**). It also uses a fuel temperature sensor as an input to help calculate fuel delivery and ignition timing. The PCM uses these sensors along with other typical engine sensors to optimize performance, emissions, and fuel economy.

Refilling a CNG vehicle is done through a traditional fuel door (**Figure 14-16**). The connector to which you attach the quick connect fitting contains a one-way check valve (**Figure 14-17**). Another one-way valve is installed between the fuel fill line and the fuel-pressure regulator to prevent any release of gas to the atmosphere through the filling connection. It is important to remember that most CNG vehicles are serviced at a dealership or specialized company by factory-certified and trained technicians.

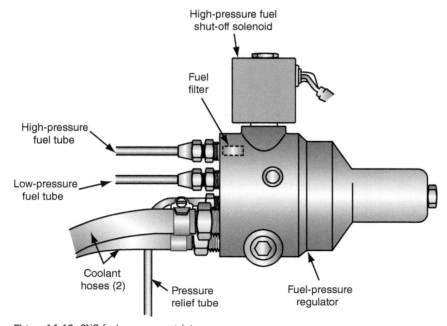

Figure 14-13 CNG fuel-pressure regulator.

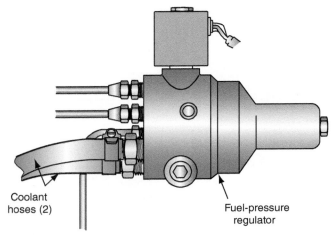

Figure 14-14 Engine coolant hoses routed to the regulator prevent icing as the pressure drops at the regulator.

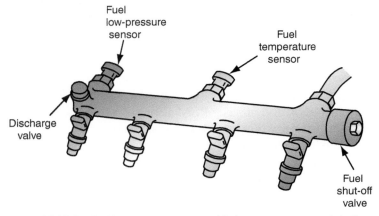

Figure 14-15 The fuel low-pressure sensor and fuel temperature sensor help the PCM determine the injector on time.

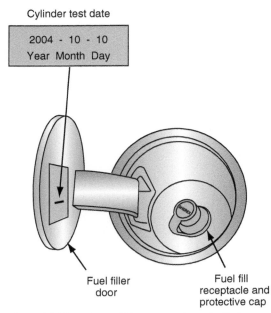

Figure 14-16 CNG tank filling receptacle and cylinder test date sticker.

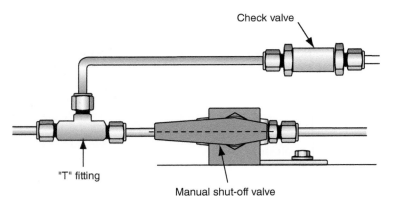

Figure 14-17 A one-way check valve prevents fuel from escaping the system during refueling.

THE HONDA CIVIC GX CNG VEHICLE

As an example of a specific dedicated CNG vehicle, we look at Honda's Civic CNG vehicle. It was the first vehicle in the United States to meet the California Air Resources Board (CARB) "Advanced Technology Partial Zero–Emissions Vehicle" (AT-PZEV) standard. This gives users emissions credits toward highway funding and EPA regulations. It also easily meets the Super-Ultra-Low-Emission Vehicle (SULEV) standard that was required of certain vehicles in some states in 2005.

Its engine is an aluminum alloy in-line four-cylinder with 1.7 L (1,590 cc) displacement. It runs a compression ratio of 12.5 to 1, and the CNG octane is rated at 130. It has an SOHC 16-valve head with variable valve timing (VTEC-E). The valves and valve seats have been modified. The pistons and connecting rods have been strengthened (**Figure 14-18**). It produces 98 lbs.-ft. of torque at 4,100 rpm. It is connected to a continuously variable transmission (CVT). It holds the equivalent of 7.5 to 8 gallons of gasoline and has a "real-world" driving range of 200 miles (320 km). It has a fuel economy rating of 31 mpg city and 34 mpg highway.

The emissions from the vehicle are particularly impressive (**Figure 14-19**). The CO, NMOG (hydrocarbon), and NOx emissions compared to the Civic gasoline Ultra-Low-Emission Vehicle (ULEV) shows the significant reductions. The CNG Civic uses two specially coated oxygen sensors to reduce the effects of hydrogen on their operation. The catalyst system includes a close-coupled catalyst and an under-floor catalyst to reduce emissions (**Figure 14-20**). Fleets are widely using Honda CNG vehicles (**Figure 14-21**).

> A **SULEV** vehicle produces an extremely low measurable amount of pollution from its tailpipe. This is a Federal government vehicle standard.

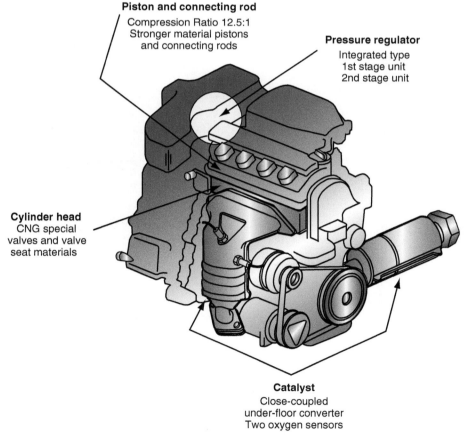

Piston and connecting rod
Compression Ratio 12.5:1
Stronger material pistons
and connecting rods

Pressure regulator
Integrated type
1st stage unit
2nd stage unit

Cylinder head
CNG special
valves and valve
seat materials

Catalyst
Close-coupled
under-floor converter
Two oxygen sensors

Figure 14-18 Engine modifications are made to the dedicated CNG Civic.

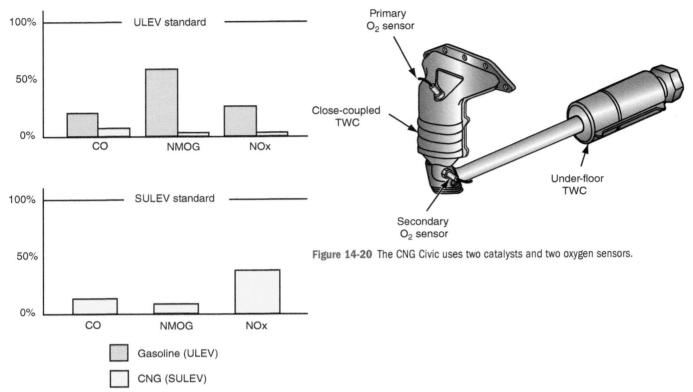

Figure 14-20 The CNG Civic uses two catalysts and two oxygen sensors.

Figure 14-19 The CNG Civic easily meets the Ultra-Low-Emission Vehicle (ULEV) and Super-Ultra-Low-Emission Vehicle (SULEV) standards. It also reduces emissions dramatically compared to its very clean Civic gasoline model.

Figure 14-21 This urban cab company uses CNG vehicles in its fleet.

ELECTRIC VEHICLES

Electric vehicles (EVs) are driven by an electric motor powered by batteries. General Motors introduced the EV1 electric car in 1996. Many manufacturers have produced some version of an electric vehicle in the 1990s for fleet use, in part to meet California's regulation that a certain percentage of fleet vehicles be zero-emission vehicles (ZEV). These vehicles are ideal for urban driving where range is short. The driving range of GM's EV1 is roughly 75 miles, depending on the type of driving. This limits the use of electric vehicles dramatically.

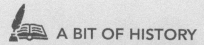

A BIT OF HISTORY

Electric vehicles have been around since the early 1800s, although they were relatively impractical until later in the century. They were at the peak of popularity in the United States during the early 1900s.

Battery technology has also limited the further development of electric vehicles as an alternative for the general public. The EV1 used 26 12-volt batteries to power the 102-kilowatt AC electric motor that drives the front wheels. The basic layout of the electric vehicle is shown in **Figure 14-22**. Fully charging a discharged set of batteries through a 220-volt outlet can take 14 hours. A Magne Charge inductive charger was developed by AC Delco to reduce the time to 3 to 4 hours. Nonetheless, until batteries that can provide longer range are developed, the production of electric vehicles for use by the general public is diminished. GM discontinued the production of the EV1 in 1999 and removed all of them from the road by 2003 because they were all originally leased. Some of the technology and experiences learned from the EV1 led to the development of today's hybrid electric vehicles (HEV).

The Chevy Volt (**Figure 14-23**) is considered an "extended range" EREV. The Gen 1 Volt has a battery range of between 25 and 50 miles while its total or extended range is up to 350 miles. In most circumstances it uses its 1.4-liter gasoline engine to power the generator. The engine can operate on demand under its more efficient rpm range. Under normal circumstances, the engine does *not* produce more energy than demanded at the time. The second generation Volt has a battery range of up to 53 miles while its total or extended range is up to 425 miles. It is more efficient using a 1.5 liter engine with direct injection and uses a wider battery charge range.

GM also has the all-electric Spark Bolt EV. The Bolt EV has a 200-mile range, 200 horsepower (hp), and 266 lbs.-ft. of torque, using a lithium-ion battery. Out of the multiplatform vehicle manufactures, the Nissan Leaf has the next closest range at 100 miles per charge. The Spark EV has an 82-mile range, 140 hp, and 327 lbs.-ft. of torque. Most of the vehicle manufacturers now offer at least one full electric model vehicle. Tesla builds electric cars exclusively that have a range of 218 to 270 miles, 287 to 713 lbs.-ft. of torque and 259- to 762-motor horsepower.

Electric vehicles and transport mechanisms are widely used in industrial applications, such as in warehouses, where emissions cannot be easily vented.

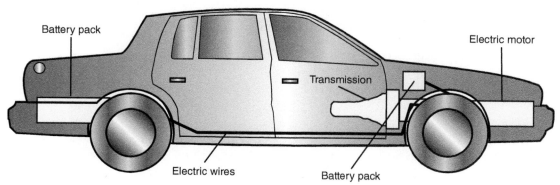

Figure 14-22 The batteries supply power to the electric motor that drives the vehicle.

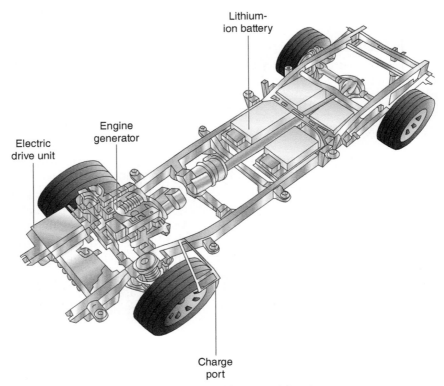

Figure 14-23 Chevy Volt's engine generator and battery pack layout.

HYBRID ELECTRIC VEHICLES

Hybrid electric vehicles use two power sources to propel the vehicle. Versions sold in the United States and Canada use an electric motor(s) and a gas or diesel engine to reduce emissions, improve fuel economy, and in some cases, increase power. The first mass-produced, commercially available hybrid vehicles were the Honda Insight and the Toyota Prius (**Figure 14-24**). The top three hybrid cars sold are the Toyota Prius, the Camry, and the Honda Civic. The problem of vehicle range has been eliminated, as the gas engine and brake system can be used to recharge the battery pack. The Honda Insight has a range of more than 600 miles on its full 10.6-gallon gas tank, depending on driving conditions. Tailpipe emissions are also greatly reduced; in 2008, the Toyota Prius earned the AT PZEV government emissions rating. The Toyota Prius continues to be a well-selling vehicle; as of

Figure 14-24 The Toyota Prius has become a very popular hybrid vehicle.

April 2016, total sales of the car reached over 5.6 million. Many people believe that until fuel cell and hydrogen-powered vehicles become a mass-produced affordable reality, HEVs will have a significant impact on market product and emissions.

AUTHOR'S NOTE I see more and more HEVs commuting on the road with me as I regularly drive to work. Friends report that they are very pleased with the performance of their Prius and the Honda Civic hybrids. I drove a Prius for work for a few years. I found it to be easy to operate and work on, plus it fit my family in there. A few of the owners with these cars have, however, complained of lower fuel mileage than the posted estimates. These fuel economy ratings are provided by the Environmental Protection Agency, not the vehicle manufacturer. Most vehicles achieve lower actual fuel economy than the posted rating. I found that there is a wide gap in the fuel economy, and it is based on driving habits and patterns. On some days the Prius I drove could get up to 65 mpg average with easy driving and slow acceleration. On other days when I was in heavy highway commuter traffic and harder acceleration, it would go down to 40 mpg. Another consideration is that HEVs are designed to get better fuel economy during city driving, when the electric motor gets more use, than during extended travel on the highway.

HEV OPERATION

The vehicle is powered by either the electric motor, the engine, or in some cases, both. An HEV system can be characterized as a **series system**, a **parallel system**, or a **dual parallel system**. In a series system, just one of the power sources can be used to drive the vehicle. The electric motor drives the vehicle, while the gasoline or diesel engine powers a generator that charges the batteries to power the motor. A parallel system, which most current models employ, can use the power of both the electric motor and the engine simultaneously to propel the vehicle. The parallel system is named as such because both power sources can apply torque to move the vehicle. The electric motor can also act in reverse, as a generator to recharge the batteries. A parallel system can also use either the electric motor or the engine exclusively to power the vehicle (**Figure 14-25**). A system is called **dual parallel** when it has two electric motors and one engine, all of which can drive the vehicle at the same time. This is employed on pickup and sport utility vehicles to date.

An AC electric motor is typically used to start the vehicle from a stop. Electric motors generate maximum torque at 0 rpm, so acceleration is acceptable even when a smaller gas engine is used. The electric motor is charged by a battery pack. During light acceleration or cruise, the electric motor may work alone to power the vehicle. On some vehicles when power is in high demand, both the electric motor and the gasoline engine are used to maximize performance. During moderate acceleration, the gasoline engine may run alone to provide the torque and horsepower required at higher speeds. The engine may also run without the electric motor when the batteries need to be charged. The gasoline engine powers the generator to charge the batteries while it is running. When the vehicle comes to a stop, both the gas engine and the electric motors are turned off to reduce emissions and consumption to zero. At the instant the throttle is pressed, the electric motor begins to spin and accelerate the vehicle. Some trucks and SUVs using a dual parallel hybrid electric vehicle have a traction motor at each axle to retain the four-wheel-drive capacity (**Figure 14-26**).

Some vehicles also employ **regenerative braking**. On these systems, the kinetic energy typically wasted during braking is converted into electrical energy to recharge the batteries. It also helps stop the vehicle, as a noticeable drag is felt while the vehicle momentum turns the motor or generator to charge the batteries.

A **series system** HEV uses the electric motor to drive the vehicle; the gas motor recharges the batteries to power the electric motor.

A **parallel system** HEV can use the electric motor and the engine to power the vehicle. It can also use just one power source at a time.

In a **dual parallel system** HEV, two electric motors and an engine are used to drive the vehicle.

Regenerative braking uses the lost kinetic energy during braking to help recharge the batteries.

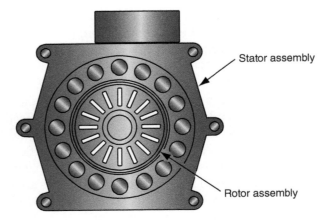

Figure 14-25 The motor is used to generate electricity as well as drive the vehicle. A rotor and stator are also used in a traditional generator.

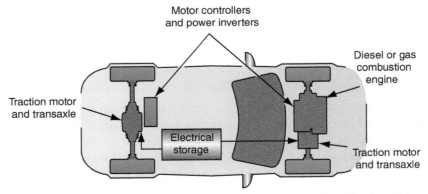

Figure 14-26 This dual parallel hybrid has a traction motor at each axle to drive the vehicle, plus a traditional gas engine.

On vehicles that use the gasoline motor and regenerative braking to recharge the batteries, external charging is never needed. This makes these vehicles much more appealing to the consumer. The consumer only needs to fill up the gas tank, like any traditional vehicle owner.

Batteries

The batteries currently being used or in prototype are our familiar lead-acid batteries, nickel-metal hydride (NiMH) batteries, and lithium-ion batteries. Each of these batteries has advantages and disadvantages over the other. Considerations for use include longevity, recharging ability, weight, and cost. The battery packs must be maintained at a relatively constant temperature, so they require a heating and cooling system (**Figure 14-27**). Manufacturers estimate that battery packs may need to be replaced after 10 years of use. The current price of a battery pack runs between $1,800 and $3,000. Some of the other electronic components can cost several thousand dollars; an inverter for a new Prius is close to $5,000. Manufacturers are offering excellent warranty programs for their batteries and hybrid components to keep the new technology popular and customers satisfied. Battery packs and HEV electronic components are generally warrantied for 10 years.

Voltages on hybrid vehicles can run up to 500 volts, so a technician must thoroughly understand the safety precautions before servicing the hybrid vehicle.

Electric Motors

Modern HEVs typically use one or more alternating current (AC) permanent magnet (PM) motors to drive the vehicle. The operation of the motor is similar in theory to that of the operation of the permanent magnet starter motors used on many of today's vehicles (**Figure 14-28**). As a matter of fact, one of the electric motors may be used to start the engine and take the place of a traditional starter. The HEV motors use high AC voltage to deliver the high-power demands of moving the vehicle. They may also be called traction motors. The motor shown in **Figure 14-29** is mounted between the engine and transmission on the flywheel.

The starting system of a hybrid electric vehicle is different from the normal starting system. The high-voltage battery pack is typically used to start the gasoline engine. The same set of electric motors that are used for propulsion are used in conjunction with a gear set to start the gasoline engine. An HEV's main computer will start the engine when the conditions are met. These conditions may be low voltage on the HEV battery pack, A/C demand, electrical demand, or acceleration demand. The starters on some modern gas and diesel engines are controlled by the computer, similar to an HEV. Special care must be taken when servicing these vehicles.

Other Hybrid Technology Components

A **power splitting device** is used to change the power from one source to the other and to combine the forces of both when necessary (**Figure 14-30**). A planetary gear set may be used to allow the electric motor to drive one member and the gasoline motor to drive another (**Figure 14-31**). Some HEVs use a continuously variable transmission (CVT) that houses some of the components and performs some of the functions of the power splitting device.

An **inverter** is used to convert the DC voltage stored by the battery pack to the AC voltage required by the electric motors. A **converter** is used to change the high-voltage battery output of 270 to 500 volts down to 12 volts, which can be used by the conventional electrical system. In most cases, there is no longer a traditional 12-volt generator because the battery pack can supply plenty of power for the vehicle accessories.

A **power splitting device** changes the mode of power delivery from one source to the other or to a combination of both sources.

A **converter** is used to change the very high battery pack voltage to 12 volts for the conventional electrical system.

An **inverter** is used to convert DC volts to AC volts.

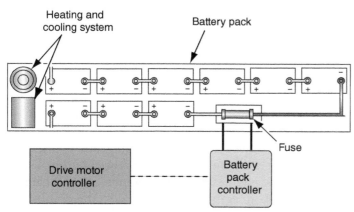

Figure 14-27 A typical 360-volt battery pack layout with a heating and cooling system.

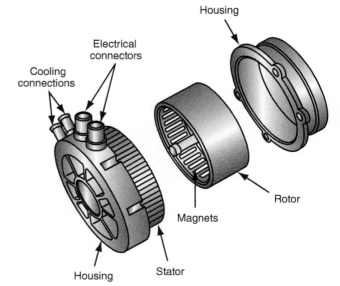

Figure 14-28 The components of a permanent magnet motor.

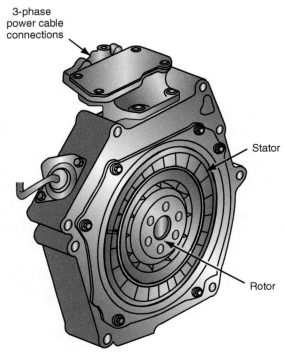

Figure 14-29 A flywheel-mounted PM motor.

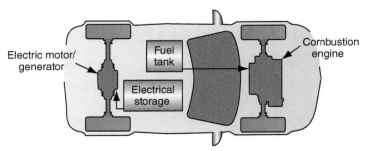

Figure 14-30 The power splitting device allows the vehicle to run on a combination of power sources or either one individually.

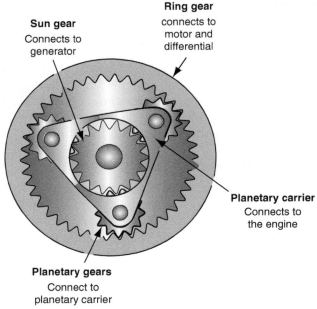

Figure 14-31 A planetary gear set uses different gear combinations to distribute power from the motor and engine.

Gasoline Engine

The gasoline engine used in an HEV may be the same as the one used in the traditional variant of the model or it may be specially fitted to be optimized for hybrid use (**Figure 14-32**). The changes to the gasoline engine are minimal in terms of operation and diagnosis. Many of the HEV gasoline engines are newer technology engines and employ variable valve timing and some changes to the intake system.

Atkinson-Cycle Engine

The **Atkinson-cycle engine** is a common gasoline internal combustion engine design used in some HEVs such as the Toyota Prius, Chevrolet Tahoe, Toyota Camry, and Ford Escape/Mercury Mariner. The Atkinson-cycle engine has a lower power output when compared to similar displacement Otto-cycle engines but also has increased efficiency. Originally, the Atkinson-cycle engine had a connecting link attached to the crankshaft that allowed all four cycles of a four-stroke cycle to occur in one crankshaft revolution. Today's Atkinson-cycle engines in HEVs refer to the principles of the cycles and the ratios, rather than having an additional linkage on the crankshaft.

The Atkinson-cycle engine actually sacrifices a little bit of power but gains 7 to 10 percent fuel mileage and further reduces emissions. Engines lose power as they try to pump the piston up against pressurized air during the compression stroke. These are

The **Atkinson-cycle engine** holds the intake valve open longer during the compression stroke to increase fuel economy.

Figure 14-32 This Honda Civic Hybrid uses a 1.3-L engine in its Integrated Motor Assist system.

the **pumping losses** that the Atkinson-cycle engine reduces. The Atkinson-cycle allows the intake valve to be open longer during the compression stroke. This action allows for more air to be pushed back into the intake manifold and less compression to occur. During the power stroke, the volume of air that was in the compression stroke has to expand to a volume that is greater. This event leads to an increase in the heat energy that is converted to mechanical energy. The mechanical efficiency of the engine is now higher, at the expense of the lower-power density of the charge during the power stroke. This event makes the volumetric ratio of the compression stroke to the power stroke less than 1:1.

The intake camshaft lobe is modified in an Atkinson-cycle engine to hold the intake valve open longer (**Figure 14-33**).

Pumping losses are wasted power from the pistons pushing up against compressed air during the compression stroke.

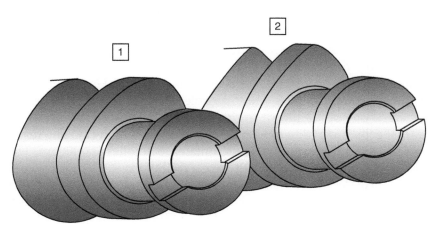

Figure 14-33 Lobe No. 1 shows the Atkinson style camshaft, compared to the traditional lobes shown in No. 2.

The **Miller-cycle engine** is similar to the Atkinson-cycle engine, but it holds the intake valve open for a longer time. This requires it to be accompanied by a supercharger; otherwise it would be a very low-power engine.

Miller-Cycle Engine

The **Miller-cycle engine** is a modification of the four-stroke cycle engine. A few manufacturers are producing prototype HEV vehicles that use a Miller-cycle gas engine. Some vehicles such as the Mazda Millennia are powered by a Miller-cycle engine. In this type of engine, a supercharger supplies highly compressed air through an intercooler into the combustion cylinders. The intake valves remain open longer compared to a conventional four-stroke cycle engine. This action prevents the upward piston movement from compressing the air-fuel mixture in the cylinder until the piston has moved one-fifth of the way upward on the compression stroke. The later valve closing allows the supercharger to force more air into the cylinder. The shorter compression stroke lowers cylinder temperatures and still gets the same (or more) compression as a standard Otto four-stroke design because the Miller design uses a supercharger to increase compression. Normally, without a supercharger, holding the intake valve open for a longer duration would result in a loss of compression and power. This operation makes the compression stroke less of a consumer of the total engine power.

By delaying the start of the compression stroke, the engine is not wasting energy by pushing the piston up on the compression stroke. As compression builds in the cylinder, the energy needed to turn the crankshaft and move the piston increases. This energy would normally be delivered by the crankshaft, which is powered by the other cylinders in the engine that are on the power stroke. Allowing for this delay to occur also allows for the crankshaft to be in a more mechanically advantageous position (past BDC).

The greater volume of air in the cylinder creates a longer combustion time, which maintains downward pressure on the piston for a longer time on the power stroke. This action increases engine torque and efficiency. The Miller-cycle principle allows an engine to produce a high horsepower and torque rating for the displacement of the engine. The Miller-cycle engine is similar to a modern Atkinson-cycle engine, except the Miller-cycle engines have a supercharger associated with their incoming air charge.

The greater volume of air in the cylinder creates a longer combustion time, which maintains downward pressure on the piston for a longer time on the power stroke. This action increases engine torque and efficiency. The Miller-cycle principle allows an engine to produce a high horsepower and torque for the CID of the engine.

Toyota Prius HEV

With more than 5 million on the road, Toyota is on its fourth-generation Toyota Prius. Its Hybrid Synergy Drive system is a very popular vehicle in many parts of North America (**Figure 14-34**). The Prius is a true hybrid, engineered from the ground up as an HEV. The newest Gen 4 uses a 1.8-L inline four-cylinder 13:1 compression Atkinson-cycle engine. Its DOHC cylinder head is fitted with 16 valves and variable valve timing, VVT-i. The engine itself delivers 95 hp at 5,000 rpm and 105 lbs.-ft. of torque at 3,600 rpm. The 600-volt/71 hp permanent magnet drive motor delivers 120 lbs.-ft. of torque. The estimated total horsepower available from the power plant is 121 hp. While less torque and horsepower than the third generation, its performance is comparable due to increased efficiency. Acceleration is from 0 mph to 60 mph (0–100 km/h) in just under 10 seconds. Some of its increased efficiency is due to a cooled exhaust gas recirculation system. Its drag coefficient is one of the lowest in the industry, helping contribute to its awarded midsize car fuel economy rating of 52 mpg combined city/highway. The fuel tank holds 11.3 gallons (43 L) and has a range of approximately 600 miles (1,000 km). The batteries never need external charging and do not have a facility for it. The Prius is also the best vehicle on the North American market in terms of emissions performance. The vehicle is classified as a SULEV vehicle and meets CARB's AT-PZEV rating. It is rated as one of the cleanest production cars on earth, producing 90 percent fewer emissions than the traditional conventional internal combustion engine. Toyota has increased production for worldwide sales from one vehicle every 8 to 10 minutes to one vehicle per minute.

Figure 14-34 Under the hood of the Toyota Prius hybrid electric vehicle, second generation. Note the orange high-voltage cables.

The Prius is coupled to an Electronically Controlled Continuously Variable Transmission (ECCVT). A small joystick mounted on the dash controls shifting. It also has a power split device within the transmission to allow the vehicle to run on either the electric motor or the gasoline engine exclusively. One of the motors (also called generators or MG1 and MG2) works as a starter and a generator, eliminating those traditional components (**Figure 14-35**). Vehicle acceleration is provided by "drive-by-wire" technology.

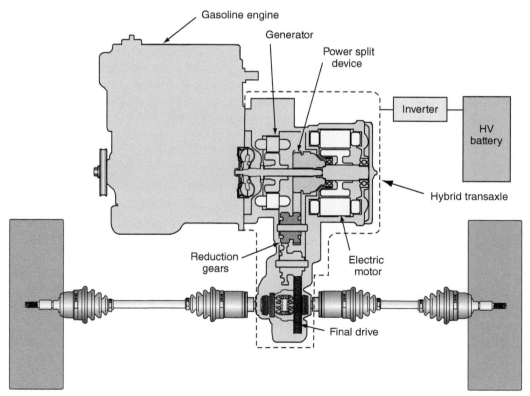

Figure 14-35 The layout of the Toyota Prius power system.

Figure 14-36 The dashboard information offers the driver an opportunity to drive with fuel economy in mind.

Three throttle pedal position sensors inform the powertrain management system about the requested power, and depending on other inputs from the engine and motor circuitry, the electronic control units respond with acceleration from an appropriate power source. The throttle plate on the gasoline engine is driven by an electric motor to the degree requested by the control module. Braking is also managed in a similar manner. The regenerative braking system captures about 30 percent of the energy normally lost in traditional braking systems.

The Prius gas engine stops when the vehicle is at a stop or at very low speeds. This further reduces fuel consumption. It is started automatically by the powertrain management system when accelerator input and other inputs indicate the need. The battery pack is a NiMH comprised of 168 1.2-volt DC cells connected in series. Twenty-eight of these modules connected in series give the Prius motors their nominal operating voltage of 201 volts (600-volt maximum) and reduce the weight of the previous battery packs used. A 56-cell lithium-ion battery pack is an available upgrade. The Prius is equipped with a dashboard that displays available battery power, instantaneous fuel economy, and a mileage bar for each 5-minute segment of driving (**Figure 14-36**).

PLUG-IN HYBRID ELECTRIC VEHICLES

The Toyota Prius is also available as a Plug-In Hybrid Electric Vehicle (PHEV). This means that the high-voltage battery pack can be charged up at home and is designed to allow the vehicle to be driven for a longer period on just electricity (battery power only—engine off). After the battery charge goes down to a certain point, the vehicle operates as a normal HEV Prius. This effectively raises the fuel economy of the vehicle by allowing it to travel farther on the same amount of gasoline. There is a trade-off by not figuring in the cost of electricity to charge up the battery pack. In some areas of the country electricity can be expensive.

The Chevy Volt is another example of a PHEV. Chevrolet's car, however, acts more like an electric car and the internal combustion engine acts more like a generator.

FUEL CELL VEHICLES

Today, there are three fuel cell electric vehicles (FCEVs) available to drive in the United States or, more specifically, in the state of California: the Honda FCX Clarity, the Hyundai Tucson Fuel Cell, and the Toyota Mirai. Except for the Toyota Mirai, they can only be leased, not purchased. Many experts believe, however, that the fuel cell will eventually become the primary power plant for the modern automobile or at least capture the advanced technology market. Many manufacturers have prototype fuel cell vehicles. An FCEV is an electric car: the battery pack is charged by the hydrogen fuel cell. The reality of mass-marketed fuel cell vehicles may be at least a decade off, however, while research and development into hydrogen conversion, fuel cell technology, and refilling infrastructure concerns are resolved. Interest and research investment is high because hydrogen is a regenerative fuel and its use could drastically reduce our consumption of nonrenewable fuel sources.

A fuel cell vehicle operates very much like an electric vehicle except that hydrogen (or another type of fuel cell) supplies electrical power to the motor(s) rather than batteries. Fuel cells electrochemically combine oxygen (from the air) and hydrogen to produce electricity. A special membrane is used to allow interaction between hydrogen and oxygen. Hydrogen will normally "explode" when introduced into the atmosphere with any electrical charge in it. Safety issues related to using hydrogen as a fuel have slowed development of the hydrogen fuel cell vehicle. Obviously, all development will place safety as the foremost issue, considering hydrogen's potential hazards.

Many other fuels are being considered to power fuel cells. Hydrogen is at the forefront because it can be regenerated, but methanol, ethanol, and gasoline are also being tested. A reformer can be used to separate the hydrogen out of the fuels, and the hydrogen feeds the fuel cell. The fuel cell produces electricity to run the electric motor that powers the vehicle. Additionally, solid fuel cells are being researched.

SUMMARY

- Federal and provincial regulations are driving manufacturers to develop more fuel-efficient, environmentally friendly vehicles.
- Automobile sources are a primary source of air pollution.
- Many areas of North America have air quality that at times is below the safe standard.
- Alternate fuel and hybrid electric vehicles can produce fewer emissions.
- Vehicles can be designed to run on alternative fuels or retrofitted to run on alternate fuels.
- Many fleet vehicles choose to run on alternative fuels.
- Propane vehicles produce lower HC and CO emissions.
- E85 is a renewable fuel source that could reduce dependence on imported oil.
- Flexible fuel vehicles can run on E85, gasoline, or a combination of both.
- Natural gas is a plentiful resource in North America.

- CNG vehicles store their fuel in tanks under pressures between 3,000 psi and 3,600 psi.
- CNG vehicles reduce emissions and can run higher compression ratios.
- CNG tank inspections are regulated by the U.S. and Canadian governments to ensure safety.
- Electric vehicles produce zero emissions but have a short driving range between recharging.
- HEVs use an electric motor and a gasoline or alternate fuel engine to power the vehicle.
- HEVs can dramatically increase fuel economy and decrease emissions.
- HEVs offer adequate power and excellent range between fill-ups.
- No external battery recharging is required on modern HEVs.
- FCEVs may be the next big development in reducing vehicle emissions and increasing fuel economy.

REVIEW QUESTIONS

Short-Answer Essays

1. Describe the reasons that manufacturers are looking for ways to reduce emissions and fuel consumption.

2. Explain the difference between a dedicated and a retrofitted alternate fuel vehicle.

3. Explain two benefits of propane as an alternative fuel.

4. What are the benefits of E85 as an automotive fuel?

5. What are some of the added or changed components on a flexible fuel vehicle?

6. Explain the basic fuel system of a CNG vehicle.

7. List three safety precautions incorporated into CNG vehicles.

8. What are the emission reduction percentages when using CNG as a fuel?

9. Describe the power flows in a parallel system HEV.

10. Explain the basic operation of a typical parallel system HEV from a stop sign to full acceleration.

Fill-in-the-Blanks

1. The _____ _____ _____ of 1990 produced regulations in the United States that required manufacturers to produce vehicles with lower and lower emissions.

2. _____ is a mix of ethanol and petroleum.

3. A _____ alternative fuel vehicle has been originally designed to run on the designated fuel.

4. CNG is typically stored at pressures between _____ and _____.

5. AT-PZEV stands for _____ _____ _____ _____.

6. The Honda Civic CNG vehicle engine has a compression ratio of _____:1.

7. A system that uses the wasted kinetic energy from braking is called a _____ _____ system.

8. An HEV that uses a gasoline engine strictly to power an electric motor is a _____ system HEV.

9. The Atkinson-cycle engine increases fuel economy by reducing the engine's _____ _____.

10. The second-generation Toyota Prius can accelerate from 0 mph to 60 mph (0–100 km/h) in roughly _____ seconds.

Multiple Choice

1. According to the U.S. EPA, mobile sources contribute roughly _____ percent of the carbon monoxide emissions to the air pollution problem.
   A. 10   C. 50
   B. 25   D. 75

2. When a conversion is made to a gas vehicle to allow it to run on an alternative fuel, it is called a _____ vehicle.
   A. HEV   C. Flexible fuel
   B. Dedicated   D. Retrofitted

3. When discussing propane, which statement is true?
   A. Propane is denser than gasoline.
   B. Propane has more available energy.
   C. Propane is stored at 2,000 psi.
   D. Propane has lower fuel consumption than gasoline.

4. When running a vehicle on propane, engine wear is greatly reduced because
   A. propane is a cleaner fuel.
   B. the engine runs on higher compression pressures.
   C. it produces lower HC emissions.
   D. the engine runs on higher fuel pressures.

5. E85 is typically made from all of the following, *except*:
   A. Corn   C. Sugarcane
   B. Wheat   D. Peanuts

6. The "greenhouse gas" that contributes to global warming is
   A. CO.
   B. CO_2.
   C. HC.
   D. NO_x.

7. Flexible fuel vehicles use a fuel sensor that detects _____ so the PCM can better calculate spark timing and fuel delivery.
   A. Fuel temperature
   B. Fuel density
   C. Alcohol content
   D. Oxygen content

8. The use of CNG vehicles would reduce _____ emissions more than the use of any other alternate fuel or HEV other than the electric vehicle.
   A. HC
   B. CO
   C. NO_x
   D. CO_2

9. Natural gas has an octane rating of
   A. 60–80.
   B. 80–105.
   C. 110–130.
   D. 120–150.

10. *Technician A* says that an HEV uses a smaller gas engine than the conventional model vehicle.
    Technician B says that some parallel HEVs can use the electric motor(s) to boost the power of the gasoline engine.
    Who is correct?
    A. A only
    B. B only
    C. Both A and B
    D. Neither A nor B

Note: **Terms are highlighted in bold**, followed by Spanish translation in color.

Advanced camshaft An advanced camshaft has the intake valve open more than the exhaust valve at TDC. It is used to improve low speed torque.

Árbol de levas de avance Un árbol de levas de avance tiene la válvula de admisión más abierta que la válvula de escape cuando se encuentra en el punto muerto superior. Se utiliza para mejorar el momento de torsión a baja velocidad.

Aerobic sealants Aerobic-type liquid gasket maker cures in the presence of oxygen to form a seal between two parts.

Obturadores aeróbicos Los obturadores aeróbicos requieren la presencia de oxígeno para secarse.

Air filter The air filter prevents dirt from entering the engine.

Filtro de aire El filtro de aire impide el ingreso de suciedad al motor.

Airflow restriction indicator An airflow restriction indicator is a meter that shows the amount of air flow past a port. This gives the technician an indicator that the air filter may be not flowing as well as it could because it is dirty.

Indicador de restricción de flujo de aire El indicador de restricción de flujo de aire es un medidor que muestra la cantidad de aire que fluye a través de un puerto. Este dato le indica al técnico que la circulación a través del filtro de aire podría no ser óptima por la presencia de suciedad en el filtro.

Alloys Alloys are mixtures of two or more metals. For example, brass is an alloy of copper and zinc.

Aleaciones Las aleaciones son combinaciones de dos o más metales. Por ejemplo, el latón es una aleación de cobre y cinc.

Anaerobic sealants Anaerobic liquid gasket maker cures in the absence of oxygen to form a seal between two parts.

Obturadores anaeróbicos Los obturadores anaeróbicos solamente se secan en ausencia de oxígeno.

Annealing A heat-treatment process used to reduce metal hardness or brittleness, relieve stresses, improve machinability, or facilitate cold working of the metal.

Esmaltación La esmaltación es un proceso de tratamiento por calor para reducir el peligro que los metales se pongan duros o agrios, relevar las tensiones, mejorar el torneo, o facilitar el trabajo en frío del metal.

Armature The movable component of the motor that consists of a conductor wound around a laminated iron core and is used to create a magnetic field.

Armadura La armadura es un componente movible del motor que consiste en un conductor enredado alrededor de un núcleo de hierro la-minado y se usa para crear un campo magnético.

Atkinson-cycle engine Holds the intake valve open longer during the compression stroke to reduce pumping losses.

Motor de ciclo Atkinson Mantiene la válvula de admisión abierta por más tiempo durante la carrera de compresión para reducir las pérdidas de bombeo.

Autoignition temperature The minimum temperature required to ignite a gas in air without a spark or flame to assist.

Temperatura de autoencendido La temperatura de autoencendido es la temperatura mínima necesaria para que un gas arda en contacto con el aire sin la ayuda de una chispa o una llama.

Back pressure Back pressure reduces an engine's volumetric efficiency.

Contrapresión La contrapresión reduce el rendimiento volumétrico de un motor.

Balance shaft A shaft driven by the crankshaft that is designed to reduce engine vibrations.

Eje de balance El eje de balance es un eje accionado por el cigüeñal que está diseñado para reducir las vibraciones del motor.

Battery terminals Battery terminals provide a means of connecting the battery plates to the vehicle's electrical system.

Bornes de la batería Los bornes de la batería proporcionan una manera de conectar las placas de la batería al sistema eléctrico del vehículo.

Bearing crush Refers to the extension of the bearing half beyond the seat.

Aplastamiento del cojinete El término aplastamiento del cojinete hace referencia a la extensión del cojinete que sobresale sobre una mitad del asiento.

Bearing inserts Bearing inserts are soft-faced bearings used to protect the crankshaft main and rod journals as well as the camshaft journals.

Insertos de casquillos Los insertos de casquillos son casquillos de cara lisa que se usan para proteger los gorrones principales y los de bielas del cigüeñal, como también los gorrones del árbol de levas.

Bearing spread Bearing spread means the distance across the outside parting edges is larger than the diameter of the bore.

Extensión del cojinete El término extensión del cojinete implica que la distancia entre los bordes exteriores es mayor que el diámetro del calibre.

Bedplate The bedplate is the lower part of a two-piece engine block. The bedplate may also be called the main girdle.

Placa de asiento La placa de asiento es la parte inferior de un bloque de motor de dos piezas. También se la denomina bancada.

Blow-by Refers to compression pressures that escape past the piston.

Fuga de gases El término fuga de gases, o *blowby*, hace referencia a las presiones de compresión que escapan a través del pistón.

Body control module (BCM) A computer used to control many systems, such as the instrument panel, radio, interior lighting, climate control, and driver information.

Módulo de control de la carrocería El módulo de control de la carrocería (BCM, por sus siglas en inglés) es una computadora que se utiliza para controlar numerosos sistemas, como el tablero de instrumentos, la radio, la iluminación interior, el control climático y la información para el conductor.

Bolt diameter Bolt diameter indicates the diameter of an imaginary line running through the center of the bolt threaded area.

Diámetro del tornillo El diámetro del tornillo indica el diámetro de una línea imaginaria que atraviesa el centro del área enroscada del tornillo.

Bolt length Bolt length indicates the distance from the underside of the bolt head to the tip of the bolt threaded end.

Longitud del tornillo La longitud del tornillo indica la distancia de la parte de abajo de la cabeza del tornillo hasta la punta del enroscado del tornillo.

Bore A bore is the diameter of a hole.

Calibre El calibre es el diámetro de un orificio.

Bottom dead center (BDC) A term used to indicate that the piston is at the very bottom of its stroke.

Punto muerto inferior El punto muerto inferior (BDC, por sus siglas en inglés) es un término utilizado para indicar que el pistón se encuentra en la posición más baja de su carrera.

Brake horsepower The usable power produced by an engine. It is measured on a dynamometer using a brake to load the engine.

Caballos de fuerza del freno Los caballos de fuerza del freno es la potencia de uso que produce un motor. Se mide sobre un dinamómetro usando un freno para cargar el motor.

Burn time The amount of time from the instant the mixture is ignited until the combustion is complete.

Tiempo de combustión El tiempo de combustión es el período que transcurre desde el instante en que la mezcla comienza a arder hasta que se completa la combustión.

Burned valves Valves that have warped and melted, leaving a groove across the valve head.

Válvulas quemadas Las válvulas quemadas son válvulas que se han deformado y fundido y, como consecuencia, presentan una ranura en la cabeza de la válvula.

Cam ground pistons A piston is not a perfect circle shape, it is more of an oval shape. Cam ground pistons are oval shaped so that they can expand to a more perfect circle shape when heated.

Pistón fresado excéntricamente Un pistón no tiene una forma circular perfecta, sino que es un poco ovalado. Los pistones fresados excéntricamente tienen una forma ovalada que les permite expandirse y adquirir una forma circular más perfecta al calentarse.

Cam-phasing Another term used for variable valve timing.

Ajuste de fase de leva Otro término utilizado para el ajuste de válvula variable.

Camshaft A shaft with eccentrically shaped lobes on it to force the opening of the valves.

Árbol de levas El árbol de levas es un eje que posee lóbulos montados excéntricamente para forzar la apertura de las válvulas.

Camshaft bearings Camshaft bearings are required in most engines as a means of reducing friction between the camshaft and the cylinder block.

Cojinetes del árbol de levas En la mayoría de los motores, los cojinetes del árbol de levas son necesarios para reducir la fricción entre el árbol de levas y el bloque de cilindros.

Camshaft lift Camshaft lift is the valve lifter movement created by the rotating action of the camshaft lobe. Typical lobe lift of original equipment in the block camshafts is between 0.240 in. and 0.280 in. (6 mm and 6.6 mm).

Elevación del árbol de levas La elevación del árbol de levas es el movimiento de la carrera de la válvula que crea la acción giratoria del lóbulo del árbol de levas. La elevación típica del lóbulo del equipo original en los árboles de levas de garrucha está entre 6 y 6.6 mm (0.240 y 0.280 pulg.).

Cam-shifting Sometimes called *cam-switching*, provides the ability for the valves to open different amounts for increased engine power and fuel economy.

Desplazamiento de leva Conocido algunas veces como cambio de leva, permite regular la apertura de las válvulas para aumentar la potencia del motor y el ahorro de combustible.

Carbon steel Steel with a specific carbon content.

Acero semiduro El acero semiduro tiene un contenido específico de carbono.

Case hardening A heating and cooling process for hardening metal. Because the case hardening is only on the surface of the component, it is usually removed during machining procedures performed at the time the engine is being rebuilt. It is not necessary to replace the case hardening in most applications.

Cimentación en caja La cimentación en caja es un proceso de calentamiento y enfriamiento para carburar o cimentar el metal. Ya que la cimentación en caja sólo es en la superfi cie del componente, generalmente se quita durante las gamas de mecanizado que se realizan al tiempo que se reconstruye el motor. No es necesario reemplazar la cimentación en caja en la mayoría de las aplicaciones.

Catalytic converter A catalytic converter reduces tailpipe emissions of carbon monoxide, unburned hydrocarbons, and nitrogen oxides.

Convertidor catalítico Un convertidor catalítico reduce las emisiones de monóxido de carbono del tubo de escape, hidrocarburos sin quemar y óxidos de nitrógeno.

C-class oil The C stands for compression ignition. In some areas the C stands for commercial engines. C-class oil is designed for commercial or diesel engines.

Aceite clase C El aceite clase C está diseñado para motores diésel o comerciales. La letra C proviene del inglés *"compression ignition"*, encendido por compresión. En el caso del aceite clase S, la letra S proviene del inglés *"spark ignition"*, encendido por chispa. No obstante, en algunos lugares la S representa el término en inglés *"service engines"*, motores de servicio, y la C, *"commercial engines"*, motores comerciales.

Ceramics A combination of nonmetallic powdered materials fired in special kilns. The end product is a new product.

Cerámica La cerámica es una combinación de materiales no metálicos pulverizados que se cuecen en hornos especiales. El resultado de este proceso es un nuevo producto final.

Cetane number The cetane number of a fuel is the measurement of how easily a diesel fuel can be ignited.

Número de cetano El número de cetano, o cetanaje, de un combustible indica la facilidad de encendido de un combustible diésel.

Chamber-in-piston A piston with a dish or depression in the top of the piston.

Pistón en cámara Un pistón en cámara es un pistón con un disco o depresión en la parte superior del pistón.

Channeling Channeling is referred to as local leakage caused by extreme temperatures developing at isolated locations on the valve face and head.

Canalización La canalización es una fuga local provocada por el desarrollo de temperaturas extremas en puntos aislados en la cabeza y la cara de la válvula.

Clearance ramp The area of the mechanical lifter camshaft from base circle to the edge comprises the clearance ramp.

Rampa de desahogo El área de la elevación mecánica del árbol de levas desde el círculo base hasta la orilla incluyen la rampa de desahogo.

Closed loop system A closed loop system uses sensors (typically an oxygen sensor) that inform the PCM about the air-fuel ratio, allowing the engine to adjust accordingly.

Sistema de circuito cerrado Un sistema de circuito cerrado utiliza sensores (típicamente un sensor de oxígeno) que informa al MCM sobre la relación combustible aire, dejando al motor para que pueda ajustarse en onsecuencia.

CNC A CNC machine uses special computer software to machine close tolerances.

CNC Un equipo CNC emplea un software especial para mecanizar tolerancias estrechas (rigurosas).

Coil An electrical device that converts low voltage from the battery into a very high voltage that is able to push a spark across the gap of the spark plug.

Bobina La bobina es un dispositivo eléctrico que convierte el bajo voltaje de la batería a un muy alto voltaje que puede empujar una chispa a través del huelgo de la bujía.

Coke Coke is a very hot-burning fuel formed when coal is heated in the absence of air. Coal becomes coke at temperatures above 1022°F(550°C). It is produced in special coke furnaces.

Coque El coque es un combustible de alta temperatura que se forma al calentar carbón en ausencia de aire. El carbón se convierte en coque a temperaturas superiores a los 1022° F(550°C). Se produce en hornos especiales para coque.

Combustion Combustion is the controlled burn created by the spark igniting the hot, compressed air-fuel mixture in the combustion chamber.

Combustión La combustión es el proceso de quemado controlado que se produce cuando una chispa enciende la mezcla caliente y comprimida de aire y combustible en la cámara de combustión.

Combustion chamber The combustion chamber is a sealed area in the engine where the burning (combustion) of the air and fuel mixture takes place.

Cámara de combustión La cámara de combustión es una zona sellada del motor donde se quema la mezcla de aire y combustible.

Composites Composites are human-made materials using two or more different components tightly bound together. The result is a material that has characteristics that neither component possesses on its own.

Compuestos Los compuestos son materiales artificiales formados por dos o más componentes estrechamente unidos. El resultado es un material con características propias que no están presentes en ninguno de los componentes.

Compressed natural gas (CNG) Lightly refined natural gas stored under high pressure.

Gas natural comprimido El gas natural comprimido (GNC o CNG, por sus siglas en inglés) es un gas natural ligeramente refinado que se almacena a alta presión.

Compression-ignition (CI) engines A Compression-ignition (CI) engine ignites the air-fuel mixture in the cylinders from the heat of compression (this engine is often fueled by diesel fuel).

Motores de encendido por compresión Un motor de encendido por compresión (CI, por sus siglas en inglés) enciende la mezcla de aire y combustible en los cilindros con el calor de la compresión (estos motores a menudo utilizan combustible diésel).

Compression ratio The compression ratio is a comparison between the volume above the piston at BDC and the volume above the piston at TDC. Compression ratio is the measure used to indicate the amount the piston compresses each intake charge.

Índice de compresión El índice de compresión es la relación entre el volumen que queda por encima del pistón cuando se halla en el punto muerto inferior y el volumen que queda por encima del pistón cuando se halla en el punto muerto superior. El índice de compresión es la medida utilizada para indicar la compresión del pistón en cada carga de admisión.

Compressions rings Compressions rings are the top rings in a piston assembly that provide cylinder sealing during the compression and power strokes.

Anillos de compresión Los anillos de compresión son los anillos superiores del conjunto del pistón, que permiten sellar los cilindros durante las carreras de compresión y de potencia.

Compressor wheel A compressor wheel is a vaned wheel mounted on the turbocharger shaft that forces air into the intake manifold.

Rueda compresora Una rueda compresora es un difusor de paletas montado en un eje turboalimentador que fuerza aire dentro del colector de entrada.

Connecting rod bearing journals Connecting rod bearing journals are offset from the crankshaft centerline to provide leverage for the piston to turn the crankshaft.

Muñones de los cojinetes de la biela Los muñones de los cojinetes de la biela están desplazados con respecto a la línea central del cigüeñal para accionar el pistón a fin de que haga girar el cigüeñal.

Connecting rod bearings Connecting rod bearings are insert-type bearings retained in the connecting rod and mounted between the large connecting rod bore and the crankshaft journal.

Cojinetes de la biela Los cojinetes de la biela son cojinetes tipo encarte que se retienen en la biela y se montan entre el orifi cio grande de la biela y el gorrón del cigüeñal.

Connecting rods The connecting rod forms a link between the piston and the crankshaft.

Bielas La biela establece una unión entre el pistón y el cigüeñal.

Converter A converter is used to change the very high battery pack voltage to 12 volts for the conventional electrical system.

Convertidor El convertidor se utiliza para cambiar el voltaje elevado del paquete de baterías a los 12 voltios del sistema eléctrico convencional.

Coolant recovery system The coolant recovery system prevents loss of coolant if the engine overheats, and it keeps air from entering the system.

Sistema de recobro de refrigerante El sistema de recobro de refrige-rante previene la pérdida de fl uido refrigerador si se sobrecalienta el motor, y no permite que entre aire al sistema.

Cooling system A cooling system controls the temperature of an engine by allowing it to heat up quickly, operate at the correct temperature, and remove excess heat.

Sistema de enfriamiento El sistema de enfriamiento controla la temperatura del motor, ya que le permite calentarse rápidamente, funcionar a la temperatura adecuada y eliminar el exceso de calor.

Core engine A used engine that is disassembled for purposes of rebuilding in large volumes.

Motor reconstruible Un motor reconstruible es un motor usado que se desmonta para su reconstrucción.

Core plugs Core plugs are used to seal the holes machined into the block to remove the sand after casting. Core plugs are often called freeze plugs. Occasionally, when coolant is not adequately protected with antifreeze, it freezes in the block; a core plug will pop out and prevent the block from cracking. People are not always that lucky, however, and that is not the purpose of core plugs.

Tapones del núcleo Los tapones del núcleo se utilizan para sellar los orificios rectificados en el bloque a fin de quitar la arena que queda después de la fundición. Los tapones del núcleo a menudo se denominan tapones o sellos del bloque del motor. A veces, cuando el refrigerante no se protege adecuadamente con un anticongelante, se congela en el bloque. En tal caso, a veces un tapón del núcleo salta para impedir que el bloque se fisure. No todos son tan afortunados y, además, ese no es el propósito de los tapones del núcleo.

Crankshaft A crankshaft is a shaft held in the engine block that converts the linear and reciprocating motion of the pistons into a nonreversing rotary motion used to turn the transmission. The crankshaft rotates in only one direction. It is not reversible.

Cigüeñal El cigüeñal es un eje presente en el bloque del motor que convierte el movimiento lineal recíproco de los pistones en un movimiento de rotación no reversible que se utiliza para hacer girar la transmisión. El cigüeñal gira solamente en una dirección. No se puede revertir.

Crankshaft counterweights Crankshaft counterweights are positioned opposite the connecting rod journals and balance the crankshaft.

Contrapesos del cigüeñal Los contrapesos del cigüeñal se colocan opuestos a los gorrones de la biela y balancean el cigüeñal.

Crankshaft throw Crankshaft throw is the measured distance between the centerline of the rod journal and the centerline of the crankshaft main journal. Crankshaft throw multiplied by two equals the piston stroke.

Codo del cigüeñal El codo del cigüeñal es la distancia entre la línea central del muñón de la biela y la línea central del muñón principal del cigüeñal. La medida del codo del cigüeñal multiplicada por dos equivale a la carrera del pistón.

Crate engine A crate engine is a new or remanufactured engine that is built with fresh components, such as bearings, rings, and lifters. You can purchase crate engines with or without cylinder heads.

Motor en enrejado Un motor en enrejado es un motor Nuevo o refabricado que está construido con nuevos componentes, tales como cojinetes, anillos y taqués.

Cycle A cycle is a complete sequence of events.

Ciclo Un ciclo es una secuencia completa de eventos.

Cylinder An engine cylinder is a round hole bored into the cylinder block.

Cilindro Un cilindro de motor es un orificio redondo taladrado en el bloque de cilindros.

Cylinder block The main structure of the engine. Most of the other engine components are attached to the block.

Bloque de cilindros El bloque de cilindros es la estructura principal del motor. Casi todos los demás componentes del motor están conectados al bloque.

Cylinder head A large object that houses most of the valvetrain and covers the combustion chamber.

Culata del cilindro La culata del cilindro es el componente de gran tamaño que aloja la mayor parte del tren de válvulas y cubre la cámara de combustión.

Dedicated alternative fuel vehicles Dedicated alternative fuel vehicles were designed to run on alternative fuel.

Vehículos de combustible alternativo dedicado Los vehículos de combustible alternativo dedicado están diseñados para funcionar con combustibles alternativos.

Detergents Gasoline additives that clean the fuel system.

Detergentes Los detergentes son aditivos que limpian el sistema de combustible.

Detonation Detonation is different from preignition. It is characterized as abnormal combustion. Detonation occurs when pressure and temperature build in gas chambers outside of the normal spark flame. This may be caused by improper octane, high engine loads, or improper combustion. Detonation is often known as "spark knock" or "pinging."

Detonación La detonación es diferente del preencendido. Se caracteriza por una combustión anormal. La detonación se produce cuando en las cámaras de combustión se acumulan presión y temperatura más allá de la chispa normal. Esto puede deberse a un octanaje erróneo, a una carga elevada del motor o a una combustión incorrecta. La detonación a menudo se denomina "pistoneo", "golpeteo" o *"pinging"*.

Diesel Any fuel that will operate in a diesel (compression ignition) engine. It is named after its inventor, Rudolf Diesel. Diesel fuel has a lower volatility rating than gasoline. It can be derived from petroleum or alternative sources such as bio-mass, natural gas, or soy beans.

Diésel Se denomina combustible diésel a cualquier combustible que funcione en un motor diésel (encendido por compresión). Lleva el nombre de su inventor, Rudolf Diesel. El combustible diésel tiene un índice de volatilidad menor que la gasolina. Puede ser derivado del petróleo o de fuentes alternativas, como la biomasa, el gas natural o la soja.

Differential The differential allows the wheels to turn at different speeds as the vehicle turns corners. It is internal to a transaxle and external to a transmission.

Diferencial El diferencial permite que las ruedas giren a velocidades diferentes cuando el vehículo gira en una esquina. Es interno para el transeje y externo para la transmisión.

Displacement Displacement is the measure of engine volume. The larger the displacement, the greater the power output.

Desplazamiento El desplazamiento es la medida del volumen del motor. A mayor desplazamiento, mayor potencia de salida.

Distributor The distributor is usually driven by the camshaft. It uses the timing of the valves to control the timing of the spark ignition.

Distribuidor El distribuidor generalmente es accionado por el árbol de levas. Utiliza la sincronización de las válvulas para controlar la sincronización del encendido por chispa.

Distributor ignition (DI) A distributor ignition (DI) system has a distributor that distributes spark to each spark plug.

Encendido con distribuidor Un sistema de encendido con distribuidor (DI, por sus siglas en inglés) posee un distribuidor que suministra la chispa a cada bujía.

Dual overhead cam (DOHC) A dual overhead cam (DOHC) engine has two camshafts mounted above each cylinder head.

Leva de disco superior dual El motor de leva de disco superior dual tiene dos árboles de levas montados sobre cada cabeza del cilindro.

Dual parallel system In a dual parallel system HEV, two electric motors and an engine are used to drive the vehicle.

Sistema doble paralelo Los vehículos híbridos eléctricos con sistema doble paralelo se impulsan con dos motores eléctricos y un motor de combustión interna.

Duration Camshaft duration is the amount of time the camshaft holds the valve open. It is measured in degrees of crankshaft rotation.

Duración La duración del árbol de levas es la cantidad de tiempo durante el cual el árbol de levas mantiene la válvula abierta. Se mide en grados de rotación del cigüeñal.

E85 E85 is an automotive fuel made of 85 percent ethanol and 15 percent petroleum.

E85 El E85 es un combustible para automóviles hecho con 85 por ciento de etanol y 15 por ciento de petróleo.

Eddy currents An eddy is a current that runs against the main current.

Corrientes de Foucault Una corriente de Foucault, o corriente torbellino, es una corriente que se opone a la corriente principal.

Efficiency A measure of a device's ability to convert energy into work.

Eficiencia La eficiencia es una medida de la capacidad que tiene un dispositivo para convertir la energía en trabajo.

Electrolysis The chemical and electrical process of decomposition that occurs when two dissimilar metals are joined in the presence of moisture. The lesser of the two metals is eaten away.

Electrólisis La electrólisis es el proceso electroquímico de descomposición que tiene lugar cuando se unen dos metales diferentes en presencia de humedad. El menor de los metales se consume.

Electronic ignition (EI) An electronic ignition system uses an electronic module to turn the coils on and off.

Encendido electrónico Un sistema de encendido electrónico usa un modulo electrónico para encender y apagar las bobinas.

Emission control system The emission control system includes various devices connected to the engine or exhaust system to reduce harmful emissions of hydrocarbons (HC), carbon monoxide (CO), and oxides of nitrogen (NOx).

Sistema de control de emisiones El sistema de control de emisiones incluye varios dispositivos conectados al motor o al sistema de escape para reducir las emisiones dañinas e hidrocarburos (HC), monóxido de carbono (CO) y óxidos de nitrógeno (NOx).

End gases Unburned gases are called end gases.

Gases finales Los gases que no se queman se denominan gases finales.

Engine An engine is defined as a device or machine that converts thermal energy into mechanical energy.

Motor Un motor se define como un dispositivo o máquina que convierte energía térmica en energía mecánica.

Engine bearings Engine bearings are two bearing halves that form a circle to fit around the crankshaft journals.

Cojinetes de eje motor Los cojinetes de eje motor son dos mitades de cojinete que forman un círculo para ajustarse alrededor de los gorrones del cigüeñal.

Engine block The engine block is the main structure of the engine that forms the combustion chamber and houses the pistons, crankshaft, and connecting rods.

Bloque del motor El bloque del motor es la estructura principal del motor que conforma la cámara de combustión y aloja los pistones, el cigüeñal y las bielas.

Engine cradle The engine cradle, which may also be called the suspension cradle, is the front frame member(s) in a unibody design. The engine is mounted to this frame, as is the suspension.

Soporte del motor El soporte del motor, al que también se le llama soporte de la suspensión neumática, es el miembro(s) del bastidor delantero en un diseño de una sola carrocería. El motor está montado en su lojamiento, tomo también la suspensión.

Engine mount An engine mount attaches the engine to the chassis and absorbs the engine vibrations.

Bancada del motor La bancada del motor se sujeta del motor al chasis y absorbe las vibraciones del motor.

Ethanol Ethanol is an alcohol that can be used as an automotive fuel.

Etanol El etanol es un alcohol que puede emplearse como combustible para automóviles.

Exhaust manifold The exhaust manifold is an exhaust collector that is mounted onto the cylinder head.

Múltiple de escape El múltiple de escape es un colector de gases de escape que está montado sobre la culata del cilindro.

Face The face is the tapered part of the section of the valve head that makes contact with the valve seat.

Cara La cara es la parte mecanizada de la sección de la cabeza de la válvula que hace contacto con el asiento de la válvula.

Ferrous metals Ferrous metals contain iron. Cast iron and steel are examples of ferrous metals. These metals are easy to identify because they will naturally attract a magnet.

Metales ferrosos Los metales ferrosos contienen hierro. El hierro fundido y el acero son metales ferrosos. Estos metales son fáciles de identificar porque atraen naturalmente a los imanes.

Fillet The fillet is the curved area between the stem and the head.

Filete El filete es el área redondeada entre el vástago y la cabeza.

Fire ring A fire ring is a metal ring surrounding the cylinder opening in a head gasket.

Anillo de encendido Un anillo de encendido es un anillo metálico circundante a la abertura del cilindro en la guarnición de la cabeza.

Flexible fuel vehicle A flexible fuel vehicle is designed to be able to run on either gasoline or a methanol or ethanol blend, M85 or E85.

Vehículo de combustible flexible Un vehículo de combustible fl exible está diseñado para poder correr ya sea con gasolina o con una mezcla de metanol o de etanol, M85 o E85.

Flexplate A flexplate is a lighter version of a flywheel and is bolted to the torque converter of an automatic transmission.

Plato flexible Un plato flexible es una versión más liviana del volante. Se atornilla al convertidor de torsión de las transmisiones automáticas.

Flywheel A heavy disk mounted at the end of the crankshaft to make the engine power pulses and vibrations more stable. It has a smooth surface to which the clutch assembly of a manual transmission is attached. This smooth surface is similar to the surface of a brake rotor.

Volante El volante es un disco pesado que se monta en el extremo del cigüeñal para hacer más estables las vibraciones y los impulsos de potencia del motor. Tiene una superficie lisa a la cual se conecta el conjunto de embrague de las transmisiones manuales. Esta superficie es similar a la de un disco de freno.

Free-wheeling engine In a free-wheeling engine, valve lift and angle prevent valve-to-piston contact if the timing belt or chain breaks. In an interference engine, if the timing belt or chain breaks or is out of phase, the valves will contact the pistons.

Motor de rueda libre En un motor de rueda libre, la elevación y el ángulo de la válvula impiden el contacto de la válvula con el pistón si se averían la cadena o la correa de distribución. Si se averían o se desfasan la cadena o la correa de distribución en un motor de interferencia, las válvulas entran en contacto con los pistones.

Friction The resistance to motion when one object moves across another object. Friction generates heat and is an excellent method of controlling mechanical action.

Fricción La fricción es la resistencia al movimiento que se produce cuando un objeto se mueve en contacto con otro. La fricción genera calor y es una manera excelente de controlar la acción mecánica.

Front main seal A lip-type seal that prevents leaks between the timing gear cover and the front of the crankshaft.

Sello principal delantero Un sello principal delantero es una junta redonda de tipo escarpia de vía que previene las fugas entre la cubierta del engranaje de distribución y el frente del cigüeñal.

Fuel injection Fuel injection systems use the PCM to control electromechanical devices, fuel injectors that deliver a closely controlled amount of fuel to the cylinders.

Por inyección Los sistemas por inyección usan el MCM para controlar los dispositivos electromecánicos, los inyectores que mandan una cantidad de combustible muy controlada a los cilindros.

Fuel system The fuel system includes the intake system that is used to bring air into the engine and the components used to deliver the fuel to the engine.

Sistema de combustible El sistema de combustible consiste en el sistema de admisión que se utiliza para introducir aire al motor y los componentes empleados para suministrar combustible al motor.

Full-filtration system A full-filtration system means all of the oil is filtered before it enters the oil galleries.

Sistema de filtración máxima Un sistema de filtración maxima significa que todo el aceite ese filtra antes de entrar a las galeras de aceite.

Gaskets Gaskets are used to prevent engine fluids or pressure from leaking between two stationary parts.

Juntas Las juntas o empaques se utilizan para impedir la fuga de fluidos o presión del motor entre dos piezas fijas.

Gasoline Gasoline has many ingredients and compounds. It is refined from crude oil and is the most common fuel used in today's engines.

Gasolina La gasolina consta de numerosos ingredientes y compuestos. Es un producto refinado a partir del petróleo crudo y es el combustible más utilizado en los motores actuales.

Glow plugs Glow plugs are threaded into the combustion chamber and use electrical current to heat the intake air on diesel engines.

Bujías incandescentes Las bujías incandescentes se encuentran roscadas en la cámara de combustión y utilizan corriente eléctrica para calentar el aire de admisión en los motores diésel.

Grade The grade of a bolt is the classification of its strength.

Calificación La calificación de un tornillo es la clasificación de su resistencia.

Grade marks Grade marks are radial lines on the bolt head.

Marcas de grado Las marcas de grado son líneas radiales situadas en la cabeza del tornillo.

Gray cast iron A type of cast iron that has a graphitic microstructure. It is a widely used material because of its tensile strength to weight ratio. It is a common consideration for engines because of their high strength and low weight requirements.

Hierro fundido gris El hierro fundido gris es un tipo de hierro que tiene microestructura grafítica. Es un material muy utilizado por la relación entre resistencia a la tracción y peso. Se lo tiene en cuenta para los motores porque estos requieren alta resistencia y bajo peso.

Gross horsepower The the maximum power that an engine can develop. It is measured without additional accessories installed (such as belt-driven items and the transmission).

Potencia bruta La potencia bruta es la máxima potencia que puede desarrollar un motor. Se mide sin accesorios adicionales (como los elementos accionados por correas o la transmisión).

Harmonic balancers Harmonic balancers are used to control and compensate for torsional vibrations.

Equilibradores armónicos Los equilibradores armónicos se utilizan para controlar y compensar las vibraciones torsionales.

HD-5 LPG HD-5 LPG is a "consumer grade" propane.

Gas licuado de petróleo HD-5 El gas licuado de petróleo (LPG, por sus siglas en inglés) HD-5 es un gas propano "para consumo".

Head The enlarged part of the valve.

Cabeza La cabeza es la parte agrandada de la válvula.

Head gasket The head gasket is used to prevent compression pressures, gases, and fluids from leaking. It is located on the connection between the cylinder head and the engine block.

Junta de culata La junta de culata se utiliza para evitar la fuga de fluidos, gases y presiones de compresión. Está ubicada en la conexión entre la culata del cilindro y el bloque del motor.

Heater core The heater core dissipates heat like a radiator. The radiated heat is used to warm the passenger compartment.

Núcleo calentador El núcleo calentador disipa el calor como un radiador. El calor del radiador se usa para calentar el compartimiento del pasajero.

Hemispherical chamber The term hemi may be used for an engine with hemispherical combustion chambers.

Cámara hemisférica El término "hemi" se puede aplicar a un motor con cámaras de combustión hemisféricas.

Honing Honing a cylinder is a machining process that prepares the surface for the installation of new piston rings.

Rectificado El rectificado de un cilindro es un proceso de maquinado que prepara la superficie para la instalación de nuevos aros o anillos de pistón.

Horsepower A measure of the rate of work.

Potencia La potencia es una medida de la tasa de trabajo.

Hybrid electric vehicle (HEV) A hybrid electric vehicle (HEV) uses an electric motor in combination with a gas, diesel, or alternate fuel engine to power the vehicle.

Vehículo híbrido eléctrico Los vehículos híbridos eléctricos se impulsan con un motor eléctrico junto con un motor a gasolina, diésel o combustible alternativo.

Ignition system The delivery of the spark used to ignite the compressed air-fuel mixture is the function of the ignition system.

Sistema de encendido La función del sistema de encendido es suministrar una chispa para encender la mezcla comprimida de aire y combustible.

Indicated horsepower The amount of horsepower the engine can theoretically produce.

Potencia disponible La potencia disponible es la cantidad de caballos de fuerza que teóricamente puede producir el motor.

Inlet check valve The inlet check valve keeps the oil filter filled at all times, so when the engine is started, an instantaneous supply of oil is available.

Válvula de retención La válvula de retención mantiene el filtro de aceite lleno en todo momento, de manera que cuando se ponga en marcha el motor, exista un suministro instantáneo de aceite.

Intake air temperature (IAT) sensor The intake air temperature sensor sends an analog voltage signal to the PCM in relation to air intake temperature.

Sensor de la temperatura del aire de entrada El sensor de la temperatura del aire de entrada envía una señal e voltaje análogo al MCM en relación con la temperatura de entrada.

Intake manifold The intake manifold connects the air inlet tubes to the cylinder head ports to equally distribute air to each cylinder.

Múltiple de admisión El múltiple de admisión conecta los tubos de admisión de aire con los puertos de las culatas de los cilindros para suministrar aire a cada cilindro de manera equitativa.

Intake manifold gasket The intake manifold gasket fits between the manifold and the cylinder head to seal the air-fuel mixture or intake air.

Junta del múltiple de admisión La junta del múltiple de admisión se encuentra entre el múltiple y la culata del cilindro, y sirve para sellar la mezcla de aire y combustible o el aire de entrada.

Integral valve seats Integral valve seats are part of the cylinder head.

Asiento de válvula integral El asiento de válvula integral forma parte de la culata del cilindro.

Intercooler The intercooler cools the intake air temperature to increase the density of the air entering the cylinders. It is also referred to as a charge air cooler.

Refrigerador El refrigerador disminuye la temperatura del aire de entrada para aumentar la densidad del aire que ingresa a los cilindros. También se lo denomina enfriador de aire cargado o intercambiador de calor.

Interference engine When a timing belt or chain breaks, the pistons and crankshaft will quickly stop moving, but the valvetrain will stop moving almost immediately. If an engine is designed so that the valves could possibly touch the piston when this happens (because a valve is fully open and the piston is at TDC), it is called an interference engine.

Motor de interferencia Cuando se averían la cadena o la correa de distribución, los pistones y el cigüeñal dejan de moverse rápidamente, pero el tren de válvulas se detiene casi de inmediato. Si el motor está diseñado para que las válvulas puedan llegar a tocar el pistón cuando esto ocurre (porque la válvula está totalmente abierta y el pistón se encuentra en el punto muerto superior), se lo denomina motor de interferencia.

Internal combustion engine Internal combustion engines burn their fuels within the engine. The power that is used as a result of burning that fuel is also developed inside the engine. In comparison, in an external combustion engine the burning of fuel occurs in an external source, or tank. The heat is then transferred to a separate component where it can be used to power the engine and move parts. Examples of external combustion engines would be steam locomotives and the Stirling engine.

Motor de combustión interna Los motores de combustión interna queman combustibles dentro del motor. La potencia que se utiliza como resultado de esa combustión también se desarrolla dentro del motor. Por el contrario, en los motores de combustión externa, el combustible se quema en una fuente externa (o tanque). Luego el calor se transfiere a otro componente donde se puede utilizar para accionar el motor y mover las piezas. Algunos ejemplos de motores de combustión externa son las locomotoras a vapor y el motor Stirling.

Inverter An inverter is used to convert DC volts to AC volts.

Invertidor Un invertidor permite convertir los voltios de CC en voltios de CA.

Journals Main journals are round machined portions on the centerline of the crankshaft where the crankshaft is held in the main bore.

Muñones Los muñones principales son partes redondeadas rectificadas ubicadas en la línea central del cigüeñal, donde se mantiene el cigüeñal en el calibre principal.

Kinetic energy The amount of energy that is currently being used or is currently working.

Energía cinética La energía cinética es la cantidad de energía que se está utilizando o se encuentra en acción en un momento determinado.

Knock sensor A knock sensor is used to detect abnormal combustion. Its data is used by the PCM to change the ignition timing.

Sensor de golpeteo El sensor de golpeteo se utiliza para detectar la combustión anormal. El módulo de control del tren de potencia utiliza estos datos para cambiar la sincronización del encendido.

Law of conservation of energy A principal law of physics that states that the total energy of an isolated system remains constant despite any internal changes.

Ley de conservación de la energía La ley de conservación de la energía es un principio de la Física que establece que la energía total de un sistema aislado permanece constante independientemente de los cambios internos que se produzcan.

Lifters Lifters are mechanical (solid) or hydraulic connections between the camshaft and the valves.

Alzaválvulas Los alzaválvulas son conexiones mecánicas (sólidas) o hidráulicas entre el árbol de levas y las válvulas.

Lip seals Lip seals are formed from synthetic rubber and have a slight raise (lip) that is the actual sealing point. The lip provides a positive seal while allowing for some lateral movement of the shaft.

Sellos con reborde Los sellos con reborde están fabricados de caucho sintético y tienen un pequeño reborde o labio, que es el punto real de sellado. El reborde proporciona un sello positivo y, al mismo tiempo, permite cierto movimiento lateral del eje.

Lobes The lobes of the camshaft push the valves open as the camshaft is rotated.

Lóbulos Los lóbulos del árbol de levas van abriendo las válvulas a medida que el árbol de levas gira.

Lobe separation angle The lobe separation angle is the average of the intake and exhaust lobe centers. It defines the amount of valve overlap. The tighter the angle (lower number of degrees), the more overlap the camshaft provides.

Ángulo de separación del lóbulo El ángulo de separación del lóbulo es el promedio de los centros de los lóbulos de entrada y de escape. Se defi ne como la cantidad de superposición de la válvula. Entre más cerrado el ángulo (menor número de grados), mayor superposición proporcionará el árbol de levas.

Locating tab The locating tab is a tab sticking out of bearing inserts that fits in a groove in the block, main bearing cap, connecting rod, or connecting rod cap to prevent insert rotation.

Aleta de posicionamiento La aleta de posicionamiento es una aleta que sale de los insertos del cojinete que cabe en una ranura del bloque, la tapa del cojinete principal, la biela o la tapa de la biela para prevenir que gire el inserto.

Long block A long block is an engine block with the cylinder head(s) and other components mounted.

Bloque largo Un bloque largo en un bloque motor con las cabezas del cilindro y otros componentes montados.

Longitudinally mounted engine A longitudinally mounted engine is placed in the vehicle lengthwise, usually front to rear.

Motor montado longitudinalmente Un motor montado longitudinalmente se coloca a lo largo del vehículo, generalmente del frente a la parte posterior.

Lubrication system The engine's lubrication system supplies oil to high-friction and high-wear locations.

Sistema de lubricación El sistema de lubricación del motor suministra aceite a las áreas de gran fricción y desgaste.

M85 M85 is an automotive fuel that contains 85 percent methanol and 15 percent gasoline. It is not commonly used anymore.

M85 El M85 es un combustible para automóviles que contiene un 85 por ciento de metanol y un 15 por ciento de gasolina. Ya no es de uso habitual.

Main-bearing journals Main-bearing journals are machined along the centerline of the crankshaft and support the weight and forces applied to the crankshaft.

Muñón del cojinete principal Los muñones de los cojinetes principales están ubicados a lo largo de la línea central del cigüeñal y soportan el peso y las fuerzas que se aplican al cigüeñal.

Main bore The main bore is a straight hole drilled through the block to support the crankshaft.

Agujero principal El agujero principal es un hoyo derecho taladrado a través del bloque para apoyar el cigüeñal.

Main caps The main caps form the other half of the main bore and bolt onto the block to form round holes for the crankshaft bearings and journals.

Tapas principales Las tapas principales forman la otra mitad del agujero principal y se sujetan con tornillos al bloque para formar agujeros redondos en los gorrones y en los cojinetes del cigüeñal.

Major thrust surface The skirt that pushes against the cylinder wall during the power stroke is the major thrust surface.

Superficie de empuje principal La superficie de empuje principal es la falda que empuja contra la pared del cilindro durante la carrera de potencia.

Margin The margin is the material between the valve face and the valve head.

Margen El margen es el material entre la cara de la válvula y la cabeza de la válvula.

Mass airflow (MAF) sensor The mass airflow (MAF) sensor measures the mass of airflow entering the engine and sends an input to the PCM reflecting this value. This is a primary input for the PCM's control of fuel delivery.

Sensor de flujo de la masa de aire El sensor de flujo de la masa de aire (MAF, por sus siglas en inglés) mide la masa de aire que ingresa al motor y envía una entrada que refleja este valor al módulo de control del tren de potencia (PCM, por sus siglas en inglés). Se trata de una entrada primordial para el control de suministro de combustible del PCM.

Mechanical efficiency The mechanical efficiency of an engine is the comparison between actual power measured at the crankshaft and the actual power developed within the cylinders. There will always be frictional losses in the crankshaft, cylinders, and connecting rods.

Rendimiento mecánico El rendimiento mecánico de un motor es la relación entre la fuerza real medida en el cigüeñal y la fuerza real desarrollada en los cilindros. Siempre existen pérdidas por fricción en el cigüeñal, los cilindros y las bielas.

Metallurgy The science of extracting metals from their ores and refining them for various uses.

Metalurgia La metalurgia es la ciencia de extraer metales de sus minerales y refi narlos para sus varios usos.

Micron A micron is a thousandth of a millimeter or about 0.0008 inch.

Micrón Un micrón es una milésima de milímetro o aproximadamente 0,0008 pulgadas.

Mid-mounted engine A mid-mounted engine is located between the rear of the passenger compartment and the rear suspension. It may also be called a mid-engine.

Motor central Un motor central se localiza entre la parte posterior del compartimiento de pasajeros y la suspensión posterior. También se le llama motor de en medio.

Miller-cycle engine Holds the intake valve open even longer during the compression stroke than the Atkinson-cycle design requiring an even higher pressure in the intake manifold which is provide by a supercharger.

Motor de ciclo Miller Mantiene abierta la válvula de admisión aún más tiempo durante la carrera de compresión que en el diseño de ciclo Atkinson, para lo cual requiere una presión todavía mayor en el múltiple de admisión, la que es proporcionada por un supercargador.

Minor thrust surface The skirt that pushes against the cylinder wall during the compression stroke is the minor thrust surface.

Superficie de empuje secundaria La superficie de empuje secundaria es la falda que empuja contra la pared del cilindro durante la carrera de compresión.

Misfire A misfire means that combustion either is incomplete or does not occur at all.

Fallo de encendido Un fallo de encendido significa que la combustión fue incompleta o no sucedió del todo.

Motor Octane Number (MON) RON and MON are octane testing numbers that are averaged together to get the advertised octane number. Each uses a different test procedure that represents different real-world driving situations.

Número de octanos del motor El número de octanos de carretera y el número de octanos del motor (RON y MON respectivamente, por sus siglas en inglés) representan números obtenidos en pruebas de octanaje que se promedian para obtener el número de octanos publicado. Cada uno utiliza un procedimiento de prueba diferente que representa diferentes condiciones de conducción reales.

Muffler A device mounted in the exhaust system behind the catalytic converter that reduces engine noise.

Mofle El mofle es un dispositivo montado en el sistema de escape atrás del convertidor catalítico que reduce el ruido del motor.

Multi-point fuel injection Multi-point fuel injection systems use one injector for each cylinder to deliver fuel.

Inyección de combustible multipunto Los sistemas de inyección de combustible multipunto utilizan un inyector para suministrar combustible a cada cilindro.

Net horsepower The amount of horsepower an engine has when it is installed in the vehicle.

Potencia neta La potencia neta es la potencia que posee un motor cuando se instala en el vehículo.

Nodular iron Nodular iron, also known as ductile iron, is a version of gray cast iron but is more flexible and elastic.

Hierro nodular El hierro nodular, también llamado hierro dúctil, es una variedad de hierro fundido gris más flexible y elástica.

Nonferrous metals Nonferrous metals contain no iron. Aluminum, magnesium, and titanium are examples of nonferrous metals. These metals will not attract a magnet.

Metales no ferrosos Los metales no ferrosos no contienen hierro. El aluminio, el magnesio y el titanio son ejemplos de metales no ferrosos. Estos metales no atraerán un imán.

Nucleate boiling The process of maintaining the overall temperature of a coolant to a level below its boiling point, but allowing the portions of the coolant actually contacting the surfaces (the nuclei) to boil into a gas.

Ebullición nucleada La ebullición nucleada es el proceso de mantener la temperatura general de un refrigerante en un nivel inferior al punto de ebullición, pero permitiendo que las porciones de refrigerante que están en contacto con las superficies (los núcleos) se evaporen.

Octane The octane number is the ratio of two test fuels that match the knock characteristic of the test fuel. The advertised octane number at a gas station is an average from RON and MON and is an indicator of the fuel's anti-knock quality. Lower octane fuels will burn faster. Higher octane fuels will burn slower.

Octano El número de octanos es una relación entre dos combustibles de ensayo que se ajusta a la característica de golpeteo del combustible de ensayo. El número de octanos publicado en las estaciones de gasolina es un promedio entre los números RON y MON, e indica la cualidad de resistencia al golpeteo del combustible. Los combustibles de menor octanaje se queman más rápido. Los combustibles de mayor octanaje se queman más lento.

Octane rating The octane rating of a fuel describes its ability to resist detonation (also known as spark knock).

Octanaje El octanaje de un combustible representa su capacidad antidetonante (o de evitar el pistoneo del motor).

Off set pin An off set pin is a piston pin that is not mounted on the vertical center of the piston.

Garrucha de desvío Una garrucha de desvío es una garrucha de pistón que no está montada en el centro vertical del pistón.

Off-square Off-square seating or tipping refers to the seat and valve stem not being properly aligned. This condition causes the valve stem to flex as the valve face is forced into the seat by spring and combustion pressures.

Fuera de escuadra El término inclinación o asiento fuera de escuadra significa que el asiento y el vástago de la válvula no están alineados correctamente. Esta situación hace que el vástago de la válvula se arquee cuando las presiones de la combustión y el muelle empujan la cara de la válvula contra el asiento.

Oil breakdown All oils will break down their additives over time. Oil breakdown usually refers to the results from prolonged high temperatures and pressures that quickly degrade the oil and do not allow it to perform its duties.

Descomposición del aceite Con el paso del tiempo, en todos los aceites se van separando los aditivos. La descomposición del aceite suele ser resultado de una exposición prolongada a temperaturas y presiones elevadas que degradan el aceite rápidamente y no permiten que cumpla su función.

Oil coking Oil coking occurs after the engine is shut down and the turbocharger is still spinning. The oil is no longer circulating, and what remains inside the turbo can literally bake into a white, hard, caked restriction in the lines. Allowing the engine to idle for a minute before shutting the engine off allows the turbo compressor to wind down and prevents this.

Carbonización del aceite La carbonización del aceite se produce cuando se apaga el motor y el turbo sobrealimentador sigue girando. El aceite deja de circular y lo que queda dentro del turbo literalmente puede transformarse en una sustancia blanca y dura que obstruye las tuberías. Al dejar que el motor marche en vacío durante un minuto antes de apagarlo, el turbocompresor puede detenerse poco a poco y evitar este problema.

Oil control rings The piston rings that control the amount of oil on the cylinder walls.

Anillos de control del aceite Los anillos de control del aceite son ani-llos de pistón que controlan la cantidad de aceite en las paredes del cilindro.

Oil cooler A heat exchanger that looks much like a small radiator.

Refrigerador del aceite El refrigerador del aceite es un intercambiador de calor que se parece mucho a un radiador pequeño.

Oil galleries Drilled passages throughout the engine that carry oil to key areas needing lubrication.

Canales de aceite Los canales de aceite son conductos perforados en el motor que llevan el aceite a las principales áreas que necesitan lubricación.

Oil pan gaskets Oil pan gaskets are used to prevent leakage from the crankcase at the connection between the oil pan and the engine block.

Juntas del colector de aceite Las juntas del colector de aceite se utilizan para impedir las fugas del cárter en la conexión entre el colector de aceite y el bloque del motor.

Oil pump The oil pump creates a vacuum so atmospheric pressures can push oil from the sump. The pump then pressurizes the oil and delivers it throughout the engine by use of galleries.

Bomba de aceite La bomba de aceite crea un vacío para que la presión atmosférica pueda empujar el aceite desde el depósito de aceite. Luego la bomba presuriza el aceite y lo envía al motor a través de canales.

Oil sludge Oil sludge buildup occurs when the oil gels and thickens due to oxidizing.

Sedimento de aceite La acumulación de sedimentos de aceite se produce cuando el aceite se gelifica y se espesa debido a la oxidación.

Open loop system An open loop system delivers fuel based on demand and has no way of detecting whether the amount sent was ideal in terms of performance, emissions, or fuel economy.

Mecanismo de bucle abierto El mecanismo de bucle abierto entrega combustible basado en la demanda y no tiene manera de detectar si la cantidad que mandó era la ideal en términos de rendimiento, emisiones o economía del combustible.

Overhead camshaft engine (OHC) If a cylinder head contains the camshaft, it is called an overhead camshaft engine. The camshaft can also be located in the cylinder block.

Motor de árbol de levas en la culata Si la culata del cilindro contiene el árbol de levas, se denomina motor de árbol de levas en la culata (OHC, por la sigla en inglés de "*OverHead Cam*"). El árbol de levas también puede estar en el bloque de cilindros.

Overhead valve (OHV) An overhead valve engine has the valves mounted in the cylinder head and the camshaft mounted in the block.

Válvula en culata Un motor de válvulas en culata (OHV, por sus siglas en inglés) tiene las válvulas montadas en la culata del cilindro y el árbol de levas montado en el bloque.

Oversize bearings Oversize bearings are thicker than standard, with the outside diameter increased in order to fit an oversize bearing bore. The inside diameter is the same as standard bearings.

Cojinetes de medidas superiores Los cojinetes de medidas superiores son más gruesos que los cojinetes estándar para aumentar el diámetro externo a fin de adaptarlo a un calibre de mayor tamaño. El diámetro interno es igual al de los cojinetes estándar.

Oxidation Oxidation occurs when hot engine oil combines with oxygen and forms carbon.

Oxidación La oxidación ocurre cuando el aceite caliente del motor se combina con oxígeno y forma carbono.

Oxidation inhibitors The formation of gum and varnish is prevented by oxidation inhibitors.

Antioxidantes Los antioxidantes evitan la formación de resinas y barniz.

Parallel system A parallel system HEV can use the electric motor and the engine to power the vehicle. It can also use just one power source at a time.

Sistema paralelo Los vehículos híbridos eléctricos con sistema paralelo se impulsan mediante un motor eléctrico y un motor de combustión interna. Pueden utilizar una sola fuente de alimentación a la vez.

Pedestal-mounted rocker arm A pedestal-mounted rocker arm is mounted on a threaded pedestal that is an integral part of the cylinder head.

Palanca de basculador montada en pedestal La palanca de basculador montada en pedestal está montada en un pedestal enroscado que es una parte integral de la cabeza del cilindro.

Pentroof combustion chambers Pentroof combustion chambers are a common design for multivalve engines using two intake valves with one or two exhaust valves and provide good S/V ratio.

Cámaras de combustión con parte superior inclinada Las cámaras de combustión con parte superior inclinada son un diseño frecuente en los motores multiválvulas que utilizan dos válvulas de admisión y una o dos válvulas de escape, y que ofrecen una buena relación superficie-volumen (S/V).

Piezoresistive sensor A piezoresistive sensor is sensitive to pressure changes. The most common use of this type of sensor is to measure the engine oil pressure.

Sensor piezorresistente El sensor piezorresistente es sensible a los cambios de presión. El uso más común para este tipo de sensor es el de medir la presión del aceite del motor.

Piston The piston is a round-shaped part that is driven up and down in the engine cylinder bore.

Pistón El pistón es una pieza redondeada que es impulsada hacia arriba y hacia abajo en el calibre del cilindro del motor.

Piston rings Piston rings are hard cut rings that fit around the piston to form a seal between the piston and the cylinder wall.

Anillos de pistón Los anillos o aros de pistón son anillos con perfiles definidos que se ubican alrededor del pistón para formar un sello entre el pistón y la pared del cilindro.

Pitch Pitch is the number of threads per inch on a bolt in the USC system.

Declive El declive es el número de roscas por pulgada en un tornillo en el sistema de medidas acostumbradas en EU.

Plastics Plastics are made from petroleum and are used in several engine components because of their light weight and heat transfer properties.

Plástico El material plástico se fabrica con petróleo y se utiliza en varios componentes del motor por sus propiedades de transferencia térmica y su peso liviano.

Pole shoes Pole shoes are made of high magnetic permeability material to help concentrate and direct the lines of force in the field assembly.

Expansiones polares Las expansiones polares están hechas con un material de alta impermeabilidad magnética que ayudan a concentrar y dirigir las isostáticas en el ensamblado de campo.

Poppet valves Poppet valves control the opening or passage by linear movement.

Válvulas de resorte Las válvulas de resorte controlan la abertura o pasaje mediante un movimiento lineal.

Positive displacement pumps Positive displacement pumps deliver the same amount of oil with every revolution, regardless of speed.

Bombas de desplazamiento positivo Las bombas de desplazamiento positivo suministran la misma cantidad de aceite en cada revolución, independientemente de la velocidad.

Potential energy The energy that is not being used at a given time, but which can be used.

Energía potencial La energía potencial es la energía que no está usándose en un tiempo específi co, pero que puede usarse.

Power splitting device A power splitting device changes the mode of power delivery from one source to the other or to a combination of both sources.

Dispositivo de división de la potencia Un dispositivo de division de la potencia cambia el modo de entregar la potencia de una fuente a otra o a una combinación de ambas fuentes.

Powertrain A powertrain includes the engine and all the components that deliver that force to the road, including the transmission and axles.

Motor El motor incluye a la máquina y todos los componentes que entregan esa fuerza en el camino, incluyendo la transmisión y los ejes.

Powertrain control module (PCM) An onboard computer that controls functions related to the powertrain, such as fuel delivery, spark timing, temperature, and shift points.

Módulo de control del tren de potencia El módulo de control del tren de potencia (PCM, por sus siglas en inglés) es una computadora de a bordo que controla funciones relacionadas con el tren de potencia, como el suministro de combustible, la sincronización del encendido, la temperatura y los puntos de cambio.

Preignition Preignition is an explosion in the combustion chamber resulting from the air-fuel mixture igniting prior to the spark being delivered from the ignition system.

Preencendido El preencendido es una detonación en la cámara de combustión que se produce cuando la mezcla de aire y combustible se enciende antes de que el sistema de encendido suministre la llama.

Prove-out circuit A prove-out circuit completes the warning light circuit to ground through the ignition switch when it is in the start position. The warning light will be on during engine cranking to indicate to the driver that the bulb is working properly.

Circuito de comprobación Un circuito de comprobación complete el circuito de luz de alerta a tierra mediante el interruptor de encendido cuando está en la posición de iniciar. La luz de alerta se encenderá durante el arranque del motor para indicarle al conductor que el foco está funcionando apropiadamente.

Pumping losses Pumping losses are wasted power from the pistons pushing up against compressed air during the compression stroke.

Saldo de bombeo Los saldos de bombeo son potencia perdida de los pistones empujando hacia el aire comprimido durante los golpes de compresión.

Pump-up Lifter pump-up occurs when excessive clearance in the valvetrain allows a valve to float. The lifter attempts to compensate for the clearance by filling with oil. The valve is unable to close, since the lifter does not leak down fast enough.

Bombear Los bombeos del empujador suceden cuando la holgura excesiva en el tren de la válvula permite que fl ote la válvula. El levantador intenta compensar por la holgura al llenarla de aceite. La válvula no puede cerrarse ya que el levantador no deja que gotee lo sufi cientemente rápido.

Pushrod A long tube that connects the camshaft to the valvetrain components in the cylinder head. This is typically not used on an OHC engine.

Varilla de empuje La varilla de empuje es un tubo largo que conecta el árbol de levas con los componentes del tren de válvulas situado en la cabeza del cilindro. Habitualmente no se utiliza en los motores OHC.

Quenching Quenching is the cooling of gases as a result of compressing them into a thin area. The quench area has a few thousandths of an inch clearance between the piston and combustion chamber. Placing the crown of the piston this close to the cooler cylinder head prevents the gases in this area from igniting prematurely.

Amortiguamiento El amortiguamiento es el enfriamiento rápido de los gases que se produce al comprimirlos en un área estrecha. El área de amortiguación es un espacio libre de unas milésimas de pulgada entre el pistón y la cámara de combustión. Al colocar la corona del pistón muy cerca de la culata del cilindro del enfriador, se evita que los gases de esta área se enciendan de manera prematura.

Radiator The radiator is a heat exchanger used to transfer heat from the engine to the air passing through it.

Radiador El radiador es un intercambiador de calor que se utiliza para transferir calor del motor al aire que lo atraviesa.

Rear main seal The rear main seal is a seal installed behind the rear main bearing on the rear main-bearing journal to prevent oil leaks.

Junta principal posterior La junta principal posterior es un sello insta-lado detrás del cojinete principal posterior en el gorrón del cojinete principal posterior para prevenir fugas de aceite.

Reciprocating A repetitive up-and-down or back-and-forth motion.

Recíproco Se llama recíproco a un movimiento repetitivo de arriba a abajo o de un lado a otro.

Reed valve A reed valve is a one-way check valve. The reed opens to allow the air-fuel mixture to enter from one direction, while closing to prevent movement in the other direction.

Válvula de lengüeta La válvula de lengüeta es una válvula de una vía. La lengüeta se abre para permitir que la mezcla de aire y combustible ingrese desde una dirección y se cierra para impedir el movimiento en la dirección contraria.

Regenerative braking Regenerative braking uses the lost kinetic energy during braking to help recharge the batteries.

Frenado regenerativo El frenado regenerativo utiliza la energía cinética perdida durante el frenado para ayudar a recargar las baterías.

Reid vapor pressure (RVP) A method of describing the volatility of the fuel. You can test the RVP to rate the fuel's volatility.

Presión de vapor Reid La presión de vapor Reid (RVP, por sus siglas en inglés) es un método que permite describir la volatilidad del combustible. Se pueden hacer pruebas de RVP para calcular la volatilidad del combustible.

Remanufactured engine An engine that has been professionally rebuilt by a manufacturer or engine supplier.

Motor re-fabricado Un motor re-fabricado es el que el fabricante o un proveedor de motores volvió a fabricar profesionalmente.

Research Octant Number (RON) RON and MON are octane testing numbers that are averaged together to get the advertised octane number. Each uses a different test procedure that represents different real-world driving situations.

Número de octanos del motor El número de octanos de carretera y el número de octanos del motor (RON y MON respectivamente, por sus siglas en inglés) representan números obtenidos en pruebas de octanaje que se promedian para obtener el número de octanos publicado. Cada uno utiliza un procedimiento de prueba diferente que representa diferentes condiciones de conducción reales.

Resonator A device mounted on the tailpipe to reduce exhaust noise.

Resonador El resonador es un dispositivo montado en el tubo de escape para reducir más el ruido de escape.

Retarded camshaft A retarded camshaft has the exhaust valve open more than the intake at TDC.

Árbol de levas atrasado Un árbol de levas atrasado tiene la válvula de escape más abierta que la válvula de admisión cuando se encuentra en el punto muerto superior.

Retrofitted A vehicle that was originally designed to run on gasoline but has been changed to run on an alternate fuel is often called retrofitted or converted.

Modificado Los vehículos diseñados originalmente para funcionar con gasolina que se modifican para funcionar con combustibles alternativos suelen llamarse modificados o convertidos.

Reverse flow cooling system In a reverse flow cooling system, coolant flows from the water pump through the cylinder heads and then through the engine block.

Sistema de enfriamiento de flujo inverso En un sistema de enfriamiento de flujo inverso, el refrigerante fluye desde la bomba de agua y atraviesa las culatas de los cilindros y luego el bloque del motor.

Ring gear The ring gear around the circumference of a flywheel or flexplate allows the starter motor drive ear to turn the engine. Some flexplates and flywheels are used as part of a sensor to determine the speed of the engine.

Engranaje anular El engranaje anular que rodea la circunferencia del volante o el plato flexible permite que el engranaje de mando del motor de arranque ponga en marcha el motor. Algunos volantes y platos flexibles forman parte de un sensor que determina la velocidad del motor.

Rocker arm A pivoting lever used to transfer the motion of the pushrod to the valve stem.

Balancín El balancín es una palanca pivotante que se utiliza para transferir el movimiento de la varilla de empuje al vástago de la válvula.

Rocker arm ratio The difference in the rocker arm dimensions from center to valve stem and center to pushrod is called the rocker arm ratio. It is a mathematical expression of the leverage being applied to the valve stem.

Relación del balancín Se denomina relación del balancín a la diferencia en las dimensiones del balancín desde el centro hasta el vástago de la válvula y desde el centro hasta la varilla de empuje. Es una expresión matemática de la fuerza de palanca que se aplica al vástago de la válvula.

Roller lifter A normal lifter with a rolling pin on the bottom to decrease friction.

Alzaválvulas de rodillo Un alzaválvulas de rodillo es un alzaválvulas normal que tiene un rodillo en la parte superior para disminuir la fricción.

Room-temperature vulcanizing (RTV) Room-temperature vulcanizing (RTV) is a type of liquid gasket maker that may be used in place of a gasket on some applications.

Vulcanización a temperatura ambiente La vulcanización a temperatura ambiente (RTV, por sus siglas en inglés) hace referencia a un tipo de sellador de motor que puede utilizarse en lugar de las juntas en algunas aplicaciones.

Roots-type supercharger The Roots-type supercharger is a positive displacement supercharger that uses a pair of lobed vanes to pump and compress the air.

Sobrealimentador tipo Roots El sobrealimentador tipo Roots es un sobrealimentador de desplazamiento positivo que utiliza un par de aletas con lóbulos para bombear y comprimir el aire.

Rotary valve A valve that rotates to cover and uncover the intake port.

Válvula rotativa Una válvula rotativa es aquella que gira para tapar y destapar el puerto de admisión.

Saddle The portion of the crankcase bore that holds the bearing half in place.

Asiento El asiento es la parte del calibre del cigüeñal que mantiene la mitad del cojinete en su lugar.

S-class oil S-class oil is used for automotive gasoline engines.

Aceite clase S El aceite clase S está diseñado para motores de automóviles a gasolina.

Sealants Sealants are commonly used to fill irregularities between the gasket and its mating surface.

Obturadores Los obturadores generalmente se utilizan para rellenar las irregularidades entre una junta y la superficie de contacto. Algunos obturadores están diseñados para reemplazar a las juntas.

Seals Seals are used to prevent leakage of fluids around a rotating part.

Sellos Los sellos se utilizan para evitar fugas de fluidos alrededor de las piezas giratorias.

Series system A series system HEV uses the electric motor to drive the vehicle; the gas motor recharges the batteries to power the electric motor.

Sistema en serie Los vehículos híbridos eléctricos con sistema en serie se impulsan con el motor eléctrico. El motor a gas recarga las baterías para alimentar el motor eléctrico.

Service sleeves Service sleeves (sometimes referred to as speedy-sleeves) press over the damaged sealing area to provide a new, smooth surface for the seal lip.

Manguitos de desgaste Los manguitos de desgaste (a veces llamados manguitos de servicio o speedi-sleeve) ejercen presión sobre el área de sellado dañada y proporcionan una nueva superficie lisa para el reborde del sello.

Shaft-mounted rocker arms Shaft-mounted rocker arms are used in a mounting system that positions all rocker arms on a common shaft mounted above the cylinder head.

Conjunto de levanta válvulas montadas en el eje El conjunto de le-vanta válvulas montadas en el eje se usan en un sistema de montaje que coloca todo el conjunto de levanta válvulas en un eje común montado encima de la cabeza del cilindro.

Short block A block assembly without the cylinder heads and some other components.

Bloque corto Un bloque corto es el ensamblado de bloque sin las cabezas del cilindro y algunos otros componentes.

Shot peening A cold-working process used to make the outer layers of a metal more compressive.

Granallado de compresión El granallado de compresión, o *shotpeening*, es un proceso de forjado en frío que se utiliza para aumentar la compresión de las capas externas de un metal.

Sintering Powder metallurgy is the manufacture of metal parts by compacting and sintering metal powders. Sintering is done by heating the metal to a temperature below its melting point in a controlled atmosphere. The metal is then pressed to increase its density.

Sinterización La pulvimetalurgia fabrica piezas metálicas mediante la compactación y la sinterización de polvos metálicos. La sinterización se realiza calentando el metal a una temperatura inferior al punto de fusión en una atmósfera controlada. Posteriormente, el metal se prensa para aumentar su densidad.

Skin effect   Skin effect is a small layer of unburned gases formed around the walls of the combustion chamber.

Efecto superficial   El efecto superficial o pelicular consiste en la formación de una pequeña capa de gases no quemados alrededor de las paredes de la cámara de combustión.

Solenoid   Is an electromechanical device changing electricity into mechanical motion such as a fuel injector.

Solenoide   Es un dispositivo electromecánico que cambia la electricidad en movimiento mecánico, como un inyector de combustible.

Spark-ignition (SI) engine   A spark-ignition (SI) engine ignites the air-fuel mixture in the combustion chamber by a spark across the spark plug electrodes (this engine is often fueled by gasoline, propane, ethanol, or natural gas).

Motor de encendido por chispa   Los motores de encendido por chispa (SI, por sus siglas en inglés) encienden la mezcla de aire y combustible en la cámara de combustión mediante una chispa entre los electrodos de la bujía (estos motores suelen funcionar con gasolina, propano, etanol o gas natural).

Splayed crankshaft   A splayed crankshaft is a crankshaft in which the connecting rod journals of two adjacent cylinders are off set on the same throw.

Cigüeñal ensanchado   Un cigüeñal ensanchado es un cigüeñal en el que los gorrones de la biela de dos cilindros adyacentes no están paralelos en el mismo alcance.

Squish area   The squish area of the combustion chamber is the area where the piston is very close to the cylinder head. The air-fuel mixture is rapidly pushed out of this area as the piston approaches TDC, causing a turbulence and forcing the mixture toward the spark plug. The squish area can also double as the quench area.

Área de compresión   El área de compresión, o squish, de la cámara de combustión es el área donde el pistón se encuentra muy cerca de la culata del cilindro. La mezcla de aire y combustible es expulsada rápidamente de esta área cuando el pistón se aproxima al punto muerto superior, lo que genera una turbulencia y empuja la mezcla hacia la bujía. El área de compresión también puede funcionar como área de amortiguación.

Stainless steel   Stainless steel is an alloy that is highly resistant to rust and corrosion.

Acero inoxidable   El acero inoxidable es una aleación que es altamente resistente a la oxidación y la corrosión.

Steel   Iron containing very low carbon (between 0.05 and 1.7 percent) is called steel.

Acero   El hierro que contiene muy bajo carbono (entre 0.05 y 1.7 por ciento) se llama acero.

Stellite   A hard facing material made from a cobalt-based material with a high chromium content.

Estelita   La estelita es un material de revestimiento duro fabricado con un material con base de cobalto y un alto contenido de cromo.

Stem necking   Stem necking is a condition of the valve stem in which the stem narrows near the bottom.

Rebajamiento del vástago   El rebajamiento del vástago es una condición del vástago de la válvula en el que el vástago se estrecha cerca del fondo.

Stratified-charge   A stratified-charge engine has two combustion chambers. A rich air-fuel mixture is supplied to a small auxiliary chamber, and the very lean air-fuel mixture is supplied to the main combustion chamber.

Carga estratificada   Un motor de carga estratificada tiene dos cámaras de combustión. Se suministra una mezcla rica de aire y combustible a una cámara auxiliar pequeña y una mezcla de aire y combustible muy pobre a la cámara de combustión principal.

Strip seal   A strip seal provides a seal between the flat surfaces on the front and back of the engine block and the intake manifold.

Junta de impermeabilización   Una junta de impermeabilización proporciona un sello entre las superfi cies planas del frente y la parte posterior del bloque motor y del colector de entrada.

Stroke   The distance that the piston travels from TDC to BDC or vice versa. The stroke is the amount of piston travel from TDC to BDC measured in inches or millimeters.

Carrera   La carrera es la distancia que recorre el pistón desde el punto muerto superior hasta el punto muerto inferior o viceversa. La carrera es la longitud de recorrido del pistón desde el punto muerto superior hasta el punto muerto inferior, expresada en pulgadas o milímetros.

Stud-mounted rocker arm   A stud-mounted rocker arm is mounted on a stud that is pressed or threaded into the cylinder head.

Balancín montado en perno   El balancín montado en perno se monta sobre un perno que está roscado o presionado contra la culata del cilindro.

SULEV   A SULEV vehicle produces an extremely low measurable amount of pollution from its tailpipe. This is a Federal government vehicle standard.

Vehículo de emisiones superultrabajas   Un vehículo de emisiones superultrabajas, o SULEV, por sus siglas en inglés, emite una cantidad extremadamente baja de contaminación por el tubo de escape. Es una norma para vehículos del Gobierno Federal de EE. UU.

Supercharger   The supercharger is a belt-driven air pump used to increase the compression pressures in the cylinders.

Sobrealimentador   El sobrealimentador es una bomba de aire impulsada por correas que se utiliza para aumentar las presiones de compresión en los cilindros.

Surface-to-volume (S/V) ratio   A high surface-to-volume ratio results in higher HC emissions. The S/V ratio is a mathematical comparison between the surface area of the combustion chamber and the volume of the combustion chamber. A typical S/V ratio is 7.5:1.

Relación superficie-volumen   Una relación superficie-volumen (S/ V) alta genera emisiones más elevadas de hidrocarburos. La relación S/V es una comparación matemática entre la superficie y el volumen de la cámara de combustión. Una relación S/V típica es 7.5:1.

Swirl chambers   Swirl chambers create an airflow that is in a horizontal direction.

Cámaras de turbulencia   Las cámaras de turbulencia crean un flujo de aire con dirección horizontal.

Tailpipe   The tailpipe conducts exhaust gases from the muffler to the rear of the vehicle.

Tubo de escape   El tubo de escape conduce los gases de emission desde el mofl e a la parte posterior del vehículo.

Tapered piston   A tapered piston is narrower at the top of the piston where it will expand when heated to become a similar diameter.

Pistón cónico   Un pistón cónico es más angosto en la parte superior; al calentarse, esta parte se expande para alcanzar un diámetro similar al de la parte inferior.

Tempering   In tempering, metal is heated to a specific temperature to reduce the brittleness of hardened carbon steel.

Templado   En el proceso de templado, el metal se calienta a una temperatura específica para reducir la fragilidad del acero al carbono endurecido.

Tensile strength Tensile strength is the metal's resistance to being pulled apart.

Resistencia a la tracción La resistencia a la tracción es la resistencia del metal a ser separado.

Tetraethyl (TEL) TEL is not used as an additive in gasoline anymore. Unleaded gasoline is readily available and is less toxic.

Tetraetilo (TEL) El tetraetilo ya no se utiliza como aditivo en la gasolina. Actualmente, se dispone de gasolina sin plomo, que es menos tóxica.

Thermal efficiency The difference between potential and actual energy developed in a fuel measured in Btus per pound or gallon.

Rendimiento térmico El rendimiento térmico es la diferencia entre la energía potencial y la energía real que se desarrolla en un combustible, expresada en BTU por libra o por galón.

Thermistor A resistor whose resistance changes in relation to changes in temperature. It is often used as a coolant temperature sensor.

Termistor El termistor es un resistor cuya resistencia cambia en relación con los cambios de temperatura. A menudo se utiliza como sensor de temperatura del refrigerante.

Thermodynamics The study of the relationship and efficiency between heat energy and mechanical energy.

Termodinámica La termodinámica es el estudio de la relación y el rendimiento de la energía térmica y la energía mecánica.

Thermostat A mechanical component that blocks or allows coolant flow to the radiator. The thermostat allows the engine to quickly reach normal operating temperatures and maintains the desired temperature.

Termostato El termostato es un componente mecánico que bloquea o permite el flujo de refrigerante hacia el radiador. El termostato permite que el motor alcance rápidamente una temperatura de funcionamiento normal y mantenga la temperatura deseada.

Thread depth The height of the thread from its base to the top of its peak.

Profundidad de la rosca La profundidad de la rosca es la altura de la rosca medida desde la base hasta la parte superior.

Thrust bearing The thrust bearing prevents the crankshaft from sliding back and forth by using flanges that rub on the side of the crankshaft journal.

Cojinete de empuje El cojinete de empuje evita que el cigüeñal se deslice de un lado a otro utilizando bridas que rozan la parte lateral del muñón del cigüeñal.

Timing chain guide A timing chain guide is a length of metal with a soft face to hold the chain in position as it rotates.

Guía de la cadena de distribución La guía de la cadena de distribución es una extensión de metal con una cara lisa que mantiene la cadena en su lugar mientras gira.

Timing chain tensioner A timing chain tensioner maintains tension on the chain or belt to prevent slapping or skipping a tooth.

Tensor de la cadena de tiempo El tensor de la cadena de tiempo mantiene la tensión en la cadena o banda para prevenir el ruido de impulso o saltearse un diente.

Top dead center (TDC) A term used to indicate that the piston is at the very top of its stroke.

Punto muerto superior El punto muerto superior (TDC, por sus siglas en inglés) es un término utilizado para indicar que el pistón se encuentra en la posición más elevada de su carrera.

Torque A rotating force around a pivot point.

Torsión La torsión es una fuerza giratoria alrededor de un punto de pivote.

Torque converter The torque converter mounts to the crankshaft through a flexplate. Its weight and fluid coupling do a fine job of maintaining a more constant crankshaft speed.

Convertidor de torsión El convertidor de torsión se monta en el cigüeñal mediante un plato flexible. Por su peso y su acoplamiento fluido es eficaz para mantener más constante la velocidad del cigüeñal.

Torque-to-yield Torque-to-yield is a stretch-type bolt that must be tightened to a specific torque and then rotated a certain number of degrees.

De tensión a rendimiento De tensión a rendimiento es un tornillo de tipo alargado que debe estar apretado a una tensión específica y luego girarlo a cierto número de grados.

Torsional vibration Torsional vibration is the result of the crankshaft twisting and snapping back during each revolution.

Vibración torsional La vibración torsional se produce debido a que el cigüeñal gira y vuelve rápidamente a su lugar en cada revolución.

Transmission The transmission is a device that uses the rotary motion and power of the crankshaft to turn the differential and drive shafts. The transmission multiplies the power output and changes the speed of the driveshaft using different gear ratios.

Transmisión La transmisión es un dispositivo que emplea el movimiento giratorio y la potencia del cigüeñal para hacer girar el diferencial y los ejes motores. La transmisión multiplica la potencia de salida y cambia la velocidad del eje motor utilizando diferentes relaciones de transmisión.

Transverse-mounted engine When an engine is mounted perpendicular to the drivetrain, it is said to be a transverse-mounted engine.

Motor transversal Cuando un motor está montado en posición perpendicular al tren de transmisión, se lo denomina motor transversal.

Tumble port Tumble port combustion chambers use a modified intake port design.

Puerto invertido Las cámaras de combustión con puerto invertido tienen un puerto de admisión con diseño modificado.

Turbine wheel A turbine wheel is a vaned wheel in a turbocharger to which exhaust pressure is supplied to provide shaft rotation.

Rodete Un rodete es una rueda de paletas en un turboalimentador en la cual la presión de salida se suministra para proporcionar la rotación del eje.

Turbo boost A term used to describe the amount of positive pressure increase of the turbocharger; for example, 10 psi (69 kPa) of boost means the air is being induced into the engine at 24.7 psi (170 kPa). Normal atmospheric pressure is 14.7 psi (101 kPa); add the 10 psi (69 kPa) to get 24.7 psi (170 kPa).

Turbosoplante Turbosoplante es un término que se usa para describir la cantidad de aumento de presión positiva del turboalimentador; por ejemplo, 10 psi (69 kPa) de soplo significa el aire que se introduce en el motor a 24.7 psi (170 kPa). La presión atmosférica normal es de 14.7 psi (101 kPa); se añaden los 10 psi (69 kPa) para obtener 24.7 psi (170 kPa).

Turbocharger A turbocharger uses the expansion of exhaust gases to rotate a fan-type wheel, which increases the pressure inside the intake manifold.

Turbo sobrealimentador El turbo sobrealimentador utiliza la expansión de los gases de escape para hacer girar una rueda similar a un ventilador y aumentar la presión dentro del múltiple de admisión.

Turbo lag Turbo lag is a short delay period before the turbocharger develops sufficient boost pressures.

Demora de respuesta La demora de respuesta es el breve período antes de que el turbo sobrealimentador desarrolle presiones de sobrealimentación suficientes.

Turbulence The movement of air inside the combustion chamber.

Turbulencia La turbulencia es el movimiento de aire dentro de la cámara de combustión.

Undersize bearings Undersize bearings have the same outside diameter as standard bearings, but the bearing material is thicker to fit an undersize crankshaft journal.

Cojinetes de medidas inferiores Los cojinetes de medidas inferiores tienen el mismo diámetro externo que los cojinetes estándar, pero el material de apoyo tiene mayor espesor para adaptarse a un muñón de cigüeñal más pequeño.

Vacuum A space in which the pressure is significantly lower than the pressure around it. In an engine, we define this as the pressure inside the engine being less than atmospheric and therefore a vacuum exists. Vacuum is considered a force and can move fluids from their rest position.

Vacío El vacío es un espacio cuya presión es significativamente inferior a la presión circundante. En un motor, el vacío se produce cuando la presión dentro del motor es inferior a la presión atmosférica. Se considera que el vacío es una fuerza capaz de poner en movimiento fluidos que se encuentran en reposo.

Valley pans Valley pans prevent the formation of deposits on the underside of the intake manifold.

Bandejas cóncavas Las bandejas cóncavas impiden la formación de depósitos en el lado inferior del múltiple de admisión.

Valve cover gaskets Valve cover gaskets seal the connection between the valve cover and the cylinder head. The gasket is not subject to pressures, but must be able to seal hot, thinning oil.

Juntas de la tapa de la válvula Las juntas de la tapa de la válvula sellan la conexión entre la tapa de la válvula y la culata del cilindro. La junta no está sometida a presión, pero debe ser capaz de impedir el paso de aceite caliente muy ligero.

Valve float Valve float is a condition that allows the valve to remain open longer than intended. It is the effect of inertia on the valve.

Flotación de la válvula La flotación de la válvula es una situación en que la válvula puede permanecer abierta por un período mayor al previsto. Es el efecto de la inercia en la válvula.

Valve guides Valve guides support and guide the valve stem through the cylinder head.

Guías de la válvula Las guías dan apoyo y guían al vástago de la válvula a través de la culata del cilindro.

Valve job A valve job includes refinishing or replacing the valves and seats and any other worn components in the valvetrain.

Trabajo en válvulas El trabajo en válvulas consiste en restaurar o reemplazar las válvulas y los asientos, así como cualquier otro componente desgastado del tren de válvulas.

Valve lift Valve lift is the total movement of the valve off of its seat, expressed in inches or millimeters.

Carrera de la válvula La carrera de la válvula es el movimiento total de la válvula fuera de su asiento, y se expresa en milímetros o en pulgadas.

Valve overlap When the intake and exhaust valves are both open at the same time an engine has valve overlap.

Superposición de válvulas Cuando las válvulas de admisión y de escape están abiertas al mismo tiempo, el motor tiene superposición de válvulas.

Valve seats Provides the mating surface for the valve face.

Asientos de las válvulas Proporcionan la superficie de contacto de la cara de la válvula.

Valve seat inserts Valve seat inserts are pressed into a machined recess in the cylinder head.

Insertos del asiento de la válvula Los insertos del asiento de la válvula están prensadas dentro de la ranura torneada de la cabeza del cilindro.

Valve seat recession Valve seat recession is the loss of metal from the valve seat, causing the seat to recede into the cylinder head.

Retroceso del asiento de la válvula El retroceso del asiento de la válvula se produce cuando el asiento de la válvula pierde metal y queda retraído en la culata del cilindro.

Valve spring A valve spring connects to the valve by a keeper and retainer, and pulls the valve back into the valve seat, thus putting the valve in a closed position where no fuel or air can pass through it.

Muelle de la válvula El muelle de la válvula está conectado a la válvula mediante una abrazadera y retén, y comprime la válvula contra el asiento de manera que quede en una posición de cierre que no permita el paso de aire o combustible.

Valve stem The valve stem guides the valve through its linear movement.

Vástago de la válvula El vástago de la válvula guía la válvula a través de su movimiento linear.

Valve timing Refers to the relationship between the rotation of the crankshaft and camshaft. That correct relationship ensures that the valves open at the correct time.

Reglaje de válvula El reglaje de válvula se refi ere a la relación entre el giro del cigüeñal y del árbol de levas. Esa correcta relación asegura que las válvulas se abren al tiempo correcto.

Valve timing mechanism The valve timing mechanism drives the camshaft in the correct timing with the crankshaft. It may use a belt, chain, or gears.

Mecanismo de reglaje de válvula El mecanismo de reglaje de válvula corre el árbol de levas en la puesta a punto correcta con el cigüeñal. Puede hacer uso de una banda, una cadena o ruedas dentadas.

Valvetrain The valvetrain is made up of the components that open and close the valves.

Tren de válvulas El tren de válvulas está formado por los componentes que abren y cierran las válvulas.

Variable cylinder displacement Allows the cancelation of individual cylinders for better fuel economy.

Cilindrada variable permite cancelar cilindros individuales para mejorar el ahorro de combustible.

Variable-length intake manifold The variable-length intake manifold has two runners of different lengths connected to each cylinder head intake port.

Colector de entrada de longitud variable El colector de entrada de longitud variable tiene dos muelas de diferentes longitudes conectadas a cada orifi cio de admisión de la cabeza del cilindro.

Variable valve timing (VVT) systems Refers to an engine that uses an adjustable valve train to vary the amount and timing of an air/fuel mixture entering or leaving a cylinder.

Sistemas de ajuste de válvula variable (VVT) Usan un tren de válvulas variable para variar el tiempo y la cantidad de aire/combustible que entra al motor.

Viscosity The measure of the resistance of the oil when placed under shear and extensional stress (the same stresses placed on oil by the crankshaft and connecting rod bearings). It is used to describe an oil's internal resistance to flow. Technicians often use viscosity to describe an oil's thickness or weight.

Viscosidad La viscosidad es la medida de la resistencia del aceite al someterlo a esfuerzo cortante y elongacional (el mismo esfuerzo al que lo someten los muñones de los cojinetes de la biela y el cigüeñal). Se utiliza para describir la resistencia interna al flujo del aceite. Los técnicos a menudo utilizan la viscosidad para describir la densidad o el peso de un aceite.

Viscosity index A measure of the change in viscosity an oil will have with changes in temperature.

Índice de viscosidad El índice de viscosidad de un aceite es la medida del cambio en la viscosidad que experimentará el aceite con las variaciones de temperatura.

Volatility Fuel volatility is the ability of the fuel to vaporize. The higher the volatility, the easier it is for the fuel to evaporate; this allows easier cold starts.

Volatilidad La volatilidad del combustible es su capacidad de evaporación. Cuanto mayor es la volatilidad, más fácil es que se evapore el combustible. Esto facilita el arranque en frío.

Volumetric efficiency The comparison between the amount of air that could enter the engine and the amount of air that actually enters the engine.

Rendimiento volumétrico El rendimiento volumétrico es la relación entre la cantidad de aire que podría ingresar al motor y la cantidad de aire que ingresa realmente.

Warning light A lamp that is illuminated to warn the driver of a possible problem or hazardous condition.

Luz de alerta La luz de alerta es una lámpara que está prendida para avisarle al conductor de un posible problema o de una condición peligrosa.

Wastegate A wastegate limits the maximum amount of turbocharger boost by directing the exhaust gases away from the turbine wheel. The wastegate valve is also referred to as a bypass valve.

Válvula de expulsión La válvula de expulsión limita la maxima cantidad de empuje del turboalimentador al desviar los gases de escape de la rueda de la turbina. La válvula de expulsión también se conoce como válvula de desviación.

Wedge chamber Wedge chamber design locates the spark plug between the valves in the widest portion of the wedge in the cylinder head.

Cámara en cuña En el diseño de cámara en cuña, la bujía se encuentra entre las válvulas en la parte más ancha de la cuña, en la culata del cilindro.

Wrist pin A wrist pin is a hardened steel pin that connects the piston to the connecting rod and allows the rod to rock back and forth as it travels with the crankshaft.

Pasador del pistón El pasador del pistón es un pasador de acero endurecido que conecta el pistón con la biela y permite que la biela se balancee de un lado a otro mientras se desplaza con el cigüeñal.

Zero lash The point at which there is no clearance or interference between components of the valvetrains.

Holgura cero La holgura cero es el punto en el que no hay holgura ni interferencia entre los componentes de los trenes de válvulas.